T0137236

Intelligent Systems Reference Library

Volume 138

Series editors

Janusz Kacprzyk, Polish Academy of Sciences, Warsaw, Poland
e-mail: kacprzyk@ibspan.waw.pl

Lakhmi C. Jain, University of Canberra, Canberra, Australia;
Bournemouth University, UK;
KES International, UK
e-mail: jainlc2002@yahoo.co.uk; jainlakhmi@gmail.com
URL: http://www.kesinternational.org/organisation.php

The aim of this series is to publish a Reference Library, including novel advances and developments in all aspects of Intelligent Systems in an easily accessible and well structured form. The series includes reference works, handbooks, compendia, textbooks, well-structured monographs, dictionaries, and encyclopedias. It contains well integrated knowledge and current information in the field of Intelligent Systems. The series covers the theory, applications, and design methods of Intelligent Systems. Virtually all disciplines such as engineering, computer science, avionics, business, e-commerce, environment, healthcare, physics and life science are included.

More information about this series at http://www.springer.com/series/8578

Urszula Stańczyk · Beata Zielosko
Lakhmi C. Jain
Editors

Advances in Feature Selection for Data and Pattern Recognition

 Springer

Editors
Urszula Stańczyk
Silesian University of Technology
Gliwice
Poland

Beata Zielosko
University of Silesia in Katowice
Sosnowiec
Poland

Lakhmi C. Jain
University of Canberra
Canberra
Australia

and

Bournemouth University
Poole
UK

and

KES International
Shoreham-by-Sea
UK

ISSN 1868-4394   ISSN 1868-4408   (electronic)
Intelligent Systems Reference Library
ISBN 978-3-319-88452-3   ISBN 978-3-319-67588-6   (eBook)
https://doi.org/10.1007/978-3-319-67588-6

This Springer imprint is published by Springer Nature
The registered company is Springer International Publishing AG
The registered company address is: Gewerbestrasse 11, 6330 Cham, Switzerland

Preface

This research book provides the reader with a selection of high-quality texts dedicated to the recent advances, developments, and research trends in the area of feature selection for data and pattern recognition.

The book can be treated as a continuation of our previous multi-authored volume as below:

Urszula Stańczyk, Lakhmi C. Jain (Eds.)
Feature Selection for Data and Pattern Recognition
Studies in Computational Intelligence vol. 584
Springer-Verlag, Germany, 2015

In particular, this second volume points to a number of advances topically subdivided into four parts:

- nature and representation of data;
- ranking and exploration of features;
- image, shape, motion, and audio detection and recognition;
- decision support systems.

The volume presents one introductory and 14 reviewed research papers, reflecting the work by 33 researchers from nine countries, namely Australia, Brazil, Canada, Germany, Hungary, Poland, Romania, Turkey, and USA.

Preparation and compilation of this monograph has been made possible by a number of people. Our warm thanks go to the commendable efforts of many institutions, teams, groups, and all individuals who have supported their laudable work. We wish to express our sincere gratitude to the contributing authors and all who helped us in the review process of the submitted manuscripts. In addition, we extend an expression of gratitude to the members of staff at Springer, for their support in making this volume possible.

Gliwice, Poland   Urszula Stańczyk
Sosnowiec, Poland   Beata Zielosko
Canberra, Australia   Lakhmi C. Jain
August 2017

Contents

Part II   Ranking and Exploration of Features

Part IV   Decision Support Systems

Editors and Contributors

About the Editors

Urszula Stańczyk received the M.Sc. degree in Computer Science, and Ph.D. degree (with honours) in technical sciences with specialisation in informatics from the Silesian University of Technology (SUT), Gliwice, Poland. She is with the Institute of Informatics, SUT. From 2003 to 2010 Editor in Chief of the "Activity Report" for the Institute, Dr. Stańczyk is a member of KES International (www.kesinternational.org), ADAA Group (http://adaa.polsl.pl/), MIR Labs (http://www.mirlabs.org/), and International Rough Set Society (http://www.roughsets.org/), a member of Program Committees for several scientific conferences, and one of the key persons responsible for establishing a series of *International Conferences on Man-Machine Interactions (ICMMI)*. She is a member of the Editorial Board of Intelligent Decision Technologies: An International Journal (http://www.iospress.nl/journal/intelligent-decision-technologies/).

Her research interests include artificial intelligence, pattern recognition and classification, neural and rough processing, feature extraction and selection, induction of decision rules, rule quality measures, stylometric processing of texts, data mining. She co-edited conference proceedings and a multi-authored monograph on feature selection, authored and co-authored a two-volume monograph on synthesis and analysis of logic circuits, academic textbooks on arithmetic of digital systems, chapters, conference papers, and journal articles focused on applications of computational intelligence techniques to stylometry.

Beata Zielosko works as an Associate Professor in the Institute of Computer Science, University of Silesia (US), in Katowice. She earned the M.Sc. degree in 2002 and received a Ph.D. in Computer Science in 2008 from this University. In January 2017, she received a habilitation from the Institute of Computer Science of the Polish Academy of Sciences. From 2009 to 2016, Dr. Zielosko worked as an Assistant Professor in the Institute Computer Science, US. From 2011 to 2013, she worked as a Senior Research Scientist in the Mathematical and Computer Sciences and Engineering Division at King Abdullah University of Science and Technology, Saudi Arabia. Dr. Zielosko is a member of International Rough Set Society (www.roughsets.org), KES International (www.kesinternational.org) and a member of Program Committees of several international scientific conferences. She is also a co-editor of international conference proceedings IJCRS 2017.

Her main areas of research are data mining, optimization of decision and association rules, greedy algorithms for decision and association rule induction, feature selection, and machine learning. Dr. Zielosko is a co-author of three research monographs published by Springer.

Dr. Lakhmi C. Jain is with the Faculty of Education, Science, Technology and Mathematics at the University of Canberra, Australia, and Bournemouth University, UK. He is a Fellow of the Institution of Engineers Australia.

Professor Jain founded the KES International for providing a professional community the opportunities for publications, knowledge exchange, cooperation, and teaming. Involving around 5000 researchers drawn from universities and companies worldwide, KES facilitates international cooperation and generates synergy in teaching and research. KES regularly provides networking opportunities for professional community through one of the largest conferences of its kind in the area of KES (www.kesinternational.org).

His interests focus on the artificial intelligence paradigms and their applications in complex systems, security, e-education, e-healthcare, unmanned air vehicles, and intelligent agents.

Contributors

João Roberto Bertini Junior School of Technology, University of Campinas, Limeira, SP, Brazil

Jerzy Błaszczyński Institute of Computing Science, Poznań University of Technology, Poznań, Poland

Mariusz Boryczka Institute of Computer Science, University of Silesia in Katowice, Sosnowiec, Poland

Krisztian Buza Knowledge Discovery and Machine Learning, Institute für Informatik III, Rheinische Friedrich-Wilhelms-Universität Bonn, Bonn, Germany

Andrzej Czyżewski Faculty of Electronics, Telecommunications and Informatics, Gdańsk University of Technology, Gdańsk, Poland

Lynn Daniel The Daniel Group, Charlotte, NC, USA

Doug Fowler The Daniel Group, Charlotte, NC, USA

Noémi Gaskó Centre for the Study of Complexity, Babeş-Bolyai University, Cluj-Napoca, Romania

Piotr Grochowalski Faculty of Mathematics and Natural Sciences, Department of Computer Science, University of Rzeszów, Rzeszów, Poland

Jerzy W. Grzymała-Busse Department of Electrical Engineering and Computer Science, University of Kansas, Lawrence, KS, USA; Department of Expert Systems and Artificial Intelligence, University of Information Technology and Management, Rzeszów, Poland

Lakhmi C. Jain University of Canberra, Canberra, Australia

Michał Lech Faculty of Electronics, Telecommunications and Informatics, Gdańsk University of Technology, Gdańsk, Poland

Rodica Ioana Lung Centre for the Study of Complexity, Babeş-Bolyai University, Cluj-Napoca, Romania

Regina Meszlényi Department of Cognitive Science, Budapest University of Technology and Economics, Budapest, Hungary; Brain Imaging Centre, Research Centre for Natural Sciences, Hungarian Academy of Science, Budapest, Hungary

Teresa Mroczek Department of Expert Systems and Artificial Intelligence, University of Information Technology and Management, Rzeszów, Poland

Maria do Carmo Nicoletti Campo Limpo Paulista School, Campo Limpo Paulista, SP, Brazil; Computer Science Department, Federal University of São Carlos, São Carlos, SP, Brazil

Agnieszka Nowak-Brzezińska Institute of Computer Science, University of Silesia in Katowice, Sosnowiec, Poland

Wiesław Paja Faculty of Mathematics and Natural Sciences, Department of Computer Science, University of Rzeszów, Rzeszów, Poland

Krzysztof Pancerz Faculty of Mathematics and Natural Sciences, Department of Computer Science, University of Rzeszów, Rzeszów, Poland

James Francis Peters Computational Intelligence Laboratory, ECE Department, University of Manitoba, Winnipeg, MB, Canada; Faculty of Arts and Sciences, Department of Mathematics, Adiyaman University, Adiyaman, Turkey

Małgorzata Przybyła-Kasperek Institute of Computer Science, University of Silesia in Katowice, Sosnowiec, Poland

Sheela Ramanna Department of Applied Computer Science, University of Winnipeg, Winnipeg, MB, Canada; Computational Intelligence Laboratory, ECE Department, University of Manitoba, Winnipeg, MB, Canada

Zbigniew W. Raś University of North Carolina, Charlotte, NC, USA; Polish-Japanese Academy of IT, Warsaw University of Technology, Warsaw, Poland

Ashmeet Singh Department of Applied Computer Science, University of Winnipeg, Winnipeg, MB, Canada

Urszula Stańczyk Institute of Informatics, Silesian University of Technology, Gliwice, Poland

Jerzy Stefanowski Institute of Computing Science, Poznań University of Technology, Poznań, Poland

Mihai Suciu Centre for the Study of Complexity, Babeş-Bolyai University, Cluj-Napoca, Romania

Piotr Szczuko Faculty of Electronics, Telecommunications and Informatics, Gdańsk University of Technology, Gdańsk, Poland

Annamária Szenkovits Centre for the Study of Complexity, Babeş-Bolyai University, Cluj-Napoca, Romania

Katarzyna A. Tarnowska University of North Carolina, Charlotte, NC, USA

Jarosław Utracki Institute of Computer Science, University of Silesia in Katowice, Sosnowiec, Poland

Beata Zielosko Institute of Computer Science, University of Silesia in Katowice, Sosnowiec, Poland

Krzysztof Żabiński Institute of Computer Science, University of Silesia in Katowice, Sosnowiec, Poland

Chapter 1
Advances in Feature Selection for Data and Pattern Recognition: An Introduction

Urszula Stańczyk, Beata Zielosko and Lakhmi C. Jain

Abstract Technological progress of the ever evolving world is connected with the need of developing methods for extracting knowledge from available data, distinguishing variables that are relevant from irrelevant, and reduction of dimensionality by selection of the most informative and important descriptors. As a result, the field of feature selection for data and pattern recognition is studied with such unceasing intensity by researchers, that it is not possible to present all facets of their investigations. The aim of this chapter is to provide a brief overview of some recent advances in the domain, presented as chapters included in this monograph.

Keywords Feature selection · Pattern recognition · Data mining

1.1 Introduction

The only constant element of the world that surrounds us is its change. Stars die and new are born. Planes take off and land. New ideas sprout up, grow, and bear fruit, their seeds starting new generations. We observe the comings and goings, births and deaths, neglect and development, as we gather experience and collect memories, moments in time that demand to be noticed and remembered.

Human brains, despite their amazing capacities, the source of all inventions, are no longer sufficient as we cannot (at least not yet) grant anyone a right to take a

U. Stańczyk (✉)
Institute of Informatics, Silesian University of Technology, Akademicka 16,
44-100 Gliwice, Poland
e-mail: urszula.stanczyk@polsl.pl

B. Zielosko
Institute of Computer Science, University of Silesia in Katowice, Będzińska 39,
41-200 Sosnowiec, Poland
e-mail: beata.zielosko@us.edu.pl

L.C. Jain
University of Canberra, Canberra, Australia
e-mail: jainlakhmi@gmail.com

© Springer International Publishing AG 2018   1
U. Stańczyk et al. (eds.), *Advances in Feature Selection for Data and Pattern Recognition*, Intelligent Systems Reference Library 138,
https://doi.org/10.1007/978-3-319-67588-6_1

direct peek at what is stored inside. And one of irresistible human drives is to share with others what we ourselves notice, experience, and feel. Mankind has invented language as means of communication, writing to pass on our thoughts to progeny and descendants, technologies and devices to help us in our daily routines, to seek answers to universal questions, and solve problems. Both human operators and machines require some instructions to perform expected tasks. Instructions need to be put in understandable terms, described with sufficient detail, yet general enough to be adapted to new situations.

A mother tells her child: *"Do not talk to strangers"*, *"take a yellow bus to school"*, *"when it rains you need an umbrella"*, *"if you want to be somebody, you need to study hard"*. In a factory the alarm bells ring when a sensor detects that the conveyor belt stops moving. In a car a reserve lights up red or orange when the gasoline level in a tank falls bellow a certain level. In a control room of a space centre the shuttle crew will not hear the announcement of count down unless all systems are declared as "go". These instructions and situations correspond to recognition of images, detection of motion, classification, distinguishing causes and effects, construction of associations, lists of conditions to be satisfied before some action can be taken.

Information about environment, considered factors and conditions are stored in some memory elements or banks, retrieved when needed and applied in situations at hand. In this era of rapid development of IT technologies we can observe unprecedented increase of collected data, with thousands and thousands of features and instances. As a result, on one side we have more and more data, on the other side, we still construct and look for appropriate methods of processing which allow us to point out which data is essential, and which useless or irrelevant, as we need access to some established means of finding what is sought and in order to do that we must be able to correctly describe it, characterise it, distinguish from other elements [15].

Fortunately, advances in many areas of science, developments in theories and practical solutions come flooding, offering new perspectives, applications, and procedures. The constant growth of available ways to treat any concept, paths to tread, forces selection as an inseparable part of any processing.

During the last few years feature selection domain has been extensively studied by many researchers in machine learning, data mining [8], statistics, pattern recognition, and other fields [11]. It has numerous applications, for example, decision support systems, customer relationship management, genomic microarray analysis, image retrieval, image and motion detection, and text categorisation [33]. It is widely acknowledged that a universal feature selection method, applicable and effective in all circumstances, does not exists, and different algorithms are appropriate for different tasks and characteristics of data. Thus for any given application area a suitable method (or algorithm) should be sought.

The main aim of feature selection is the removal of features that are not informative, i.e., irrelevant or redundant in order to reduce dimensionality, discover knowledge, and explore stored data [1]. The selection can be achieved by ranking of variables according to some criterion or by retrieving a minimum subset of features that satisfy some level of classification accuracy. The evaluation of feature selection technique or algorithm can be measured by the number of selected features, performance of learning model, and computation time [19].

Apart from the point of view of pattern recognition tasks, feature selection is also important with regard to knowledge representation [17]. It is always preferable to construct a data model which allows for simpler representation of knowledge stored in the data and better understanding of described concepts.

This book is devoted to recent advances in the field of feature selection for data and pattern recognition. There are countless ideas and also those waiting to be discovered, validated and brought to light. However, due to space restriction, we can only include a sample of research in this field. The book that we deliver to a reader consists of 14 chapters divided into four parts, described in the next section.

1.2 Chapters of the Book

Apart from this introduction, there are 14 chapters included in the book, grouped into four parts. In the following list short descriptions for all chapters are provided.

Part I Nature and Representation of Data

Chapter 2 is devoted to discretisation [10, 13]. When the entire domain of a numerical attribute is mapped into a single interval, such numerical attribute is reduced during discretisation. The problem considered in the chapter is how such reduction of data sets affects the error rate measured by the C4.5 decision tree [26] generation system using cross-validation. The experiments on 15 numerical data sets show that for a Dominant Attribute discretisation method the error rate is significantly larger for the reduced data sets. However, decision trees generated from the reduced data sets are significantly simpler than the decision trees generated from the original data sets.

Chapter 3 presents extensions of under-sampling bagging ensemble classifiers for class imbalanced data [6]. There is proposed a two phase approach, called Actively Balanced Bagging [5], which aims to improve recognition of minority and majority classes with respect to other extensions of bagging [7]. Its key idea consists in additional improving of an under-sampling bagging classifier by updating in the second phase the bootstrap samples with a limited number of examples selected according to an active learning strategy. The results of an experimental evaluation of Actively Balanced Bagging show that this approach improves predictions of the two different baseline variants of under-sampling bagging. The other experiments demonstrate the differentiated influence of four active selection strategies on the final results and the role of tuning main parameters of the ensemble.

Chapter 4 addresses recently proposed supervised machine learning algorithm which is heavily supported by the construction of an attribute-based decision graph (AbDG) structure, for representing, in a condensed way, the training set associated with a learning task [4]. Such structure has been successfully used for the purposes of classification and imputation in both stationary and non-stationary environments [3]. The chapter provides the motivations and main technicalities involved in the process of constructing AbDGs, as well as stresses some of the strengths of this graph-based structure, such as robustness and low computational costs associated to both training and memory use.

Chapter 5 focuses on extensions of dynamic programming approach for optimisation of rules relative to length, which is important for knowledge representation [29]. "Classical" optimising dynamic programming approach allows to obtain rules with the minimum length using the idea of partitioning a decision table into subtables. Basing on the constructed directed acyclic graph, sets of rules with the minimum length can be described [21]. However, for larger data sets the size of the graph can be huge. In the proposed modification not the complete graph is constructed but its part. Only one attribute with the minimum number of values is considered, and for the rest of attributes only the most frequent value of each attribute is taken into account. The aim of the research was to find a modification of an algorithm for graph construction, which allows to obtain values of rule lengths close to optimal, but for the smaller graph than in "classical" case.

Part II Ranking and Exploration of Features

Chapter 6 describes an overview of reasons for using ranking feature selection methods and the main general classes of this kind of algorithms, with definitions of some background issues [30]. There are presented selected algorithms based on random forests and rough sets, and a newly implemented method, called Generational Feature Elimination (GFE) is introduced. This method is based on feature occurrences at given levels inside decision trees created in subsequent generations. Detailed information about its particular properties, and results of performance with comparison to other presented methods, are also included. Experiments were performed on real-life data sets as well as on an artificial benchmark data set [16].

Chapter 7 addresses ranking as a strategy used for estimating relevance or importance of available characteristic features. Depending on applied methodology, variables are assessed individually or as subsets, by some statistics referring to information theory, machine learning algorithms, or specialised procedures that execute systematic search through the feature space. The information about importance of attributes can be used in the pre-processing step of initial data preparation, to remove irrelevant or superfluous elements. It can also be employed in post-processing, for optimisation of already constructed classifiers [31]. The chapter describes research on the latter approach, involving filtering inferred decision rules while exploiting ranking positions and scores of features [32]. The optimised rule classifiers were applied in the domain of stylometric analysis of texts for the task of binary authorship attribution.

Chapter 8 discusses the use of a method for attribute selection in a dispersed decision-making system. Dispersed knowledge is understood to be the knowledge that is stored in the form of several decision tables. Different methods for solving the problem of classification based on dispersed knowledge are considered. In the first method, a static structure of the system is used. In more advanced techniques, a dynamic structure is applied [25]. Different types of dynamic structures are analyzed: a dynamic structure with disjoint clusters, a dynamic structure with inseparable clusters, and a dynamic structure with negotiations. A method for attribute selection, which is based on the rough set theory [24], is used in all of the described methods. The results obtained for five data sets from the UCI Repository are compared and some conclusions are drawn.

Chapter 9 contains the study of knowledge representation in rule-based knowledge bases. Feature selection [14] is discussed as a part of mining knowledge bases from a knowledge engineer's and from a domain expert's perspective. The former point of view is usually aimed at completeness analysis, consistency of the knowledge base and detection of redundancy and unusual rules, while in the latter case rules are explored with regard to their optimization, improved interpretation and a way to improve the quality of knowledge recorded in the rules. In this sense, exploration of rules, in order to select the most important knowledge, is based in a great extent on the analysis of similarities across the rules and their clusters. Building the representatives for created clusters of rules bases on the analysis of the premises of rules and then selection of the best descriptive ones [22]. Thus this approach can be treated as a feature selection process.

Part III Image, Shape, Motion, and Audio Detection and Recognition

Chapter 10 explores recent advances in brain imaging technology, coupled with large-scale brain research projects, such as the BRAIN initiative in the U.S. and the European Human Brain Project, as they allow to capture brain activity in unprecedented detail. In principle, the observed data is expected to substantially shape the knowledge about brain activity, which includes the development of new biomarkers of brain disorders. However, due to the high dimensionality selection of relevant features is one of the most important analytic tasks [18]. In the chapter, the feature selection is considered from the point of view of classification tasks related to functional magnetic resonance imaging (fMRI) data [20]. Furthermore, an empirical comparison of conventional LASSO-based feature selection is presented along with a novel feature selection approach designed for fMRI data based on a simple genetic algorithm.

Chapter 11 introduces the notion of classes of shapes that have descriptive proximity to each other in planar digital 2D image object shape detection [23]. A finite planar shape is a planar region with a boundary and a nonempty interior. The research is focused on the triangulation of image object shapes [2], resulting in maximal nerve complexes from which shape contours and shape interiors can be detected and described. A maximal nerve complex is a collection of filled triangles that have a vertex in common. The basic approach is to decompose any planar region containing an image object shape into these triangles in such a way that they cover either part or all of a shape. After that, an unknown shape can be compared with a known shape by comparing the measurable areas covering both known and unknown shapes. Each known shape with a known triangulation belongs to a class of shapes that is used to classify unknown triangulated shapes.

Chapter 12 presents an experimental study of several methods for real motion and motion intent classification (rest/upper/lower limbs motion, and rest/left/right hand motion). Firstly, EEG recordings segmentation and feature extraction are presented [35]. Then, 5 classifiers (Naïve Bayes, Decision Trees, Random Forest, Nearest-Neighbors, Rough Set classifier) are trained and tested using examples from an open database. Feature subsets are selected for consecutive classification experiments, reducing the number of required EEG electrodes [34]. Methods comparison and obtained results are given, and a study of features feeding the classifiers is provided. Differences among participating subjects and accuracies for real and imaginary motion are discussed.

Chapter 13 is an extension of the work presented where the problem of classifying audio signals using a supervised tolerance class learning algorithm (TCL) based on tolerance near sets was first proposed [27]. In the tolerance near set method (TNS) [37], tolerance classes are directly induced from the data set using a tolerance level and a distance function. The TNS method lends itself to applications where features are real-valued such as image data, audio and video signal data. Extensive experimentation with different audio-video data sets was performed to provide insights into the strengths and weaknesses of the TCL algorithm compared to granular (fuzzy and rough) and classical machine learning algorithms.

Part IV Decision Support Systems

Chapter 14 overviews an application area of recommendations for customer loyalty improvement, which has become a very popular and important topic area in today's business decision problems. Major machine learning techniques used to develop knowledge-based recommender system, such as decision reducts, classification, clustering, action rules [28], are described. Next, visualization techniques [12] used for the implemented interactive decision support system are presented. The experimental results on the customer dataset illustrate the correlation between classification features and the decision feature called the promoter score and how these help to understand changes in customer sentiment.

Chapter 15 presents a discussion on an alternative attempt to manage the grids that are in intelligent buildings such as central heating, heat recovery ventilation or air conditioning for energy cost minimization [36]. It includes a review and explanation of the existing methodology and smart management system. A suggested matrix-like grid that includes methods for achieving the expected minimization goals is also presented. Common techniques are limited to central management using fuzzy-logic drivers, and redefining of the model is used to achieve the best possible solution with a surplus of extra energy. In a modified structure enhanced with a matrix-like grid different ant colony optimisation techniques [9] with an evolutionary or aggressive approach are taken into consideration.

1.3 Concluding Remarks

Feature selection methods and approaches are focused on reduction of dimensionality, removal of irrelevant data, increase of classification accuracy, and improvement of comprehensibility and interpretability of resulting solutions. However, due to the constant increase of size of stored, processed, and explored data, the problem poses a challenge to many existing feature selection methodologies with respect to efficiency and effectiveness, and causes the need for modifications and extensions of algorithms and development of new approaches.

It is not possible to present in this book all extensive efforts in the field of feature selection research, however we try to "touch" at least some of them. The aim of this chapter is to provide a brief overview of selected topics, given as chapters included in this monograph.

References

1. Abraham, A., Falcón, R., Bello, R. (eds.): Rough Set Theory: A True Landmark in Data Analysis, Studies in Computational Intelligence, vol. 174. Springer, Heidelberg (2009)
2. Ahmad, M., Peters, J.: Delta complexes in digital images. Approximating image object shapes, 1–18 (2017). arXiv:170604549v1
3. Bertini Jr., J.R., Nicoletti, M.C., Zhao, L.: An embedded imputation method via attribute-based decision graphs. Expert Syst. Appl. **57**(C), 159–177 (2016)
4. Bi, W., Kwok, J.: Multi-label classification on tree- and dag-structured hierarchies. In: Getoor, L., Scheffer, T. (eds.) Proceedings of the 28th International Conference on Machine Learning (ICML-11), pp. 17–24. ACM, New York, NY, USA (2011)
5. Błaszczyński, J., Stefanowski, J.: Actively balanced bagging for imbalanced data. In: Kryszkiewicz, M., Appice, A., Ślęzak, D., Rybiński, H., Skowron, A., Raś, Z.W. (eds.) Foundations of Intelligent Systems - 23rd International Symposium, ISMIS 2017, Warsaw, Poland, June 26–29, 2017, Proceedings. Lecture Notes in Computer Science, vol. 10352, pp. 271–281. Springer, Cham (2017)
6. Branco, P., Torgo, L., Ribeiro, R.P.: A survey of predictive modeling on imbalanced domains. ACM Comput. Surv. **49**(2), 31:1–31:50 (2016)
7. Breiman, L.: Bagging predictors. Mach. Learn. **24**(2), 123–140 (1996)
8. Deuntsch, I., Gediga, G.: Rough set data analysis: a road to noninvasive knowledge discovery. Mathoδos Publishers, Bangor (2000)
9. Dorigo, M., Gambardella, L.: Ant colony system: a cooperative learning approach to the traveling salesman problem. IEEE Trans. Evolut. Comput. **1**(1), 53–66 (1997)
10. Fayyad, U.M., Irani, K.B.: On the handling of continuous-valued attributes in decision tree generation. Mach. Learn. **8**(1), 87–102 (1992)

11. Fiesler, E., Beale, R.: Handbook of Neural Computation. Oxford University Press, Oxford (1997)
12. Goodwin, S., Dykes, J., Slingsby, A., Turkay, C.: Visualizing multiple variables across scale and geography. IEEE Trans. Visual Comput. Graphics **22**(1), 599–608 (2016)
13. Grzymała-Busse, J.W.: Data reduction: discretization of numerical attributes. In: Klösgen, W., Zytkow, J.M. (eds.) Handbook of Data Mining and Knowledge Discovery, pp. 218–225. Oxford University Press Inc., New York (2002)
14. Guyon, I.: An introduction to variable and feature selection. J. Mach. Learn. Res. **3**, 1157–1182 (2003)
15. Guyon, I., Gunn, S., Nikravesh, M., Zadeh, L. (eds.): Feature Extraction: Foundations and Applications. Springer, Heidelberg (2006)
16. Guyon, I., Gunn, S.R., Asa, B., Dror, G.: Result analysis of the NIPS 2003 feature selection challenge. In: Proceedings of the 17th International Conference on Neural Information Processing Systems, pp. 545–552 (2004)
17. Jensen, R., Shen, Q.: Computational Intelligence and Feature Selection. Wiley, Hoboken (2008)
18. Kharrat, A., Halima, M.B., Ayed, M.B.: MRI brain tumor classification using support vector machines and meta-heuristic method. In: 15th International Conference on Intelligent Systems Design and Applications, ISDA 2015, Marrakech, Morocco, December 14–16, 2015, pp. 446–451. IEEE (2015)
19. Liu, H., Motoda, H.: Computational Methods of Feature Selection. Chapman & Hall/CRC, Boca Raton (2008)
20. Meszlényi, R., Peska, L., Gál, V., Vidnyánszky, Z., Buza, K.: Classification of fMRI data using dynamic time warping based functional connectivity analysis. In: 2016 24th European Conference on Signal Processing (EUSIPCO), pp. 245–249. IEEE (2016)
21. Moshkov, M., Zielosko, B.: Combinatorial Machine Learning - A Rough Set Approach, Studies in Computational Intelligence, vol. 360. Springer, Heidelberg (2011)
22. Nowak-Brzezińska, A.: Mining rule-based knowledge bases inspired by rough set theory. Fundamenta Informaticae **148**, 35–50 (2016)
23. Opelt, A., Pinz, A., Zisserman, A.: Learning an alphabet of shape and appearance for multi-class object detection. Int. J. Comput. Vis. **80**(1), 16–44 (2008)
24. Pawlak, Z.: Rough sets. Int. J. Comput. Inf. Sci. **11**, 341–356 (1982)
25. Przybyła-Kasperek, M., Wakulicz-Deja, A.: A dispersed decision-making system - the use of negotiations during the dynamic generation of a system's structure. Inf. Sci. **288**(C), 194–219 (2014)
26. Quinlan, J.R.: C4.5: Programs for Machine Learning. Morgan Kaufmann Publishers Inc., San Francisco (1993)
27. Ramanna, S., Singh, A.: Tolerance-based approach to audio signal classification. In: Khoury, R., Drummond, C. (eds.) Advances in Artificial Intelligence: 29th Canadian Conference on Artificial Intelligence, Canadian AI 2016, Victoria, BC, Canada, May 31–June 3, 2016, Proceedings, pp. 83–88. Springer, Cham (2016)
28. Raś, Z.W., Dardzinska, A.: From data to classification rules and actions. Int. J. Intell. Syst. **26**(6), 572–590 (2011)
29. Rissanen, J.: Modeling by shortest data description. Automatica **14**(5), 465–471 (1978)
30. Rudnicki, W.R., Wrzesień, M., Paja, W.: All relevant feature selection methods and applications. In: Stańczyk, U., Jain, L.C. (eds.) Feature Selection for Data and Pattern Recognition, pp. 11–28. Springer, Heidelberg (2015)
31. Stańczyk, U.: Selection of decision rules based on attribute ranking. J. Intell. Fuzzy Syst. **29**(2), 899–915 (2015)
32. Stańczyk, U.: Weighting and pruning of decision rules by attributes and attribute rankings. In: Czachórski, T.., Gelenbe, E.., Grochla, K., Lent, R. (eds.) Proceedings of the 31st International Symposium on Computer and Information Sciences, Communications in Computer and Information Science, vol. 659, pp. 106–114. Springer, Cracow (2016)
33. Stańczyk, U., Jain, L. (eds.): Feature Selection for Data and Pattern Recognition, Studies in Computational Intelligence, vol. 584. Springer, Heidelberg (2015)

34. Szczuko, P.: Real and imaginary motion classification based on rough set analysis of EEG signals for multimedia applications. Multimed. Tools Appl. (2017)
35. Tadel, F., Baillet, S., Mosher, J.C., Pantazis, D., Leahy, R.M.: Brainstorm: a user-friendly application for MEG/EEG analysis. Intell. Neuroscience **2011**(8), 8:1–8:13 (2011)
36. Utracki, J.: Building management system—artificial intelligence elements in ambient living driving and ant programming for energy saving—alternative approach. In: Piętka, E., Badura, P., Kawa, J., Wieclawek, W. (eds.) 5th International Conference on Information Technologies in Medicine, ITIB 2016 Kamień Śląski, Poland, June 20–22, 2016 Proceedings, vol. 2, pp. 109–120. Springer, Cham (2016)
37. Wolski, M.: Toward foundations of near sets: (pre-)sheaf theoretic approach. Math. Comput. Sci. **7**(1), 125–136 (2013)

Part I
Nature and Representation of Data

Chapter 2
Attribute Selection Based on Reduction of Numerical Attributes During Discretization

Jerzy W. Grzymała-Busse and Teresa Mroczek

Abstract Some numerical attributes may be reduced during discretization. It happens when a discretized attribute has only one interval, i.e., the entire domain of a numerical attribute is mapped into a single interval. The problem is how such reduction of data sets affects the error rate measured by the C4.5 decision tree generation system using ten-fold cross-validation. Our experiments on 15 numerical data sets show that for a Dominant Attribute discretization method the error rate is significantly larger (5% significance level, two-tailed test) for the reduced data sets. However, decision trees generated from the reduced data sets are significantly simpler than the decision trees generated from the original data sets.

Keywords Dominant attribute discretization · Multiple scanning discretization · C4.5 Decision tree generation · Conditional entropy

2.1 Introduction

Discretization based on conditional entropy of the concept given the attribute (feature) is considered to be one of the most successful discretization techniques [1–9, 11, 12, 15–17, 19–22]. During discretization of data sets with numerical attributes some attributes may be reduced, since the entire domain of the numerical attribute is

J.W. Grzymała-Busse (✉)
Department of Electrical Engineering and Computer Science, University of Kansas, 66045 Lawrence, KS, USA
e-mail: jerzy@ku.edu

J.W. Grzymała-Busse · T. Mroczek
Department of Expert Systems and Artificial Intelligence, University of Information Technology and Management, 35-225 Rzeszów, Poland
e-mail: tmroczek@wsiz.rzeszow.pl

mapped into a single interval. A new numerical data set may be created by removing attributes from the original numerical data set indicated by single-intervals. Such data sets are called reduced. Our main objective is to compare quality of numerical data sets, original and reduced, using the C4.5 decision tree generation system. To the best of our knowledge, no similar research was ever conducted.

We conducted a series of experiments on 15 data sets with numerical attributes. All data sets were discretized by the Dominant Attribute discretization method [12, 14]. In Dominant Attribute method, first the best attribute is selected (it is called the Dominant Attribute), a then for this attribute the best cutpoint is chosen. In both cases, the criterion of quality is the minimum of corresponding conditional entropy. New, reduced data sets were created. For pairs of numerical data sets: original and reduced, the ten-fold cross-validation was conducted using C4.5 decision tree generation system. Our results show that the error rate is significantly larger (5% significance level, two-tailed test) for the reduced data sets. However, decision trees generated from the reduced data sets are significantly simpler than the decision trees generated from the original data sets. Complexity of decision trees is measured by the depth and size.

2.2 Dominant Attribute Discretization

An example of a data set with numerical attributes is presented in Table 2.1. In this table all cases are described by variables called *attributes* and one variable called a *decision*. The set of all attributes is denoted by A. The decision is denoted by d. The set of all cases is denoted by U. In Table 2.1 the attributes are *Length*, *Width*, *Height* and *Weight*, while the decision is *Quality*. Additionally, $U = \{1, 2, 3, 4, 5, 6, 7, 8\}$. A *concept* is the set of all cases with the same decision value. In Table 2.1 there are three concepts, $\{1, 2, 3\}$, $\{4, 5\}$ and $\{6, 7, 8\}$.

Table 2.1 A numerical data set

Case	Attributes				Decision
	Length	Height	Width	Weight	Quality
1	4.7	1.8	1.7	1.7	High
2	4.5	1.4	1.8	0.9	High
3	4.7	1.8	1.9	1.3	High
4	4.5	1.8	1.7	1.3	Medium
5	4.3	1.6	1.9	1.7	Medium
6	4.3	1.6	1.7	1.3	Low
7	4.5	1.6	1.9	0.9	Low
8	4.5	1.4	1.8	1.3	Low

Let a be a numerical attribute, let p be the smallest value of a and let q be the largest value of a. During discretization, the domain $[p, q]$ of the attribute a is divided into the set of k intervals,

$$\{[a_{i_0}, a_{i_1}), [a_{i_1}, a_{i_2}), \ldots, [a_{i_{k-2}}, a_{i_{k-1}}), [a_{i_{k-1}}, a_{i_k}]\},$$

where $a_{i_0} = p$, $a_{i_k} = q$, and $a_{i_l} < a_{i_{l+1}}$ for $l = 0, 1, \ldots, k - 1$. The numbers a_{i_1}, $a_{i_2}, \ldots, a_{i_{k-1}}$ are called *cut-points*. Such intervals are denoted by

$$a_{i_0} \ldots a_{i_1}, a_{i_1} \ldots a_{i_2}, \ldots, a_{i_{k-2}} \ldots a_{i_{k-1}}, a_{i_{k-1}} \ldots a_{i_k}.$$

For any nonempty subset B of the set A of all attributes, an *indiscernibility* relation $IND(B)$ is defined, for any $x, y \in U$, in the following way

$$(x, y) \in IND(B) \text{ if and only if } a(x) = a(y) \text{ for any } a \in B, \tag{2.1}$$

where $a(x)$ denotes the value of the attribute $a \in A$ for the case $x \in U$. The relation $IND(B)$ is an equivalence relation. The equivalence classes of $IND(B)$ are denoted by $[x]_B$.

A partition on U is the set of all equivalence classes of $IND(B)$ and is denoted by B^*. Sets from $\{d\}^*$ are concepts. For example, for Table 2.1, if $B = \{Length\}$, $B^* = \{\{1, 3\}, \{2, 4, 7, 8\}, \{5, 6\}\}$ and $\{d\}^* = \{\{1, 2, 3\}, \{4, 5\}, \{6, 7, 8\}\}$. A data set is consistent if $A^* \leq \{d\}^*$, i.e., if for each set X from A^* there exists set Y from $\{d\}^*$ such that $X \subseteq Y$. For the data set from Table 2.1, each set from A^* is a singleton, so this data set is consistent.

We quote the Dominant Attribute discretization algorithm [12, 14]. The first task is sorting of the attribute domain. Potential cut-points are selected as means of two consecutive numbers from the sorted attribute domain. For example, for *Length* there are two potential cut-points: 4.4 and 4.6.

Let S be a subset of the set U. An entropy $H_S(a)$ of an attribute a, with the values $a_1, a_2, \ldots, a_n$ is defined as follows

$$-\sum_{i=1}^{n} p(a_i) \cdot \log p(a_i), \tag{2.2}$$

where $p(a_i)$ is a probability (relative frequency) of the value a_i of the attribute a, $a_1, a_2, \ldots, a_n$ are all values of a in the set S, logarithms are binary, and $i = 1, 2, \ldots, n$.

A conditional entropy for the decision d given an attribute a, denoted by $H_S(d|a)$ is

$$\sum_{i=1}^{n} p(a_i) \cdot \sum_{j=1}^{m} p(d_j|a_i) \cdot \log p(d_j|a_i), \tag{2.3}$$

$p(d_j|a_i)$ is the conditional probability of the value d_j of the decision d given the value a_i of a and $d_1, d_2, \ldots, d_m$ are all values of d in the set S. The main ideas of Dominant Attribute discretization algorithm are presented below.

Procedure Dominant Attribute

Input: a set U of cases, a set A of attributes, a set $\{d\}^*$ of concepts
Output: a discretized data set
$\{A\}^* := \{U\}$;
$\{B\}^* := \emptyset$;
while $\{A\}^* \not\leq \{d\}^*$ **do**
$\quad$ X := SelectBlock($\{A\}^*$);
$\quad$ a := BestAttribute(X);
$\quad$ c := BestCutPoint(X, a);
$\quad$ $\{S_1, S_2\}$:= Split(X, c);
$\quad$ $\{B\}^* := \{B\}^* \cup \{S_1, S_2\}$;
$\quad$ $\{A\}^* := \{B\}^*$;
end

In the Dominant Attribute discretization method, initially we need to select the dominant attribute, defined as an attribute with the smallest entropy $H_S(a)$. The process of computing of $H_U(Length)$ is illustrated in Fig. 2.1. In the Figs. 2.1 and 2.2 l stands for low, m for medium, and h for high, where {low, medium, high} is the domain of *Quality*.

$$H_U(Length) = \frac{1}{4}\left(-\frac{1}{2} \cdot \log\frac{1}{2}\right) 2 + \frac{1}{2}\left(\left(-\frac{1}{4} \cdot \log\frac{1}{4}\right) 2 - \frac{1}{2} \cdot \log\frac{1}{2}\right) + \frac{1}{4} \cdot 0 = 1.$$

Similarly, we compute remaining three conditional entropies: $H_U(Height) \approx 0.940$, $H_U(Width) \approx 1.439$ and $H_U(Weight) = 1.25$. We select *Height* since its entropy is the smallest.

Let a be an attribute and q be a cut-point of the attribute a that splits the set S into two subsets, S_1 and S_2. The conditional entropy $H_S(d|a, q)$ is defined as follows

$$\frac{|S_1|}{|S|}H_{S_1}(d|a) + \frac{|S_2|}{|S|}H_{S_2}(d|a), \tag{2.4}$$

Fig. 2.1 Computing conditional entropy $H_U(Length)$

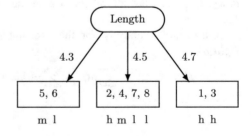

Fig. 2.2 Computing
conditional entropy
$H_U(Quality|Height, 1.5)$

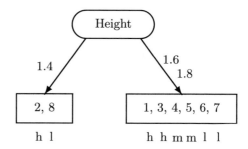

where $|X|$ denotes the cardinality of the set X. The cut-point q for which the conditional entropy $H_S(d|a, q)$ has the smallest value is selected as the best cut-point.

Thus, the next step is to find the best cutpoint for *Height*. There are two candidates: 1.5 and 1.7. We need to compute two conditional entropies, namely $H_U(Quality|Height, 1.5)$ and $H_U(Quality|Height, 1.7)$. Computing the former entropy is illustrated in Fig. 2.2. $H_U(Quality|Height, 1.5) = \frac{1}{4}(-\frac{1}{2} \cdot \log \frac{1}{2})2 + \frac{3}{4}(-\frac{1}{3} \cdot \log \frac{1}{3})3 \approx 1.439$.

Similarly, $H_U(Quality|Height, 1.7) \approx 1.201$. We select the cut-point with the smaller entropy, i.e., 1.7.

After any selection of a new cut-point we test whether discretization is completed, i.e., if the data set with discretized attributes is consistent. So far, we discretized only one attribute, *Height*. Remaining, not yet discretized attributes, have values $p..q$, where p is the smallest attribute value and q is the largest attribute value. The corresponding table is presented in Table 2.4. For Table 2.4, $A^* = \{\{1, 3, 4\}, \{2, 5, 6, 7, 8\}\}$, so $A^* \not\le \{d\}^*$, the data set from Table 2.4 needs more discretization. The cut-point 1.7 of *Height* splits the original data set from Table 2.1 into two smaller subtables, the former with the cases 1, 3 and 4 and the latter with the cases 2, 5, 6, 7 and 8. The former subtable is presented as Table 2.2, the latter as Table 2.3. The remaining computing is conducted by recursion. We find the best attribute for Table 2.2, then the best cut-point, and we check whether the currently dicretized data set is consistent. If not, we find the best attribute for Table 2.3, the best cut-point, and we check again whether the currently dicretized data set is consistent. If not, we compute new numerical data sets that result from current cut-points, and again, compute the best attribute, the best cut-point, and check whether the currently discretized data set is consistent.

Table 2.2 Numerical data set restricted to $\{1, 3, 4\}$

Case	Attributes				Decision
	Length	Height	Width	Weight	Quality
1	4.6..4.7	1.7..1.8	1.7..1.9	1.5..1.7	High
3	4.6..4.7	1.7..1.8	1.7..1.9	0.9..1.5	High
4	4.3..4.6	1.7..1.8	1.7..1.9	0.9..1.5	Medium

Table 2.3 A numerical data set restricted to {2, 5, 6, 7, 8}

	Attributes				Decision
Case	Length	Height	Width	Weight	Quality
2	4.3..4.6	1.4..1.7	1.7..1.9	0.9..1.5	High
5	4.3..4.6	1.4..1.7	1.7..1.9	1.5..1.7	Medium
6	4.3..4.6	1.4..1.7	1.7..1.9	0.9..1.5	Low
7	4.3..4.6	1.4..1.7	1.7..1.9	0.9..1.5	Low
8	4.3..4.6	1.4..1.7	1.7..1.9	0.9..1.5	Low

Table 2.4 Numerical data set with discretized *Height*

	Attributes				Decision
Case	Length	Height	Width	Weight	Quality
1	4.3..4.7	1.7..1.8	1.7..1.9	0.9..1.7	High
2	4.3..4.7	1.4..1.7	1.7..1.9	0.9..1.7	High
3	4.3..4.7	1.7..1.8	1.7..1.9	0.9..1.7	High
4	4.3..4.7	1.7..1.8	1.7..1.9	0.9..1.7	Medium
5	4.3..4.7	1.4..1.7	1.7..1.9	0.9..1.7	Medium
6	4.3..4.7	1.4..1.7	1.7..1.9	0.9..1.7	Low
7	4.3..4.7	1.4..1.7	1.7..1.9	0.9..1.7	Low
8	4.3..4.7	1.4..1.7	1.7..1.9	0.9..1.7	Low

Table 2.5 Completely discretized data set

	Attributes				Decision
Case	Length	Height	Width	Weight	Quality
1	4.6..4.7	1.7..1.8	1.7..1.9	1.5..1.7	High
2	4.3..4.6	1.4..1.5	1.7..1.9	0.9..1.1	High
3	4.6..4.7	1.7..1.8	1.7..1.9	1.1..1.5	High
4	4.3..4.6	1.7..1.8	1.7..1.9	1.1..1.5	Medium
5	4.3..4.6	1.5..1.7	1.7..1.9	1.5..1.7	Medium
6	4.3..4.6	1.5..1.7	1.7..1.9	1.1..1.5	Low
7	4.3..4.6	1.5..1.7	1.7..1.9	0.9..1.1	Low
8	4.3..4.6	1.4..1.5	1.7..1.9	1.1..1.5	Low

Our finally discretized data set, presented in Table 2.5, is consistent. The last step is an attempt to merge successive intervals. Such attempt is successful if a new discretized data set is still consistent. It is not difficult to see that all cut-points are necessary. For example, if we remove cut-point 4.6 for *Length*, cases 3 and 4 will be indistinguishable, while these two cases belong to different concept.

Note that the discretized data set, presented in Table 2.5, has four attributes and five cut-points. One of attributes, *Width*, is redundant. Thus, the reduced attribute set consists of three attributes: *Length*, *Height* and *Weight*.

2.3 Multiple Scanning Discretization

The Multiple Scanning discretization is also based on conditional entropy. However, this method uses a different strategy. The entire attribute set is scanned t times, where t is a parameter called the total number of scans. During every scan the best cut-point is computed for all attributes. The parameter t is provided by the user. If t is too small, the discretized data set is not consistent and Dominant Attribute method is used.

Procedure Multiple Scanning

Input: a set U of cases, a set A of attributes, a set $\{d\}^*$ of concepts, a number of scans t
Output: a discretized data set
$\{A\}^* := \{U\}$;
foreach *scan := 1 to t* **do**
    if $\{A\}^* \leq \{d\}^*$ **then**
        | break;
    end
    $C := \emptyset$;
    foreach $a \in A$ **do**
        cut_point := BestCutPointMS($\{A\}^*$, a);
        $C := C \cup \{$cut_point$\}$;
    end
    $\{A\}^* := $ Split($\{A\}^*$, C);
end

The main ideas of the Multiple Scanning algorithm are presented above. For details see [10, 11, 13, 14]. Obviously, for the same data set, data sets discretized by the Dominant Attribute and Multiple Scanning methods are, in general, different. We consider the Multiple Scanning method as auxiliary one.

Since during every scan all attributes are discretized, usually the discretized data set has all original attributes. The only chance to eliminate some attributes is during the last step, i.e., merging successive intervals.

2.4 Experiments

Our experiments were conducted on 15 data sets with numerical attributes presented in Table 2.6. All of these data sets are accessible at the University of California at Irvine *Machine Learning Repository*. First we discretized all data sets using Dominant Attribute method. Then we identified data sets with single interval attributes, i.e., discretized values in which the entire domain of a numerical attribute was mapped into a single interval. For two data sets, *Abalone* and *Iris*, no single interval attributes were discovered, so these two data sets were removed from further experiments. For

Table 2.6 Data sets

Data set	Cases	Number of attributes	Concepts
Abalone	4,177	8	29
Australian	690	14	2
Bankruptcy	66	5	2
Bupa	345	6	2
Connectionist Bench	208	60	2
Echocardiogram	74	7	2
E coli	336	8	8
Glass	214	9	6
Image Segmentation	210	19	7
Ionoshere	351	34	2
Iris	150	4	3
Pima	768	8	2
Wave	512	21	3
Wine Recognition	178	13	3
Yeast	1,484	8	9

any data set with single interval attributes, a new data set with numerical attributes was created by removing single interval attributes from the original, numerical data set. Such data sets are called *reduced*.

Reduced data sets are presented in Table 2.7. As it was observed in Sect. 2.3, Multiple Scanning discretization seldom produces reduced data sets. In our experiments, Multiple Scanning produced reduced data sets only for three original data sets: *Connectionist Bench, Image Segmentation* and *Ionosphere*, so during analysis of experimental results Multiple Scanning was ignored.

All numerical data sets, original and reduced, were subjected to single ten-fold cross-validation using C4.5 system [18]. The system C4.5 was selected as well-known classifier. Note that C4.5 has an internal discretization method similar to Dominant Discretization algorithm. Error rates computed by C4.5 and ten-fold cross-validation are presented in Table 2.8. For our results we used the Wilcoxon matched-pairs two-tailed test with 5% significance level. We conclude that the error rate is significantly larger for reduced data sets. An additional argument for better quality of original data sets was reported in [10, 11, 13, 14]. Multiple Scanning discretization method was better than other discretization methods since in the majority of data sets discretized by Multiple Scanning all discretized attributes have more than one interval.

Table 2.7   Reduced data sets

Data set	Number of single-interval attributes for data sets reduced by	
	Dominant Attribute	Multiple Scanning
Abalone	0	0
Australian	9	0
Bankruptcy	3	0
Bupa	2	0
Connectionist Bench	57	8
Echocardiogram	3	0
E coli	3	0
Glass	3	0
Image Segmentation	15	1
Ionoshere	29	1
Iris	0	0
Pima	2	0
Wave	15	0
Wine Recognition	9	0
Yeast	3	0

Table 2.8   C4.5 error rate, data sets reduced by dominant attribute

Name	Original data set	Reduced data set
Australian	16.09	15.36
Bankruptcy	6.06	12.12
Bupa	35.36	35.94
Connectionist Bench	25.96	25.96
Echocardiogram	28.38	44.59
E coli	17.86	19.35
Glass	33.18	33.18
Image Segmentation	12.38	10.48
Ionoshere	10.54	11.97
Pima	25.13	26.95
Wave	26.37	32.42
Wine Recognition	8.99	8.99
Yeast	44.41	48.45

We compared complexity of decision trees generated by C4.5 from original and reduced data sets as well. Results are presented in Tables 2.9 and 2.10. The tree depth is the number of edges on the longest path between the root and any leaf. The tree size is the total number of nodes of the tree. These numbers are reported by the C4.5

Table 2.9 C4.5 tree depth, data sets reduced by dominant attribute

Name	Original data set	Reduced data set
Australian	11	5
Bankruptcy	1	3
Bupa	8	8
Connectionist Bench	7	2
Echocardiogram	4	3
E coli	9	7
Glass	8	8
Image Segmentation	8	7
Ionoshere	11	10
Pima	9	7
Wave	10	9
Wine Recognition	3	3
Yeast	22	20

Table 2.10 C4.5 tree size, data sets reduced by dominant attribute

Name	Original data set	Reduced data set
Australian	63	11
Bankruptcy	3	7
Bupa	51	33
Connectionist Bench	35	5
Echocardiogram	9	7
E coli	43	37
Glass	45	45
Image Segmentation	25	25
Ionoshere	35	25
Pima	43	35
Wave	85	63
Wine Recognition	9	9
Yeast	371	411

system. Using the same Wilcoxon test we conclude that decision trees generated from reduced trees are simpler. The depth of decision trees is smaller for reduced data sets with significance level 5%. On the other hand, the size of decision trees is smaller with significance level 10%.

2.5 Conclusions

Let us recall that our main objective is to compare quality of numerical data sets, original and reduced, using the C4.5 decision tree generation system. Our experiments on 15 numerical data sets show that for a Dominant Attribute discretization method the error rate computed by C4.5 and ten-fold cross-validation is significantly larger (5% significance level, two-tailed test) for the reduced data sets than for the original data sets. However, decision trees generated from the reduced data sets are significantly simpler than the decision trees generated from the original data sets. Thus, if our top priority is accuracy, the original data sets should be used. On the other hand, if all what we want is simplicity we should use reduced data sets.

References

1. Blajdo, P., Grzymala-Busse, J.W., Hippe, Z.S., Knap, M., Mroczek, T. L., Piatek : A comparison of six approaches to discretization—a rough set perspective. In: Proceedings of the Rough Sets and Knowledge Technology Conference, pp. 31–38 (2008)
2. Bruni, R., Bianchi, G.: Effective classification using a small training set based on discretization and statistical analysis. IEEE Trans. Knowl. Data Eng. **27**(9), 2349–2361 (2015)
3. Chmielewski, M.R., Grzymala-Busse, J.W.: Global discretization of continuous attributes as preprocessing for machine learning. Int. J. Approx. Reason. **15**(4), 319–331 (1996)
4. Clarke, E.J., Barton, B.A.: Entropy and MDL discretization of continuous variables for bayesian belief networks. Int. J. Intell. Syst. **15**, 61–92 (2000)
5. de Sa, C.R., Soares, C., Knobbe, A.: Entropy-based discretization methods for ranking data. Inf. Sci. **329**, 921–936 (2016)
6. Elomaa, T., Rousu, J.: General and efficient multisplitting of numerical attributes. Mach. Learn. **36**, 201–244 (1999)
7. Fayyad, U.M., Irani, K.B.: Multi-interval discretization of continuous-valued attributes for classification learning. In: Proceedings of the Thirteenth International Conference on Artificial Intelligence, pp. 1022–1027 (1993)
8. Fayyad, U.M., Irani, K.B.: On the handling of continuous-valued attributes in decision tree generation. Mach. Learn. **8**, 87–102 (1992)
9. Garcia, S., Luengo, J., Sáez, J.A., Lopez, V., Herrera, F.: A survey of discretization techniques: taxonomy and empirical analysis in supervised learning. IEEE Trans. Knowl. Data Eng. **25**(4), 734–750 (2013)
10. Grzymala-Busse, J.W., Mroczek, T.: A comparison of two approaches to discretization: multiple scanning and C4.5. In: Proceedings of the 6-th International Conference on Pattern Recognition and Machine Learning, pp. 44–53 (2015)
11. Grzymala-Busse, J.W.: A multiple scanning strategy for entropy based discretization. In: Proceedings of the 18th International Symposium on Methodologies for Intelligent Systems, pp. 25–34 (2009)
12. Grzymala-Busse, J.W.: Discretization of numerical attributes. In: Kloesgen, W., Zytkow, J. (eds.) Handbook of Data Mining and Knowledge Discovery, pp. 218–225. Oxford University Press, New York, NY (2002)
13. Grzymala-Busse, J.W.: Discretization based on entropy and multiple scanning. Entropy **15**, 1486–1502 (2013)
14. Grzymala-Busse, J.W., Mroczek, T.: A comparison of four approaches to discretization based on entropy. Entropy **18**, 1–11 (2016)

15. Jiang, F., Sui, Y.: A novel approach for discretization of continuous attributes in rough set theory. Knowl. Based Syst. **73**, 324–334 (2015)
16. Kohavi, R., Sahami, M.: Error-based and entropy-based discretization of continuous features. In: Proceedings of the Second International Conference on Knowledge Discovery and Data Mining, pp. 114–119 (1996)
17. Nguyen, H.S., Nguyen, S.H.: Discretization methods in data mining. In: Polkowski, L., Skowron, A. (eds.) Rough Sets in Knowledge Discovery 1: Methodology and Applications, pp. 451–482. Physica-Verlag, Heidelberg (1998)
18. Quinlan, J.R.: C4.5: Programs for Machine Learning. Morgan Kaufmann Publishers, San Mateo (1993)
19. Rahman, M.D., Islam, M.Z.: Discretization of continuous attributes through low frequency numerical values and attribute interdependency. Expert Syst. Appl. **45**, 410–423 (2016)
20. Sang, Y., Qi, H., Li, K., Jin, Y., Yan, D., Gao, S.: An effective discretization method for disposing high-dimensional data. Inf. Sci. **270**, 73–91 (2014)
21. Stefanowski, J.: Handling continuous attributes in discovery of strong decision 0 rules. In: Proceedings of the First Conference on Rough Sets and Current Trends in Computing, pp. 394–401 (1998)
22. Stefanowski, J.: Algorithms of Decision Rule Induction in Data Mining. Poznań University of Technology Press, Poznań, Poland (2001)

Chapter 3
Improving Bagging Ensembles for Class Imbalanced Data by Active Learning

Jerzy Błaszczyński and Jerzy Stefanowski

Abstract Extensions of under-sampling bagging ensemble classifiers for class imbalanced data are considered. We propose a two phase approach, called Actively Balanced Bagging, which aims to improve recognition of minority and majority classes with respect to so far proposed extensions of bagging. Its key idea consists in additional improving of an under-sampling bagging classifier (learned in the first phase) by updating in the second phase the bootstrap samples with a limited number of examples selected according to an active learning strategy. The results of an experimental evaluation of Actively Balanced Bagging show that this approach improves predictions of the two different baseline variants of under-sampling bagging. The other experiments demonstrate the differentiated influence of four active selection strategies on the final results and the role of tuning main parameters of the ensemble.

Keywords Class imbalance · Active learning · Bagging ensembles · Under-sampling

3.1 Introduction

Supervised learning of classifiers from class imbalanced data is still a challenging task in machine learning and pattern recognition. Class imbalanced data sets are characterized by uneven cardinalities of classes. One of the classes, usually called a *minority class* and being of key importance in a given problem, contains significantly less learning examples than other majority classes. Class imbalance occurs in many real-world application fields, such as: medical data analysis, fraud detection,

J. Błaszczyński (✉) · J. Stefanowski
Institute of Computing Science, Poznań University of Technology, Piotrowo 2,
60-965 Poznań, Poland
e-mail: jerzy.blaszczynski@cs.put.poznan.pl

J. Stefanowski
e-mail: jerzy.stefanowski@cs.put.poznan.pl

© Springer International Publishing AG 2018
U. Stańczyk et al. (eds.), *Advances in Feature Selection for Data and Pattern Recognition*, Intelligent Systems Reference Library 138,
https://doi.org/10.1007/978-3-319-67588-6_3

technical diagnostics, image recognition or text categorization. More information about them can be found in [24, 54, 57].

If imbalance in the class distribution is severe, i.e., some classes are strongly under-represented, standard learning algorithms do not work properly. Constructed classifiers may have difficulties, in some cases they may be even completely unable, to classify correctly new instances from the minority class. Such behaviour have been demonstrated in several experimental studies such as [23, 29, 39].

Several approaches to improve classifiers for imbalanced data have been proposed [11, 25, 54]. They are usually categorized as: classifier-independent *pre-processing* methods or *modifications of algorithms* for learning particular classifiers. Methods within the first category try to re-balance the class distribution inside the training data by either adding examples to the minority class (*over-sampling*) or by removing examples from the majority class (*under-sampling*). The other category of algorithm level methods involves specific solutions dedicated to improving a given classifier. Specialized ensembles are among the most effective methods within this category [40].

Besides developing new approaches, some researchers attempt to better understand the nature of the imbalance data and key properties of its underlying distribution, which makes the class imbalanced problem difficult to be handled. They have shown, that so called, *data difficulty factors* hinder the learning performance of classification algorithms [22, 29, 41, 53]. The data difficulty factors are related to characteristics of class distribution, such as decomposition of the class into rare sub-concepts, overlapping between classes or presence of rare minority examples inside the majority class regions. It has been shown that some classifiers and data pre-processing methods are more sensitive to some of these difficulty factors than others [45, 52].

Napierała et al. have shown that several data difficulty factors may be approximated by analyzing the content of the minority example neighbourhood and modeling several types of data difficulties [45]. Moreover, in our previous works [6, 7] it has been observed, that neighbourhood analysis of minority examples may be used to change the distribution of examples in bootstrap samples of ensembles. The resulting extensions of bagging ensembles are cable to significantly improve classification performance on imbalanced data sets. The interest in studying extensions of bagging ensembles is justified by recent promising experimental results of their comparison against other classifiers dedicated to imbalanced data [6, 7, 33, 37].

Nevertheless, a research question could be posed, whether it is still possible to improve performance of these ensembles. In experimental studies, such as [5, 7, 37], it has been shown that the best proposals of extending bagging by under-sampling may improve the minority class recognition at the cost of strong decrease of recognition of majority class examples. We claim that it would be more beneficial to construct an ensemble providing a good trade-off between performance in both classes instead.

To address this research question we plan to consider a quite different perspective of extending bagging ensembles than it is present in the current solutions, which mainly modify the generation of bootstrap samples. Here, we propose instead a two phase approach. First, we start with construction of an ensemble classifier according to one of under-sampling extensions designed for imbalanced data. Then, we modify bootstrap samples, constructed in the first phase, by adding a limited number of learning examples, which are important to improve performance in both classes. To perform this kind of an example selection we follow inspiration coming from the *active learning* paradigm [2]. This type of learning is commonly used in the semi-supervised framework to update the classifier learned on labeled part of data by selecting the most informative examples from the pool of unlabeled ones. Active learning can also be considered to filter examples from the fully labeled data sets [2]. In this way, active strategies have been already applied to imbalanced learning, although these attempts are still quite limited, see Sect. 3.3.

In this chapter we will discuss a new perspective of using active learning to select examples while extending under-sampling bagging ensembles. We call the proposed extension *Actively Balanced Bagging* (ABBag) [8].

In the first phase of the approach, ABBag is constructed with previously proposed algorithms for generating under-sampling bagging extensions for imbalanced data. In the experiments we will consider two different efficient algorithms, namely Exactly Balanced Bagging (EBBag) [13], and Neighbourhood Balanced Bagging (NBBag) [7]. Then, in the second phase the ensemble classifier will be integrated with the active selection of examples. In ABBag this strategy exploits the decision margin of component classifiers in ensemble votes, which is more typical for the active learning. Since, contrary to typical active learning setting, we are dealing with fully labeled data, errors of component classifiers in ensemble will be taken into account as well. Moreover, following experiences from the previous research on data difficulty factors, the neighbourhood analysis of the examples will be also explored. All these elements could be integrated in different way, which leads us to consider four versions of the active selection strategies.

The preliminary idea of ABBag was presented in our earlier conference paper [8]. In this chapter, we discuss it in more details and put in the context of other related approaches. The next contributions include carrying out a comprehensive experimental study of ABBag usefulness and its comparison against the baseline versions of under-sampling extensions of bagging for imbalanced data. Furthermore, we experimentally study properties of ABBag with respect to different active selection strategies and tuning its parameters.

The chapter is organized as follows. The next section summarizes the most related research on improving classifiers learned from class imbalanced data. The following Sect. 3.3, discusses use of active learning in class imbalanced problems. Ensembles specialized for imbalanced data are described in Sect. 3.4. The *Actively Balanced Bagging* (ABBag) is presented in Sect. 3.5. The results of experimental evaluation of ABBag are given in Sect. 3.6. The final section draws conclusions.

3.2   Improving Classifiers Learned from Imbalanced Data

In this section we discuss concepts, which are the most related to our proposal. For more comprehensive reviews of specialized methods for class imbalanced data the reader could refer to, e.g., [11, 25, 54]. In this chapter, we consider only a typical binary definition of the class imbalance problem, where the selected minority class is distinguished from a single majority class. This formulation is justified by focusing our interest on the most important class and its real-life semantics [24, 54]. Recently some researchers study more complex scenarios with multiple minority classes, see e.g., reviews in [49, 56].

3.2.1   Nature of Imbalanced Data

In some problems characterized by high class imbalance, standard classifiers have been found to be accurate, see e.g., [3]. In particular, it has been found that, when there is a good separation (e.g., linear) between classes, the minority class may be sufficiently recognized regardless of the high *global imbalance ratio* between classes [46]. The global imbalance ratio is usually expressed as either $N_{min}:N_{maj}$ or $\frac{N_{min}}{N}$, where N_{maj}, N_{min}, N are the number of majority, minority, and total number of examples in the data set, respectively.

Some researches have shown that the global class imbalance ratio is not nec-essarily the only, or even the main, problem causing the decrease of classification performance [22, 32, 41, 42, 47, 51]. These researchers have drawn attention to other characteristics of example distributions in the attribute space called *data com-plexity* or *data difficulty factors*. Although these factors should affect learning also in balanced domains, when they occur *together* with class imbalance, then the dete-rioration of classification performance is amplified and affects mostly the minority class. The main data difficulty factors are: decomposition the minority class into rare sub-concepts, overlapping between classes, and presence of outliers, rare instances, or noise.

The influence of *class decomposition* has been noticed by Japkowicz et al. [29, 32]. They experimental showed that the degradation of classification performance has resulted from decomposing the minority class into many sub-parts containing very few examples, rather than from changing the global imbalance ratio. They have also argued that the minority class often does not form a compact homogeneous distribu-tion of the single concept, but is scattered into many smaller sub-clusters surrounded by majority examples. Such sub-clusters are referred to *small disjuncts*, which are harder to learn and cause more classification errors than larger sub-concepts.

Other factors related to the class distribution are linked to high *overlapping* between regions of minority and majority class examples in the attribute space. This difficulty factor has already been recognized as particularly important for stan-dard, balanced, classification problems, however, its role is more influential for the minority class. For instance, a series of experimental studies of popular classifiers on synthetic data have pointed out that increasing overlapping has been more influential than changing the class imbalance ratio [22, 47]. The authors of [22] have also shown that the local imbalance ratio inside the overlapping region is more influential than the global ratio.

Yet another data difficulty factor which causes degradation of classifier performance on imbalanced data is the presence of minority examples inside distributions of the majority class. Experiments presented in a study by Napierała et al. [42] have shown that single minority examples located inside the majority class regions cannot be always treated as noise since their proper treatment by informed pre-processing may lead to improvement of classifiers. In more recent papers [45, 46], they have distinguished between safe and unsafe examples. *Safe examples* are the ones located in homogeneous regions populated by examples from one class only. Other examples are *unsafe* and they are more difficult to learn from. Unsafe examples were further categorized into *borderline* (placed close to the decision boundary between classes), *rare cases* (isolated groups of few examples located deeper inside the opposite class), and *outliers*.

The same authors have introduced an approach [45] to automatically identify the aforementioned types of examples in real world data sets by analyzing class labels of examples in the local neighbourhood of a considered example. Depending on the number of examples from the majority class in the local neighbourhood of the given minority example, we can evaluate whether this example could be safe or unsafe (difficult) to be learned.

3.2.2 Evaluation of Classifiers on Imbalanced Data

Class imbalance constitutes difficulty not only during construction of a classifier but also when one evaluates classifier performance. The overall classification accuracy is not a good criterion characterizing classifier performance, in this type of problem, as it is dominated by the better recognition of the majority class which compensates the lower recognition of the minority class [30, 34]. Therefore, other measures defined for binary classification are considered, where typically the class label of the minority class is called positive and the class label of the majority class is negative. The performance of the classifiers is presented in a binary confusion matrix as in Table 3.1.

Table 3.1 Confusion matrix for the classifier evaluation

	Predicted Positive	Predicted Negative
True Positive	TP	FN
True Negative	FP	TN

One may construct basic metrics concerning recognition of the positive (minority) and negative (majority) classes from the confusion matrix:

$$Sensitivity \ = \ Recall = \frac{TP}{TP + FN}, \tag{3.1}$$

$$Specificity = \frac{TN}{FP + TN}, \tag{3.2}$$

$$Precision = \frac{TP}{TP + FP}. \tag{3.3}$$

Some more elaborated measures may also be considered (please see e.g., overviews of the measures presented in [25, 30]).

As the class imbalance task invariably involves a trade off between false positives FP and false negatives FN, to control both, some single-class measures are commonly considered in pairs, e.g., *Sensitivity* and *Specificity* or *Sensitivity* and *Precision*. These single-class measures are often aggregated to form further measures [28, 30]. The two admittedly most popular aggregations are the following:

$$\text{G-mean} = \sqrt{Sensitivity \cdot Specificity}, \tag{3.4}$$

$$\text{F-measure} = \frac{(1 + \beta) \cdot Precision \cdot Recall}{\beta \cdot Precision + Recall}. \tag{3.5}$$

The F-measure combines *Recall* (*Sensitivity*) and *Precision* as a weighted harmonic mean, with the β parameter ($\beta > 0$) as the relative weight. It is most commonly used with $\beta = 1$. This measure is exclusively concerned with the positive (minority) class. Following inspiration from its original use in the information retrieval context, *Recall* is a recognition rate of examples originally from the positive class while precision assesses to what extent the classifier was correct in classifying examples as positive that were actually positive. Unfortunately it is dependent to the class imbalance ratio.

The most popular alternative, G-mean, was introduced in [34] as a geometric mean of *Sensitivity* and *Specificity*. It has a straightforward interpretation since it takes into account the relative balance of the classifier performance in both positive class and negative class. An important, useful property of the G-mean is that it is independent of the distribution of examples between classes. As both classes have equal importance in this formula, various further modifications to prioritize the positive class, like the adjusted geometric mean, have been postulated (for their overview see [30]).

The aforementioned measures are based on single point evaluation of classifiers with purely deterministic predictions. In case of scoring classifiers, several authors use the *ROC (Receiver Operating Characteristics) curve* analysis. The quality of the classifier performance is reflected by the area under a ROC curve (so called AUC measure). Alternative proposals include Precision Recall Curves or other special cost curves (see their review in [25, 30]).

3.2.3 Main Approaches to Improve Classifiers for Imbalanced Data

The class imbalance problem has received growing research interest in the last decade and several specialized methods have been proposed. Please see [11, 24, 25, 54] for reviews of these methods, which are usually categorized in two groups:

- Classifier-independent methods that rely on transforming the original data to change the distribution of classes, e.g., by re-sampling.
- Modifications of either a learning phase of the algorithm, classification strategies, construction of specialized ensembles or adaptation of cost sensitive learning.

The first group include data *pre-processing methods*. The simplest data pre-processing (re-sampling) techniques are: *random over-sampling*, which replicates examples from the minority class, and *random under-sampling*, which randomly eliminates examples from the majority classes until a required degree of balance between classes is reached. *Focused* (also called *informed*) *methods* attempt to take into account the internal characteristics of regions around minority class examples. Popular examples of such methods are: OSS [34], NCR [38], SMOTE [14] and some extensions these methods: see e.g., [11]. Moreover, some hybrid methods integrating over-sampling of selected minority class examples with removing the most harmful majority class examples have been also proposed, see e.g., SPIDER [42, 51].

The other group includes many quite specialized methods based on different principles. For instance, some authors changed search strategies, evaluation criteria or parameters in the internal optimization of the learning algorithm - see e.g., extensions of induction of decision tress with the Hellinger distance or the asymmetric entropy [16], or reformulation of the optimization task in generalized versions of SVM [24]. The final prediction technique can be also revised, for instance authors of [23] have modified conflict strategies with rules to give more chance for minority rules. Finally, other researchers adapt the imbalance problem to cost sensitive learning. For a more comprehensive discussion of various methods for modifying algorithm refer to [24, 25].

The neighbourhood analysis has been also used to modify pre-processing methods, see extensions of SMOTE or over-sampling [9], rule induction algorithm BARCID [43] or ensembles [7].

3.3 Active Learning

Active learning is a research paradigm in which the learning algorithm is able to select examples used for its training. Traditionally, this methodology has been applied interactively with respect to unlabeled data. Please refer to the following survey for a review of different active strategies in semi-supervised learning perspective [50]. The goal of active learning, in this traditional view, is to minimize costs, i.e., time, effort, and other resources related to inquiring for class labels needed to update / train classifier.

Nevertheless, active learning may also be applied when class labels are known. The goal is then to select the best examples for training. Such definition of a goal is particularly appealing to learning from imbalanced data, where one has a limited number of examples from the minority class and too high number of examples from the majority class. Thus, a specialized selection of the best examples from majority class may be solved by an active approach. The recent survey [11] clearly demonstrates an increasing interest in applying active learning strategies to imbalanced data.

In pool-based active learning, which is of our interest here, one starts with a given pool (i.e., a set) of examples. The classifier is first built on examples from the pool. Then one queries these examples outside the pool that are considered to be potentially the most useful to update the classifier. The main problem for active learning strategy is computing the *utility* of examples outside the pool. Various definitions of *utility* have already been considered in the literature [48]. Uncertainty sampling and query-by-committee are the two most frequently applied solutions.

Uncertainty sampling queries examples one by one, at each step, selecting the one for which the current classifier is the most uncertain while predicting the class. For imbalanced data, it has been applied together with support vector machines (SVM) classifiers. In such a case, uncertainty is defined simply as a distance to the margin of SVM classifier. Ertkin et al. have started this direction and proposed an active learning with early stopping with online SVM [17]. These authors have also considered an adaptive over-sampling algorithm VIRTUAL, which is able to generate synthetic minority class examples [18]. Another method, also based on uncertainty sampling, has been proposed by Zięba and Tomczak. This proposal consists in an ensemble of boosted SVMs. Base SVM classifiers are trained iteratively on examples identified by an extended margin created in previous iteration [60].

Query by committee (QBC) [1], on the other hand, queries examples, again, one by one, at each step selecting the one for which a committee of classifiers disagrees the most. The committee may be formed in different ways, e.g., by sampling hypotheses from the version space, or through bagging ensembles [48]. Yang and Ma have proposed a random subspace ensemble for class imbalance problem that makes use of QBC [59]. More precisely, they calculate the margin between two highest membership probabilities for the two most likely classes predicted by the ensemble.

The idea of QBC have also been considered by Napierala and Stefanowski in argument based rule learning for imbalanced data, where it selects the most difficult examples to be annotated [44]. The annotated examples are further handled in generalized rule induction. The experimental results of [44] show that this approach significantly improved recognition of both classes (minority and majority) in particular for rare cases and outliers.

Other strategies to compute utility of examples were also considered. For example, Certainty-Based Active Learning (CBAL) algorithm has been proposed for imbalanced data [20]. In CBAL, neighbourhoods are explored incrementally to select examples for training. The importance of an example is measured within the neighbourhood. In this way, certain, and uncertain areas are constructed and then used to select the best example. A hybrid algorithm has been also proposed for on-line active learning with imbalanced classes [19]. This algorithm switches between different selection strategies: uncertainty, density, certainty, and sparsity.

All of the algorithms mentioned this far query only one example at time. However, querying more examples, in a batch, may reduce the labeling effort and computation time. One does not need to rebuild the classifier after each query. On the other hand, batch querying introduces additional challenges, like diversity of batch [10]. To best of our knowledge no batch querying active learning algorithm has been proposed for class imbalanced data.

3.4 Ensembles Specialized for Imbalanced Data

Specialized extensions of ensembles of classifiers are among the most efficient currently known approaches to improve recognition of the minority class in imbalanced setting. These extensions may be categorized differently. The taxonomy proposed by Galar et al. in [21] distinguishes between *cost-sensitive* approaches vs. integrations with *data pre-processing*. The first group covers mainly cost-minimizing techniques combined with boosting ensembles, e.g., AdaCost, AdaC or RareBoost. The second group of approaches is divided into three sub-categories: Boosting-based, Bagging-based or Hybrid depending on the type of ensemble technique which is integrated into the schema for learning component classifiers and their aggregation. Liu et al. categorize the ensembles for class imbalance into bagging-like, boosting-based methods or hybrid ensembles depending on their relation to standard approaches [40].

Since the most of related works [4, 6, 21, 33, 36] indicate superior performance of bagging extensions versus the other types ensembles (e.g., boosting), we focus our consideration, in this study, on bagging ensembles.

Bagging [12] classifier, proposed by Breiman, is an ensemble of m_{bag} base (component) classifiers constructed by the same algorithm from m_{bag} *bootstrap samples* drawn from the original training set. The predictions of component classifiers are combined to form the final decision as the result of the equal weight majority voting. The key concept in bagging is *bootstrap* aggregation, where the training set, called a bootstrap, for each component classifier is constructed by random uniform sampling examples from the original training set. Usually the size of each bootstrap is equal to the size of the original training set and examples are drawn with replacement.

Algorithm 3.1: Bagging scheme

Input : LS training set; TS testing set; CLA learning algorithm;
           m_{bag} number of bootstrap samples;
Output: C^* final classifier

1 *Learning phase*;
2 **for** $i := 1$ *to* m_{bag} **do**
3 | $S_i :=$ bootstrap sample {sample examples with replacement} ;
4 | $C_i :=$ CLA (S_i) {generate a component classifier} ;
5 **end**

6 *Classification phase*;
7 **foreach** y *in* TS **do**
8 | $C^*(x) :=$ combination of predictions $C_i(x)$, where $i = 1, \ldots, m_{bag}$
9 | {prediction for example x results from majority voting C_i} ;
10 **end**

Since bootstrap sampling, in the standard version, will not change drastically the class distribution in constructed bootstrap samples, they will be still biased toward the majority class. Thus, most of proposals to adapt/extend bagging to class imbalance overcome this drawback by applying pre-processing techniques, which change the balance between classes in each bootstrap sample. Usually they construct bootstrap samples with the same, or similar, cardinalities of both minority and majority classes.

In *under-sampling* bagging approaches the number of the majority class examples in each bootstrap sample is randomly reduced to the cardinality of the minority class (N_{min}). In the simplest proposal, called *Exactly Balanced Bagging* (EBBag), while constructing training bootstrap sample, the entire minority class is copied and combined with randomly chosen subsets of the majority class to exactly balance cardinalities between classes [13].

While such under-sampling bagging strategies seem to be intuitive and work efficiently in some studies, Hido et al. [26] observed that they do not truly reflect the philosophy of bagging and could be still improved. In the original bagging the class distribution of each sampled subset varies according to the binomial distribution while in the above under-sampling bagging strategy each subset has the same class ratio as the desired balanced distribution. In *Roughly Balanced Bagging* (RBBag) the numbers of instances for both classes are determined in a different way by equalizing the sampling probability for each class. The number of minority examples (S_{min}) in each bootstrap is set to the size of the minority class N_{min} in the original data. In contrast, the number of majority examples is decided probabilistically according to the negative binomial distribution, whose parameters are the number of minority examples (N_{min}) and the probability of success equal to 0.5. In this approach only the size of the majority examples (S_{maj}) varies, and the number of examples in the minority class is kept constant since it is small. Finally, component classifiers are induced by the same learning algorithm from each i-th bootstrap sample ($S_{min}^i \cup S_{maj}^i$) and their predictions form the final decision with the equal weight majority voting.

Yet another approach has been considered in *Neighbourhood Balanced Bagging* (NBBag), which is based on different principles than aforementioned under-sampling bagging ensembles. Instead of using uniform sampling, in NBBag, probability of an example being drawn into the bootstrap is modified according to the class distribution in his neighbourhood [7]. NBBag shifts sampling probability toward unsafe examples located in difficult to learn sub-regions of the minority class. To perform this type of sampling weights are assigned to the learning examples. The weight of minority example is defined as: $w = 0.5 \times \left(\frac{(N'_{min})^{\psi}}{k} + 1 \right)$ where N'_{min} is the number of majority examples among k nearest neighbours of the example, and ψ is a scaling factor. Setting $\psi = 1$ causes a linear amplification of an example weight together with an increase of unsafeness, and setting ψ to values higher than 1 results in an exponential amplification. Each majority example is assigned a constant weight $w = 0.5 \times \frac{N_{maj}}{N_{min}}$, where N_{maj} is the number of majority class examples in the training set and N_{min} is the number of minority class examples in the training set. Then sampling is performed according to the distribution of weights. In this sampling, probability of an example being drawn to the bootstrap sample is reflected by its weight.

Another way to overcome class imbalance in a bootstrap sample consists in performing *over-sampling* of the minority class before training a component classifier. In this way, the number of minority examples is increased in each sample (e.g., by a random replication), while the majority class is not reduced as in under-sampling bagging. This idea was realized in many ways as authors considered several kinds of integrations with different over-sampling techniques. Some of these ways are also focused on increasing diversity of bootstrap samples. *OverBagging* is the simplest version which applies a simplest random over-sampling to transform each training bootstrap sample. S_{maj} of minority class examples is sampled with replacement to exactly balance the cardinality of the minority and the majority class in each sample. Majority examples are sampled with replacement as in the original bagging. An over-sampling variant of Neighbourhood Balanced Bagging (NBBag) has also been proposed [7]. In this variant, weights of examples are calculated in the same way as for under-sampling NBBag.

Finally, Lango et al. have proposed to integrate a random selection of attributes (following inspirations of [27, 35]) into Roughly Balanced Bagging [36]. Then the same authors have introduced a generalization of RBBag for multiple imbalanced classes, which exploits the multinomial distribution to estimate cardinalities of class examples in bootstrap samples [37].

3.5 Active Selection of Examples in Under-Sampling Bagging

Although a number of interesting under-sampling extensions of bagging ensembles, for class imbalanced data, have been recently proposed, the prediction improvement brought by these extensions may come with a decrease of recognition of majority class examples. Thus, we identify a need for better learning a trade-of between performance in both classes. Then an open research problem is how to achieve this balance of performance in both classes.

In this study we want to take a different perspective than in current proposals. These proposals mainly make use of various modifications of sampling examples to bootstraps (usually oriented toward balancing bootstrap) and then construct an ensemble, in a standard way, by learning component classifiers in one step and aggregating their predictions according to the majority voting (please see details of bagging and its variants in [12, 35]).

More precisely, we want to consider another hypothesis: given an already good technique of constructing under-sampling bagging, could one perform an additional step of updating its bootstraps by selecting a limited number of remaining learning examples, which could be useful for improving the trade-off between recognizing minority and majority classes.

Our proposed approach, called *Actively Balanced Bagging* (ABBag) [8], is composed of two phases. The first phase consists in learning an ensemble classifier by one of approaches for constructing under-sampling extensions of bagging. Although one can choose any good performing extension, we will further consider quite simple, yet effective one: Exactly Balanced Bagging (EBBag) [21], and more complex one based on other principles: Neighbourhood Balanced Bagging NBBag [7]. The current literature, such as [7, 33, 36], contains several experimental studies, which have clearly demonstrated that both these ensembles, and Roughly Balanced Bagging [26], are the best ensembles and they also out-performed single classifiers for difficult imbalanced data. Furthermore their modifications of sampling examples are based on completely different principles which is an additional argument to better verify the usefulness of the proposed active selection strategies in ABBag. For more information on constructing the EBBag and NBBag ensembles the reader may refer to Sect. 3.4.

The second phase includes an *active selection of examples*. It includes:

1. An iterative modification of bootstrap samples, constructed in the first phase, by adding selected examples from the training set;
2. Re-learning of component classifiers on modified bootstraps. The examples selected in (1) are added to bootstraps in *batches*, i.e., small portions of learning examples.

The proposed active selection of examples can be seen as a variant of Query-by-committee (QBC) approach [1]. As discussed in the previous sections QBC uses a decision margin, or simply a measure of disagreement between members of the committee, to select the examples. Although QBC has been already successfully applied in active learning of ensemble classifiers in [10]. It has been observed that QBC does not take into account global (i.e., concerning the whole training set) properties of examples distribution, and in result, it can focus too much on selecting outliers and sparse regions [10]. Therefore, we need to adapt this strategy for imbalanced data, which are commonly affected by data difficulty factors.

Furthermore, one should remember that selecting one single example at a time is a standard strategy in active learning [50]. Contrary, in ABBag we promote, in each iteration, to select small batches of examples instead of single example. We motivate the batch selection by a potential reduction of computation time, as well as, an increase of diversity of examples in the batch. As it was observed in [10], a greedy selection of single example with respect to a single criterion, typical for active strategy, where highest utility/uncertainty measure is taken into account [50] does not provide desired diversity. In our view, giving chance for random drawing also some slightly sub-optimal examples besides the best ones may result in a higher diversity of new bootstraps and increased diversity of re-learned component classifiers.

We address the above mentioned issues twofold. First, and foremost, the proposed active selection of examples considers multiple factors to determine the usefulness of an example to be selected. More precisely, they are following:

1. Decision margin of component classifiers, and a prediction error of the single component classifier (which is a modification of QBC).
2. Factors specific to imbalanced data, which reflect more global (i.e., concerning the whole training set) and/or local (i.e., concerning example neighbourhood) class distribution of examples.
3. Additionally we use a specific variant of rejection sampling to enforce diversity within the batch through extra randomization.

The algorithm for learning ABBag ensemble is presented as a pseudocode Algorithm 3.2. It starts with training set LS, and m_{bag} bootstrap samples S and results in constructing an under-sampling extension of bagging in the first phase (lines 2–4). Moreover, it makes use of initial balancing weights w, which are calculated in accordance with the under-sampling bagging extension, used in this phase. These initial *balancing weights* w allow us to direct sampling toward more difficult to learn examples. In case of EBBag, balancing weights w reflect only the global imbalance of an example in the training set. In case of NBBag, balancing weights w expresses both global and local imbalance of an example in the training set. In the end of the first phase, component classifiers are generated from each of bootstraps S (line 3).

In the second phase, the active selection of examples is performed between lines 5–13. All bootstraps from S are iteratively (m_{al} times) enlarged by adding batches, and new component classifiers are re-learned.

In each iteration, new weights w' of examples are calculated according to weights update method um (which is described in the next paragraph), and then they are sorted (lines 7–8). Each bootstrap is enhanced by n_{al} examples selected randomly with the rejection sampling according to $\alpha = w'(x_{n_{al}}) + \varepsilon$, i.e., n_{al} random examples with weights w' higher than α are selected (lines 9–10). The parameter ε introduces additional (after α) level of randomness into the sampling. Finally, new component classifier C_i is learned resulting in new ensemble classifier C (line 11).

We consider here four different weights update methods. The simplest method, called *margin* (m),[1] is substituting the initial weights of examples with a decision margin between component classifiers in C. For a given testing example it is defined

---

[1]For simplicity margin will be denoted as m - in particular in experiments see Tables 3.3, 3.4, 3.5 and 3.6; further introduced weight update methods will be denoted analogously.

Algorithm 3.2: Actively Balanced Bagging Algorithm

Input : LS training set; TS testing set; CLA component classifier learning algorithm; m_{bag}
number of bootstrap samples; S bootstrap samples; w weights of examples from
LS; um weights update method; m_{al} number of active learning iterations; n_{al}
maximum size of active learning batch

Output: C ensemble classifier

1 *Learning phase*;
2 **for** $i := 1$ *to* m_{bag} **do**
3 $\quad\mid\quad$ $C_i :=$ CLA (S_i) {generate a component classifier} ;
4 **end**
5 **for** $l := 1$ *to* m_{al} **do**
6 $\quad\mid\quad$ **for** $i := 1$ *to* m_{bag} **do**
7 $\quad\mid\quad\quad\mid\quad$ $w' :=$ updateWeights $(w, C,$ um$)$ {update weights used in sampling} ;
8 $\quad\mid\quad\quad\mid\quad$ sort all x with respect to $w'(x)$, so that $w'(x_1) \geq w'(x_2) \geq \ldots \geq w'(x_n)$;
9 $\quad\mid\quad\quad\mid\quad$ $S'_i :=$ random sample from $x_1, x_2, \ldots, x_{n_{al}}$ according to w' {rejection sampling from
10 $\quad\mid\quad\quad\mid\quad$ $\quad$ top n_{al} x sorted according to w'; $\alpha = w'(x_{n_{al}})$ } ;
11 $\quad\mid\quad\quad\mid\quad$ $S_i := S_i \cup S'_i$;
12 $\quad\mid\quad\quad\mid\quad$ $C_i :=$ CLA (S_i) {re-train a new component classifier} ;
13 $\quad\mid\quad$ **end**
14 **end**

15 *Classification phase*;
16 **foreach** x *in* TS **do**
17 $\quad\mid\quad$ $C(x) :=$ majority vote of $C_i(x)$, where $i = 1, \ldots, m_{bag}$ {the class
assignment for object x is a combination of predictions of component classifiers C_i} ;
18 **end**

---

as: *margin* or $m = 1 - \left| \frac{V_{maj} - V_{min}}{m_{bag}} \right|$, where V_{maj} is a number of votes for majority class
and V_{min} is number of votes for minority class. As the margin may not be directly
reflecting the characteristic of imbalanced data (indeed under-sampling somehow
should reduce bias of the classifiers) we consider combining it with additional factors.
This leads to three variants of weights update methods. In the first extension, called,
margin and weight (*mw*), new weight w' is a product of margin m and initial balancing
weight w. We reduce the influence of w in subsequent iterations of active example
selection, as l is increasing. The reason for this reduction of influence is that we
expect margin m to improve (i.e., better reflect the usefulness of examples) with
subsequent iterations, and thus initial weights w becoming less important. More
precisely, $mw = m \times w^{\left(\frac{m_{al} - l}{m_{al}} \right)}$.

Both considered so far weights update methods produce bootstrap samples which,
in the same iteration l, differ only according to randomization introduced by the
rejection sampling, i.e., weights w' are the same for each i. That is why, we consider
yet another modification of methods m and mw, which makes w', and, consequently,
each bootstrap dependent on performance of the corresponding component classifier.
These two new update methods: *margin and component error* (*mce*), and *margin,
weight and component error* (*mwce*) are defined, respectfully, as follows: $mce =
m + 1_e \times w$, and $mwce = mw + 1_e \times w$. In this notation, 1_e is an indicator function

Table 3.2 Characteristics of benchmark real-world data sets

Data set	# examples	# attributes	Minority class	IR [%]
abalone	4177	8	0-4 16-29	11.47
breast-cancer	286	9	recurrence-events	2.36
car	1728	6	good	24.04
cleveland	303	13	3	7.66
cmc	1473	9	2	3.42
ecoli	336	7	imU	8.60
haberman	306	4	2	2.78
hepatitis	155	19	1	3.84
scrotal-pain	201	13	positive	2.41
solar-flare	1066	12	f	23.79
transfusion	748	4	1	3.20
vehicle	846	18	van	3.25
yeast	1484	8	ME2	28.10

defined so that $1_e = 1$ when a component classifier is making a prediction error on example, and $1_e = 0$ otherwise.

3.6 Experimental Evaluation

In this section we will carry out experiments designed to provide better understanding of the classification performance of Actively Balanced Bagging. The following two aims of these experiments are considered. First, we want to check to what extent the predictive performance of Actively Balanced Bagging can be improved in comparison to under-sampling extensions of bagging. For this part of experiments we choose two quite efficient, in classification performance, extensions of bagging: Exactly Balanced Bagging (EBBag) [13], and Neighbourhood Balanced Bagging (NBBag) [7]. Then, the second aim of experiments is to compare different variants of proposed active selection methods, which result in different versions of ABBag. Moreover, the sensitivity analysis of tuning basic parameters of ABBag is carried out.

3.6.1 Experimental Setup

The considered Actively Balanced Bagging ensembles are evaluated with respect to averaged performance in both minority and majority classes. That is why we consider G-mean measure, introduced in Sect. 3.2.2, since we want to find a good trade-off between recognition in both classes.

Similarly to our previous studies [6–8, 45] we will focus in our experiments on 13 benchmark real-world class imbalanced data sets. In this way, we include in this study data sets which have been often analyzed in many experimental studies with imbalanced data. This should make it easier to compare the achieved performances to the best results reported in the literature. The characteristics is of these data sets are presented in Table 3.2. The data sets represent different sizes, imbalance ratios (denoted by IR), domains and have both continuous and nominal attributes. Taking into account results presented in [45] some of data sets should be easier to learn for standard classifiers while most of them constitute different degrees of difficulties. More precisely, such data as vehicle and car, are easier ones as many minority class examples may categorized as safe ones. On the other hand, data sets breast cancer, clevaland, ecoli contain many borderline examples, while the remaining data sets could be estimated as the most difficult one as they additionally contain many rare cases or outliers.

Nearly all of benchmark real-world data sets were taken from the UCI repository.[2] One data set includes a medical problem and it was also used in our earlier works of on class imbalance.[3] In data sets with more than one majority class, they are aggregated into one class to have only binary problems, which is also typically done in other studies presented in the literature.

Furthermore, we include in this study a few synthetic data sets with a priori known (i.e., designed) data distribution. To this end, we applied a specialized generator for imbalanced data [58] and we produced two different types of data sets. In these data sets, examples from the minority class are generated randomly inside predefined spheres and majority class examples are randomly distributed in an area surrounding them. We consider two configurations of minority class spheres, called according to the shape they form: paw and flower, respectively. In both data sets the global imbalance ratio IR is equal to 7, and the total cardinality of examples are 1200 for paw and 1500 for flower always with three attributes. The minority class is decomposed into 3 or 5 sub-parts. Moreover, each of this data set has been generated with a different number of potentially unsafe examples. This fact is denoted by four numbers included in the name of data set. For instance, flower5-3d-30-40-15-15 represents flower with minority class that contains approximately 30% of safe examples, 40% inside the class overlapping (i.e., boundary), 15% rare and 15% outliers.

3.6.2   Results of Experiments

We conducted our experiments in two variants of constructing ensembles. In the first variant, standard EBBag or under-sampling NBBag was used. In the second variant,

---

[2]http://www.ics.uci.edu/mlearn/MLRepository.html.
[3]We are grateful to prof. W. Michalowski and the MET Research Group from the University of Ottawa for providing us an access to scrotal-pain data set.

the size of each of the classes in bootstrap samples was further reduced to 50% of the size of the minority class in the training set. Active selection parameters, used in the second phase, m_{al}, and n_{al} were chosen in a way, which enables the bootstrap samples constructed in ABBag to excess the size of standard under-sampling bootstrap by a factor not higher than two. The size of ensembles m_{bag}, in accordance with previous experiments [7], was always fixed to 50. We used WEKA[4] implementation of J48 decision tree as the component classifier in all of considered ensembles. We set under-sampling NBBag parameters to the same values as we have already used in [7]. All measures are estimated by a stratified 10-fold cross-validation repeated five times to improve repeatability of observed results.

In Tables 3.3, 3.4, 3.5 and 3.6, we present values of G-mean for all considered variants of ABBag on all considered real-world and synthetic data sets. Note that in Tables 3.3 and 3.4, we present results of active balancing of 50% under-sampling EBBag, and 50% under-sampling NBBag, respectively. Moreover, in Tables 3.5 and 3.6, we present results of active balancing of standard under-sampling EBBag, and standard under-sampling NBBag, respectively. The last row of each of Tables 3.3, 3.4, 3.5 and 3.6, contains average ranks calculated as in the Friedman test [30]. The interpretation of average rank is that the lower the value, the better the classifier.

The first, general conclusion resulting from our experiments is that ABBag performs better than under-sampling extensions of bagging, both: EBBag, and NBBag. Let us treat EBBag, and NBBag as baselines in Tables 3.3, 3.4, 3.5 and 3.6, respectively. The observed improvements of G-mean are statistically significant regardless of the considered version of ABBag. More precisely, each actively balanced EBBag has the lower average rank than the baseline EBBag, and, similarly, each actively balanced NBBag has lower average rank than the baseline NBBag. Moreover, Friedman tests result in p-values $\ll 0.00001$ in all of comparisons of both EBBag (Tables 3.3 and 3.5) NBBag (Tables 3.4 and 3.6). According to Nemenyi post-hoc test, critical difference CD between average ranks in our comparison is around 1.272. The observed difference between average ranks of each actively balanced EBBag and the baseline EBBag is thus higher than calculated CD. We can state that ABBag improves significantly classification performance over base line EBBag. An analogous observation holds for each actively balanced NBBag and the baseline NBBag. We can conclude this part of experiments by stating that ABBag is able to improve baseline classifier regardless of the setting.

However, th observed improvement of G-mean introduced by ABBag depends on the data set. Usually improvements for easier to learn data sets, like car, are smaller than these observed for the other data sets. The most apparent (and consistent) improvements are noted for the hardest to learn versions of synthetic data sets. In both cases of flower5-3d-10-20-35-35, and paw3-3d-10-20-35-35 application of baseline versions of EBBag, and NBBag gives values of G-mean equal to 0. These results are remarkably improved by all considered versions of ABBag.

---

[4]Eibe Frank, Mark A. Hall, and Ian H. Witten (2016). The WEKA Workbench. Online Appendix for "Data Mining: Practical Machine Learning Tools and Techniques", Morgan Kaufmann, Fourth Edition, 2016.

Table 3.3 G-mean of actively balanced 50% under-sampling EBBag

Data set	EBBag	m-EBBag	mce-EBBag	mw-EBBag	mwce-EBBag
abalone	79.486	79.486	80.056	79.486	79.603
breast-cancer	57.144	59.628	60.471	60.640	60.900
car	96.513	97.785	98.359	97.806	98.359
cleveland	70.818	73.154	70.818	73.672	70.818
cmc	64.203	65.146	64.572	64.771	64.687
ecoli	87.836	88.870	89.259	88.926	88.638
flower5-3d-10-20-35-35	0.000	55.055	54.251	52.653	54.046
flower5-3d-100-0-0-0	92.315	93.415	94.570	93.501	94.812
flower5-3d-30-40-15-15	77.248	77.995	78.502	78.022	78.423
flower5-3d-30-70-0-0	91.105	91.764	93.729	92.019	93.993
flower5-3d-50-50-0-0	91.966	92.470	93.972	92.317	93.834
haberman	62.908	65.355	65.916	67.299	65.520
hepatitis	78.561	79.132	80.125	79.208	80.079
paw3-3d-10-20-35-35	0.000	51.152	51.148	52.318	50.836
paw3-3d-100-0-0-0	90.857	93.020	94.001	93.011	94.391
paw3-3d-30-40-15-15	74.872	76.277	78.241	76.546	77.544
paw3-3d-30-70-0-0	88.545	90.510	91.410	90.927	91.106
paw3-3d-50-50-0-0	91.424	92.087	92.825	92.038	93.537
scrotal-pain	72.838	73.572	73.574	73.692	72.838
solar-flare	82.048	83.126	83.064	83.013	83.064
transfusion	66.812	67.929	66.812	67.448	66.812
vehicle	95.506	95.840	97.120	96.010	97.120
yeast	82.658	84.026	84.818	85.337	84.984
average rank	4.848	3.087	2.065	2.652	2.348

Now we move to examination of the influence of the proposed active modifications, i.e., weights update methods in the active selection of examples, on the classification performance. We make the following observations. First, if we consider actively balanced 50% under-sampling EBBag, *margin and component error* weights update method, thus (*mce*-EBBag), has the best average rank and the best value of median calculated for all G-mean results in Table 3.3. The second best performing weight update method in this comparison is *margin, weight and component error*, and thus (*mwce*-EBBag). These observations are, however, not statistically significant according to the critical difference and results of Wilcoxon test for a selected pair of classifiers. Results of actively balanced standard under-sampling EBBag, presented in Table 3.5, provide more considerable distinction. The best weights update method in this case is *margin, weight and component error*, and thus (*mwce*-EBBag). Moreover the observed difference in both average rank and *p*-value in Wilcoxon test allows us to state that *mwce*-EBBag is significantly better performing than *m*-EBBag.

Table 3.4 G-mean of actively balanced 50% under-sampling NBBag

Data set	NBBag	m-NBBag	mce-NBBag	mw-NBBag	$mwce$-NBBag
abalone	78.297	79.384	79.264	79.034	79.339
breast-cancer	56.521	62.559	61.957	60.106	62.316
car	93.918	95.698	97.816	97.405	98.182
cleveland	74.275	76.131	75.000	78.000	74.275
cmc	63.944	64.969	64.390	64.969	64.807
ecoli	88.056	89.326	88.846	89.412	89.139
flower5-3d-10-20-35-35	0.000	51.940	51.302	52.069	52.105
flower5-3d-100-0-0-0	89.851	92.844	94.465	93.427	94.869
flower5-3d-30-40-15-15	73.869	77.081	77.266	76.614	77.484
flower5-3d-30-70-0-0	88.513	92.055	93.814	91.903	93.975
flower5-3d-50-50-0-0	89.164	92.227	93.969	91.734	94.107
haberman	58.618	65.165	65.087	65.068	65.386
hepatitis	79.632	79.632	80.449	80.778	80.270
paw3-3d-10-20-35-35	0.000	50.644	49.804	43.549	52.541
paw3-3d-100-0-0-0	85.165	91.889	94.000	92.596	93.584
paw3-3d-30-40-15-15	29.157	75.499	75.000	74.818	70.950
paw3-3d-30-70-0-0	82.767	89.944	91.631	90.739	91.331
paw3-3d-50-50-0-0	84.787	91.327	92.079	91.826	92.474
scrotal-pain	74.471	75.180	75.625	76.507	75.363
solar-flare	82.275	85.049	83.620	83.954	83.233
transfusion	65.259	65.816	65.351	65.259	65.509
vehicle	94.304	96.720	97.733	96.513	97.498
yeast	82.450	84.454	84.887	83.549	85.005
average rank	4.935	2.652	2.435	2.957	2.022

Second best performing weight update method in this case is *margin and component error* weights update method, thus (*mce*-EBBag). Differences between this method and the other weights update methods are, however, again not statistically significant.

Similar observations are valid for actively balanced NBBag. In this case, the best weights update method according to average rank is *margin, weight and component error*, and thus (*mwce*-NBBag), regardless of the variant of the fist phase of the active selection, as seen in Tables 3.4 and 3.6. On the other hand, *margin and component error* weights update method, and thus (*mce*-NBBag), has the best value of median when 50% under-sampling *mce*-NBBag is considered. However, this observation is not statistically significant. So are all the observed differences among different weights update methods in active selection for NBBag.

To sum up observations noted so far, we should report that all different factors that we take into account in weights update methods: margin of classifiers in ensemble, weight of example, and component error are important for improving ABBag

Table 3.5 G-mean of actively balanced under-sampling EBBag

Data set	EBBag	m-EBBag	mce-EBBag	mw-EBBag	mwce-EBBag
abalone	76.927	77.740	77.940	77.740	77.874
breast-cancer	57.979	58.488	58.316	58.632	59.305
car	97.611	97.760	98.100	97.780	98.185
cleveland	68.207	68.207	68.207	68.207	68.207
cmc	62.242	62.552	62.242	62.280	62.242
ecoli	87.677	87.677	87.677	87.677	87.677
flower5-3d-10-20-35-35	0.000	43.713	54.854	51.077	54.419
flower5-3d-100-0-0-0	92.601	93.506	94.860	93.399	95.007
flower5-3d-30-40-15-15	77.749	78.027	78.654	77.977	78.979
flower5-3d-30-70-0-0	91.614	92.165	93.629	91.994	93.544
flower5-3d-50-50-0-0	91.338	92.651	94.622	92.451	94.469
haberman	60.673	64.379	64.327	65.124	65.251
hepatitis	74.217	77.505	78.631	76.937	76.688
paw3-3d-10-20-35-35	0.000	41.322	53.729	37.655	52.275
paw3-3d-100-0-0-0	92.507	93.811	94.233	93.765	94.329
paw3-3d-30-40-15-15	76.756	76.756	78.846	76.756	78.839
paw3-3d-30-70-0-0	89.362	90.206	91.069	91.443	90.990
paw3-3d-50-50-0-0	92.107	92.107	92.445	92.107	92.757
scrotal-pain	74.258	74.258	74.258	74.549	74.258
solar-flare	84.444	84.444	84.444	84.444	84.654
transfusion	64.078	65.492	67.534	65.911	67.589
vehicle	95.117	96.327	96.771	96.476	96.849
yeast	81.689	82.248	84.234	83.933	84.541
average rank	4.565	3.283	2.174	3.087	1.891

performance. Moreover, two weight update methods tend to give better results than the others. These are: *margin and component error (mce)*, and *margin, weight and component error (mwce)*. These results may be interpreted as an indication that proposed active selection of examples is able to make good use of the known labels of minority and majority examples, since weight, and component error factors are important.

In the next step of our experiments, we tested the influence of ε parameter on the performance of ABBag. ε controls the level of randomness in active selection of examples (see Sect. 3.3 for details). The results of this analysis favour small values of ε, which means that, in our setting, it is better to select the best (or almost the best, to be more precise) examples into active learning batches. This result may be partially explained by a relatively small size of batches used in the experimental evaluation. For small batches it should be important to select as good examples as possible

Table 3.6   G-mean of actively balanced under-sampling NBBag

Data set	NBBag	m-NBBag	mce-NBBag	mw-NBBag	$mwce$-NBBag
abalone	78.714	79.308	79.291	79.317	79.460
breast-cancer	58.691	62.752	62.698	62.191	62.501
car	96.200	97.518	97.847	97.775	98.801
cleveland	73.004	73.004	73.931	74.170	74.776
cmc	65.128	65.128	65.128	65.365	65.128
ecoli	88.581	88.581	88.581	88.581	88.581
flower5-3d-10-20-35-35	0.000	51.952	51.527	52.800	51.073
flower5-3d-100-0-0-0	92.373	93.594	94.481	93.437	94.683
flower5-3d-30-40-15-15	76.914	78.080	77.913	77.570	78.196
flower5-3d-30-70-0-0	91.120	92.297	93.490	92.141	94.112
flower5-3d-50-50-0-0	92.003	93.126	93.209	92.889	94.322
haberman	64.128	65.101	65.251	66.059	65.590
hepatitis	78.017	78.078	79.665	79.269	80.739
paw3-3d-10-20-35-35	0.000	50.916	49.912	51.239	52.142
paw3-3d-100-0-0-0	90.122	92.792	93.141	93.190	94.388
paw3-3d-30-40-15-15	63.966	76.440	75.945	76.990	77.057
paw3-3d-30-70-0-0	87.208	90.072	90.966	90.871	91.116
paw3-3d-50-50-0-0	91.317	92.105	92.295	91.582	92.984
scrotal-pain	73.205	75.023	74.812	75.636	74.891
solar-flare	83.435	84.731	83.574	84.015	84.286
transfusion	65.226	65.943	66.239	65.226	65.226
vehicle	95.339	96.776	97.463	96.759	97.401
yeast	84.226	84.780	85.580	85.067	85.247
average rank	4.783	3.000	2.630	2.783	1.804

while with an increase of the size of batch, more diversity in batches resulting from selection of more sub-optimal examples, should be also found to be useful.

We finish the presented experimental evaluation with an analysis of influence of parameter m_{al}, and parameter n_{al} on the classification performance of ABBag. Parameter n_{al} is, in the results presented further on, the size of active learning batch determined as the percentage of the size of minority class. Parameter m_{al} is the number of performed active learning iterations. We restrict ourselves in this analysis to *margin, weight and component error* (*mwce*) weight update method, since it gave the best results for majority of considered variants. Moreover, we will only present results of an analysis of some harder to learn data sets since they provide more information about behaviour of ABBag. Results for other data sets demonstrate usually the same tendencies as for the chosen data sets, though these tendencies might be less visible. We present changes in values of G-mean for four different data sets: cleveland, flower5-3d-10-20-35-35, paw3-3d-10-20-35-35,

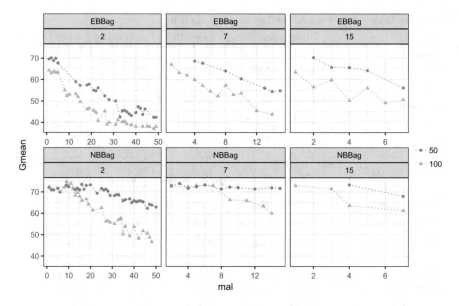

Fig. 3.1 `cleveland` - influence of m_{al} and n_{al} on G-mean of actively balanced (*mwce*) 50% under-sampling, and standard under-sampling of EBBag and NBBags

and `yeast` in the following figures: Figs. 3.1, 3.2, 3.3, and 3.4, respectively. In each of these figures one can compare G-mean performance of actively balanced EBBag ensemble, and actively balanced NBBag ensemble (in a trellis of smaller figures). Three different plots are presented, in each of figures, for each of ensembles, according to tested value of $n_{al} = \{2, 7, 15\}$. On each of plots relation between m_{al} and G-mean is presented for ensembles resulting from 50% under-sampling, and standard (i.e., 100%) under-sampling performed at the first phase of ABBag.

A common tendency for majority of plots representing G-mean performance of *mwce* ABBag on real-world data sets (here represented by data sets: `cleveland` in Fig. 3.1, and `yeast` in Fig. 3.4) is that, regardless of other parameters, one can observe an increase of G-mean performance for initial active balancing iterations (i.e., small values of m_{al}), followed by stabilization or decrease of performance for the further iterations. A decrease of performance is more visible for further iterations of actively balanced EBBag. Thus, the tendency observed for real-word data sets is in line with our motivation for proposing ABBag expressed by an intuition that it should suffice to perform a limited number of small updates of bootstrap by actively selected examples to sufficiently improve performance of under-sampling extensions of bagging.

A different tendency is visible on plots representing G-mean performance of *mwce* ABBag on hard to learn synthetic data sets that we analyzed (represented by data sets: `flower5-3d-10-20-35-35` in Fig. 3.2, and `paw3-3d-10-20-35-35` in Fig. 3.3). One can observe an almost permanent increase of actively balanced

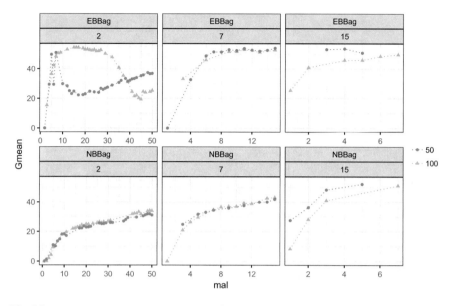

Fig. 3.2 `flower5-3d-10-20-35-35`—influence of m_{al} and n_{al} on G-mean of actively balanced (*mwce*) 50% under-sampling, and standard under-sampling of EBBag and NBBag

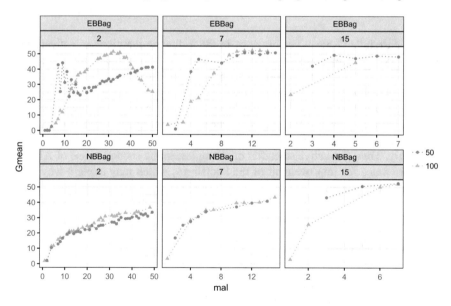

Fig. 3.3 `paw3-3d-10-20-35-35`—influence of m_{al} and n_{al} on G-mean of actively balanced (*mwce*) 50% under-sampling, and standard under-sampling of EBBag and NBBag

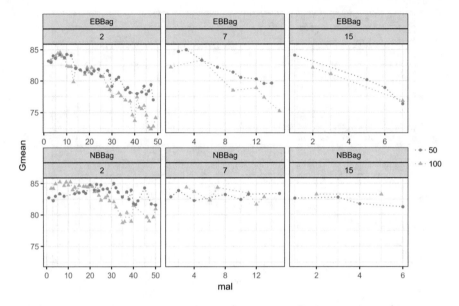

Fig. 3.4 yeast—influence of m_{al} and n_{al} on G-mean of actively balanced (*mwce*) 50% under-sampling, and standard under-sampling of EBBag and NBBag

NBBag ensemble G-mean performance in the following iterations of actively balancing procedure, regardless of other parameters. For active balancing EBBag this tendency is more similar to performance on real-world data sets. After an increase of performance observed for initial iterations, comes a decrease (e.g., see 50% under-sampling EBBag for $m_{al} = 2$ in Fig. 3.3) or stabilization (e.g., see under-sampling EBBag for $m_{al} > 2$ in the same Fig. 3.3). One should take into account, however, that these are really hard to learn data sets, for which base-line classifiers were performing poorly. Thus, the active selection of examples, in the second phase of ABBag, had more place for improvement of the bootstraps, which may explain why more iterations were needed.

3.7 Conclusions

The main aim of this chapter has been focused on the attempts to improve classification performance of the under-sampling extensions of the bagging ensembles to better address class imbalanced data. The current extensions are mainly based on modifying the example distributions inside bootstrap samples (bags). In our proposal we have decided to additionally add a limited number of learning examples coming from outside the given bag. As a result of such small adjusting bootstrap bags, the final ensemble classifier should better balance recognition of examples from minority and majority classes. To achieve this aim, we have introduced a two phase approach, which we have called Actively Balanced Bagging (ABBag).

The key idea, in this approach, is to first learn an under-sampling bagging ensemble and, then, to carry out steps of updating bootstraps by small batches composed of a limited number of actively selected learning examples. These batches are drawn by one of proposed active learning strategies, with one of the example weights. Their definition takes into account a decision margin of ensemble votes for the classified instance, balancing of the example class distribution in its neighbourhood, and prediction errors of component classifiers. To best of our knowledge such a combined approach has not been considered yet.

The results of experiments have demonstrated that:

- The new proposed ABBag ensemble with the active selection of examples has improved G-mean performance of two baseline under-sampling bagging Exactly Balanced Bagging (EBBag) and Nearest Balanced Bagging (NBBag), which are already known to be very good ensemble classifiers for imbalanced data and out-performs several other classifiers (see, e.g., earlier experimental results in [7]).
- Another observation resulting from the presented experiments is that an active selection strategy performs best when it integrates the ensemble disagreement, i.e. the decision margin (which is typically applied in the standard active learning such as QBC) with information on class distribution in imbalanced data and prediction errors of component classifiers. The best performing selection strategy is *mwce* for both EBBag and NBBag classifiers.
- A more detailed analysis of ABBag performance on harder to learn data sets allowed us to observe that it is usually better to add a small number of examples in batches to obtain the best classification performance.

Finally, an open problem for further research, related to the main topic of this book, concerns dealing with a higher number of attributes in the proposed ensemble. Highly dimensional imbalanced data sets are still not sufficiently investigated in the current research (please see, e.g., discussions in [11, 15]). In case of ensembles, it is possible to look for other solutions than typical attribute selection in a pre-processing step or in a special wrapper such as, e.g., proposed in [31]. A quite natural modification could be applying random subspace methods (such as Ho's proposal [27]) while constructing bootstraps. It has already been applied in extensions of Roughly Balanced Bagging [37].

However, our new proposal exploits modeling of the neighbourhood for minority class examples. It is based on distances between examples, e.g. with Heterogeneous Value Difference Metric. As it has been recently shown by Tomasev's research [55] on, so called, *hubness*, a k-nearest neighbourhood constructed on highly dimensional data may suffer from the curse of dimensionality and such metrics are not sufficient. Therefore, the current proposal of ABBag should be be rather applied to datasets with a smaller or medium number of attributes. The prospect extensions of ABBag for a larger number of attributes should be constructed with different techniques to estimate the example neighbourhoods.

Acknowledgements The research was supported from Poznań University of Technology Statutory Funds.

References

1. Abe, N., Mamitsuka, H.: Query learning strategies using boosting and bagging. In: Proceedings of 15th International Conference on Machine Learning, pp. 1–10 (2004)
2. Aggarwal, C., X., K., Gu, Q., Han, J., Yu, P.: Data Classification: Algorithms and Applications. Active learning: A survey, pp. 571–606. CRC Press (2015)
3. Batista, G., Prati, R., Monard, M.: A study of the behavior of several methods for balancing machine learning training data. ACM SIGKDD Explor. Newsl. **6**, 20–29 (2004). https://doi. org/10.1145/1007730.1007735
4. Błaszczyński, J., Deckert, M., Stefanowski, J., Wilk, S.: Integrating selective preprocessing of imbalanced data with Ivotes ensemble. In: Proceedings of 7th International Conference RSCTC 2010, LNAI, vol. 6086, pp. 148–157. Springer (2010)
5. Błaszczyński, J., Lango, M.: Diversity analysis on imbalanced data using neighbourhood and roughly balanced bagging ensembles. In: Proceedings ICAISC 2016, LNCS, vol. 9692, pp. 552–562. Springer (2016)
6. Błaszczyński, J., Stefanowski, J., Idkowiak, L.: Extending bagging for imbalanced data. In: Proc. of the 8th CORES 2013, Springer Series on Advances in Intelligent Systems and Computing, vol. 226, pp. 226–269 (2013)
7. Błaszczyński, J., Stefanowski, J.: Neighbourhood sampling in bagging for imbalanced data. Neurocomputing **150A**, 184–203 (2015)
8. Błaszczyński, J., Stefanowski, J.: Actively Balanced Bagging for Imbalanced Data. In: Proceedings ISMIS 2017, Springer LNAI, vol. 10352, pp. 271–281 (2017)
9. Błaszczyński, J., Stefanowski, J.: Local data characteristics in learning classifiers from imbalanced data. In: J. Kacprzyk, L. Rutkowski, A. Gaweda, G. Yen (eds.) Advances in Data Analysis with Computational Intelligence Methods, Studies in Computational Intelligence. p. 738. Springer (2017). doi:https://doi.org/10.1007/978-3-319-67946-4_2 (to appear)
10. Borisov, A., Tuv, E., Runger, G.: Active Batch Learning with Stochastic Query-by-Forest (SQBF). Work. Act. Learn. Exp. Des. JMLR **16**, 59–69 (2011)
11. Branco, P., Torgo, L., Ribeiro, R.: A survey of predictive modeling under imbalanced distributions. ACM Comput. Surv. **49**(2), 31 (2016). https://doi.org/10.1145/2907070
12. Breiman, L.: Bagging predictors. Mach. Learn. **24**(2), 123–140 (1996). https://doi.org/10.1007/ BF00058655
13. Chang, E.: Statistical learning for effective visual information retrieval. In: Proceedings of ICIP 2003, pp. 609–612 (2003). doi:https://doi.org/10.1109/ICIP.2003.1247318
14. Chawla, N., Bowyer, K., Hall, L., Kegelmeyer, W.: SMOTE: Synthetic Minority Over-Sampling Technique. J. Artif. Intell. Res. **16**, 341–378 (2002)
15. Chen, X., Wasikowski, M.: FAST: A ROC–based feature selection metric for small samples and imbalanced data classification problems. In: Proceedings of the 14th ACM SIGKDD, pp. 124–133 (2008). doi:https://doi.org/10.1145/1401890.1401910
16. Cieslak, D., Chawla, N.: Learning decision trees for unbalanced data. In: D. et al. (ed.) Proceedings of the ECML PKDD 2008, Part I, LNAI, vol. 5211, pp. 241–256. Springer (2008). doi:https://doi.org/10.1007/978-3-540-87479-9_34
17. Ertekin, S., Huang, J., Bottou, L., Giles, C.: Learning on the border: Active learning in imbalanced data classification. In: Proceedings ACM Conference on Information and Knowledge Management, pp. 127–136 (2007). doi:https://doi.org/10.1145/1321440.1321461
18. Ertekin, S.: Adaptive oversampling for imbalanced data classification. Inf. Sci. Syst. **264**, 261–269 (2013)

19. Ferdowsi, Z., Ghani, R., Settimi, R.: Online Active Learning with Imbalanced Classes. In: Proceedings IEEE 13th International Conference on Data Mining, pp. 1043–1048 (2013)
20. Fu, J., Lee, S.: Certainty-based Active Learning for Sampling Imbalanced Datasets. Neuro-computing **119**, 350–358 (2013). https://doi.org/10.1016/j.neucom.2013.03.023
21. Galar, M., Fernandez, A., Barrenechea, E., Bustince, H.: Herrera, F.: A review on ensembles for the class imbalance problem: bagging-, boosting-, and hybrid-based approaches. IEEE Trans. Syst. Man Cybern. C **99**, 1–22 (2011)
22. Garcia, V., Sanchez, J., Mollineda, R.: An empirical study of the behaviour of classifiers on imbalanced and overlapped data sets. In: Proceedings of Progress in Pattern Recognition, Image Analysis and Applications, LNCS, vol. 4756, pp. 397–406. Springer (2007)
23. Grzymala-Busse, J., Stefanowski, J., Wilk, S.: A comparison of two approaches to data mining from imbalanced data. J. Intell. Manuf. **16**, 565–574 (2005). https://doi.org/10.1007/s10845-005-4362-2
24. He H. Yungian, M.: Imbalanced Learning. Foundations, Algorithms and Applications. IEEE - Wiley (2013)
25. He, H., Garcia, E.: Learning from imbalanced data. IEEE Trans. Data Knowl. Eng. **21**, 1263–1284 (2009). https://doi.org/10.1109/TKDE.2008.239
26. Hido, S., Kashima, H.: Roughly balanced bagging for imbalance data. Stat. Anal. Data Min. **2**(5–6), 412–426 (2009)
27. Ho, T.: The random subspace method for constructing decision forests. Pattern Anal. Mach. Intell. **20**(8), 832–844 (1998)
28. Hu, B., Dong, W.: A study on cost behaviors of binary classification measures in class-imbalanced problems. CoRR abs/1403.7100 (2014)
29. Japkowicz, N., Stephen, S.: Class imbalance problem: a systematic study. Intell. Data Anal. J. **6**(5), 429–450 (2002)
30. Japkowicz, N.: Shah, Mohak: Evaluating Learning Algorithms: A Classification Perspective. Cambridge University Press (2011). doi:https://doi.org/10.1017/CBO9780511921803
31. Jelonek, J., Stefanowski, J.: Feature subset selection for classification of histological images. Artif. Intell. Med. **9**, 227–239 (1997). https://doi.org/10.1016/S0933-3657(96)00375-2
32. Jo, T., Japkowicz, N.: Class imbalances versus small disjuncts. ACM SIGKDD Explor. Newsl. **6**(1), 40–49 (2004). https://doi.org/10.1145/1007730.1007737
33. Khoshgoftaar, T., Van Hulse, J., Napolitano, A.: Comparing boosting and bagging techniques with noisy and imbalanced data. IEEE Trans. Syst. Man Cybern. Part A **41**(3), 552–568 (2011). https://doi.org/10.1109/TSMCA.2010.2084081
34. Kubat, M., Matwin, S.: Addressing the curse of imbalanced training sets: one-side selection. In: Proceedings of the 14th International Conference on Machine Learning ICML-1997, pp. 179–186 (1997)
35. Kuncheva, L.: Combining Pattern Classifiers. Methods and Algorithms, 2nd edn. Wiley (2014)
36. Lango, M., Stefanowski, J.: The usefulness of roughly balanced bagging for complex and high-dimensional imbalanced data. In: Proceedings of International ECML PKDD Workshop on New Frontiers in Mining Complex Patterns NFmC, LNAI, vol. 9607, pp. 94–107, Springer (2015)
37. Lango, M., Stefanowski, J.: Multi-class and feature selection extensions of Roughly Balanced Bagging for imbalanced data. J. Intell. Inf. Syst. (to appear). doi:https://doi.org/10.1007/s10844-017-0446-7
38. Laurikkala, J.: Improving identification of difficult small classes by balancing class distribution. Tech. Rep. A-2001-2, University of Tampere (2001). doi:https://doi.org/10.1007/3-540-48229-6_9
39. Lewis, D., Catlett, J.: Heterogeneous uncertainty sampling for supervised learning. In: Proceedings of 11th International Conference on Machine Learning, pp. 148–156 (1994)
40. Liu, A., Zhu, Z.: Ensemble methods for class imbalance learning. In: Y.M. He H. (ed.) Imbalanced Learning. Foundations, Algorithms and Applications, pp. 61–82. Wiley (2013). doi:https://doi.org/10.1002/9781118646106.ch4

41. Lopez, V., Fernandez, A., Garcia, S., Palade, V., Herrera, F.: An Insight into Classification with Imbalanced Data: Empirical Results and Current Trends on Using Data Intrinsic Characteristics. Inf. Sci. **257**, 113–141 (2014)
42. Napierała, K., Stefanowski, J., Wilk, S.: Learning from imbalanced data in presence of noisy and borderline examples. In: Proceedings of 7th International Conference RSCTC 2010, LNAI, vol. 6086, pp. 158–167. Springer (2010). doi:https://doi.org/10.1007/978-3-642-13529-3_18
43. Napierała, K., Stefanowski, J.: BRACID: A comprehensive approach to learning rules from imbalanced data. J. Intell. Inf. Syst. **39**, 335–373 (2012). https://doi.org/10.1007/s10844-011-0193-0
44. Napierała, K., Stefanowski, J.: Addressing imbalanced data with argument based rule learning. Expert Syst. Appl. **42**, 9468–9481 (2015). https://doi.org/10.1016/j.eswa.2015.07.076
45. Napierała, K., Stefanowski, J.: Types of minority class examples and their influence on learning classifiers from imbalanced data. J. Intell. Inf. Syst. **46**, 563–597 (2016). https://doi.org/10.1007/s10844-015-0368-1
46. Napierała, K.: Improving rule classifiers for imbalanced data. Ph.D. thesis, Poznań University of Technology (2013)
47. Prati, R., Batista, G., Monard, M.: Class imbalance versus class overlapping: an analysis of a learning system behavior. In: Proceedings 3rd Mexican International Conference on Artificial Intelligence, pp. 312–321 (2004)
48. Ramirez-Loaiza, M., Sharma, M., Kumar, G., Bilgic, M.: Active learning: An empirical study of common baselines. Data Min. Knowl. Discov. **31**, 287–313 (2017). https://doi.org/10.1007/s10618-016-0469-7
49. Seaz, J., Krawczyk, B., Woźniak, M.: Analyzing the oversampling of different classes and types in multi-class imbalanced data. Pattern Recognit **57**, 164–178 (2016). https://doi.org/10.1016/j.atcog.2016.03.012
50. Settles, B.: Active learning literature survey. Tech. Rep. 1648, University of Wisconsin-Madison (2009)
51. Stefanowski, J., Wilk, S.: Selective pre-processing of imbalanced data for improving classification performance. In: Proceedings of the 10th International Conference DaWaK. LNCS, vol. 5182, pp. 283–292. Springer (2008). doi:https://doi.org/10.1007/978-3-540-85836-2_27
52. Stefanowski, J.: Overlapping, rare examples and class decomposition in learning classifiers from imbalanced data. In: S. Ramanna, L.C. Jain, R.J. Howlett (eds.) Emerging Paradigms in Machine Learning, vol. 13, pp. 277–306. Springer (2013). doi:https://doi.org/10.1007/978-3-642-28699-5_11
53. Stefanowski, J.: Dealing with data difficulty factors while learning from imbalanced data. In: J. Mielniczuk, S. Matwin (eds.) Challenges in Computational Statistics and Data Mining, pp. 333–363. Springer (2016). doi:https://doi.org/10.1007/978-3-319-18781-5_17
54. Sun, Y., Wong, A., Kamel, M.: Classification of imbalanced data: a review. Int. J.Pattern Recognit Artif. Intell. **23**(4), 687–719 (2009). https://doi.org/10.1142/S0218001409007326
55. Tomasev, N., Mladenic, D.: Class imbalance and the curse of minority hubs. Knowl. Based Syst. **53**, 157–172 (2013)
56. Wang, S., Yao, X.: Mutliclass imbalance problems: analysis and potential solutions. IEEE Trans. Syst. Man Cybern. Part B **42**(4), 1119–1130 (2012). https://doi.org/10.1109/TSMCB.2012.2187280
57. Weiss, G.: Mining with rarity: A unifying framework. ACM SIGKDD Explor. Newsl. **6**(1), 7–19 (2004). https://doi.org/10.1145/1007730.1007734
58. Wojciechowski, S., Wilk, S.: Difficulty factors and preprocessing in imbalanced data sets: an experimental study on artificial data. Found. Comput. Decis. Sci. **42**(2), 149–176 (2017)
59. Yang, Y., Ma, G.: Ensemble-based active learning for class imbalance problem. J. Biomed. Sci. Eng. **3**(10), 1022–1029 (2010). https://doi.org/10.4236/jbise.2010.310133
60. Zięba, M., Tomczak, J.: Boosted SVM with active learning strategy for imbalanced data. Soft Comput. **19**(12), 3357–3368 (2015). https://doi.org/10.1007/s00500-014-1407-5

Chapter 4
Attribute-Based Decision Graphs and Their Roles in Machine Learning Related Tasks

João Roberto Bertini Junior and Maria do Carmo Nicoletti

Abstract Recently, new supervised machine learning algorithm has been proposed which is heavily supported by the construction of an attribute-based decision graph (AbDG) structure, for representing, in a condensed way, the training set associated with a learning task. Such structure has been successfully used for the purposes of classification and imputation in both, stationary and non-stationary environments. This chapter provides a detailed presentation of the motivations and main technicalities involved in the process of constructing AbDGs, as well as stresses some of the strengths of this graph-based structure, such as robustness and low computational costs associated with both, training and memory use. Given a training set, a collection of algorithms for constructing a weighted graph (i.e., an AbDG) based on such data is presented. The chapter describes in details algorithms involved in creating the set of vertices, the set of edges and, also, assigning labels to vertices and weights to edges. Ad-hoc algorithms for using AbDGs for both, classification or imputation purposes, are also addressed.

Keywords Attribute-based decision graphs · Graph-based data representation · Imputation of missing data · Data classification

J. R. Bertini Junior (✉)
School of Technology, University of Campinas, R. Paschoal Marmo 1888,
Jd. Nova Itália, Limeira, SP 13484-332, Brazil
e-mail: bertini@ft.unicamp.br

M. do Carmo Nicoletti
Campo Limpo Paulista School, R. Guatemala 167, Campo Limpo Paulista,
SP 13231-230, Brazil
e-mail: carmo@cc.faccamp.br

M. do Carmo Nicoletti
Computer Science Department, Federal University of São Carlos,
Rodovia Washington Luís s/n, São Carlos, SP 13565-905, Brazil

© Springer International Publishing AG 2018　　　　　　　　　　　　　53
U. Stańczyk et al. (eds.), *Advances in Feature Selection for Data and Pattern Recognition*, Intelligent Systems Reference Library 138,
https://doi.org/10.1007/978-3-319-67588-6_4

4.1 Introduction

In the Machine Learning (ML) area, two broad groups of algorithms can be considered, referred to as supervised and unsupervised algorithms. Supervised algorithms use a particular information associated to each training instance, referred to as *class*; such algorithms induce knowledge representations which are conventionally known as classifiers. Usually unsupervised algorithms do not require the *class* information; most of them can be characterized as clustering algorithms which, given as input a set of training instances induce, as output, a set of disjoint sets of such instances (i.e., a clustering). In an unsupervised context, the inductive process can be viewed as the provider of some sort of organization to the given data; the concept of similarity (or dissimilarity) is used to guide the grouping of similar instances [33].

In the context of supervised automatic learning as well as of the several supervised methods focused on this chapter, there are mainly two kinds of data which can be organized as graph-based structures. Data that naturally reflect a graph structure are the so-called relational data [27], and the commonly available data, usually described as vectors of attribute values, referred to as vector-based data. Lately, there has been an increasing number of machine learning tasks addressed by graph-based approaches (see e.g. [1, 11]). Graph-based approaches have been adopted in supervised classification tasks in works such as [10, 25].

Also, the capability of a graph-based structure to model data distribution has been explored in the context of unsupervised learning, involving clustering tasks [33]. There is an emphasis on graph-based representation in both, unsupervised and semi-supervised learning environments, as the basic structure to model knowledge, particularly in semi-supervised tasks, such as transduction and induction [12, 16]. Reference [38] describes a semi-supervised learning framework based on graph embedding. Within the complex network theory [31], for instance, large data sets can be clustered using a community detection algorithm, such as in [19, 28, 37]. In [23, 24] the graph-based relational learning (GBRL) is discussed as a subarea of graph-based data mining (GBDM), which conceptually differs from logic-based relational learning, implemented by, for example, inductive logic programming algorithms [29, 30]. As pointed out in [23], GBDM algorithms tend to focus on finding frequent subgraphs i.e., subgraphs in the data whose number of instances (they represent) are above some minimum support; this is distinct from a few GBRL developed systems which, typically, involve more than just considering the frequency of the pattern in the data, such as the Subdue [15] and the GBI [39].

So far in the literature, research work having focus on the process of graph construction, for representing a particular training set of vector-based data, has not yet attracted the deserved scientific community attention; this is particularly odd, taking into account the crucial role that data representation plays in any automatic learning process [40]. The way a training data is represented has a deep impact on its further use by any learning algorithm. Although one can find several works where graphs have been used as structures for representing training sets, the many ways of using graphs' representational potentialities have not been completely explored yet.

Invariably, most of the available graph-based learning algorithms are restricted, considering they employ only one out of a few algorithms for creating the graph that represents a given training set. Also, the graph construction methods surveyed so far always represent each training instance as a vertex and, then, define edges between vertices as a way of representing some sort of similarity between them. Graphs constructed in this way are referred to as *data graphs*. Among the most popular methods for constructing graphs are those that construct KNN graphs and ϵ-graphs [14, 17], as well as fully connected weighted graphs, where weights are defined by a given function, such as the Gaussian, as in [41]. Regardless of being a suitable solution for data mining related problems, these kind of graphs are still tied to their inherent advantages and disadvantages.

As already mentioned, generally methods for constructing graphs rely on some *ad-hoc* concept of neighborhood, which commonly gives rise to local structures within the data set; the global data structure is then left to be addressed by the learning algorithm [22]. Proposals contemplating new types of graph-based structures for representing data and, also, algorithms for dealing with them, will certainly contribute for further progress in areas such as data mining and automatic learning. As pointed out in [3], graph-based representations of vector-based data are capable of modelling arbitrary local data configurations, enabling the learning algorithm best capture the underlying data distribution.

A new data structure for storing relevant information about training data sets was proposed in [5] and is referred to as *Attribute-based Data Graph (AbDG)*; an AbDG models a given vector-based training data set as a weighted graph. The main motivation for proposing the AbDG was to devise a data structure compact and easily manageable, able to represent and condense all information present in a given training data set, which could also be used as the source of information for classification tasks. This chapter addresses and reviews Attribute-based Data Graph (AbDG) as well as the two task-oriented subjacent algorithms associated with AbDGs; the first that constructs the graph representing a giving training set of vector-based instances and, the second, that uses the graph for classification purposes.

Besides the Introduction section, this chapter is organized into six more sections. Section 4.2 introduces the main technicalities involved for the establishment of the concept of Attribute-based Decision Graph, focusing on the proposal of two possible graph structures namely a *p-partite*, in Sect. 4.2.1.1, and a *complete p-partite*, in Sect. 4.2.1.2, where p refers to the number of attributes that describe a vector-based training instance. Formal notation is introduced and the processes involved in the AbDG construction are described. The section first discusses the construction of the vertex set, given a vector-based data set and, then, the two possible strategies for inserting edges connecting vertices. Once the construction of the graph-structure is finished, the process that assigns labels to vertices and its counterpart, that assigns weights to edges, complete and finalize the graph construction process. Section 4.3 details the process of using the information embedded in an AbDG for classification purposes and, for that, defines the classification process as some sort of graph matching process between the AbDG and one of its subgraphs i.e., the one defined by the instance to be classified. Section 4.4 presents a numerical example of the induction of an AbDG aiming at a classification task and the use of the induced AbDG in the processes of classifying a new unclassified instance. Section 4.5 reviews the convenience of using AbDGs when learning from data sets with absent attribute values. Section 4.6 approaches the process of constructing an AbDG as a search task conducted by a genetic algorithm (GA) and, finally, Sect. 4.7 resumes the main contributions of the AbDG approach, highlighting some of its main advantages and the technicalities involved in the construction/use of such structure. The section ends with some new insights for continuing the research work with focus on AbDGs.

4.2 The Attribute-Based Decision Graph Structure (AbDG) for Representing Training Data

The AbDG is a graph-based structure proposed in [5] and extended in [9], aiming at modelling data described as vectors of attribute values; such structure has been later employed in several machine learning related tasks, such as those presented in [5–8].

4.2.1 Constructing an AbDG

Consider a training data set X, where each instance $\mathbf{x} = (x_1, \ldots, x_p, c)$ in X is a p-dimensional data vector of features followed by a class label $c \in \{\omega_1, \ldots \omega_M\}$, representing one among M classes. The process that constructs the AbDG for representing X initially focuses on the construction of the set of vertices, then on the construction of the set of edges and finally, on assigning labels to vertices and weights to edges of the induced graph, turning it into a weighted labeled data graph representing X.

As mentioned before, given a data set X having N p-dimensional instances, most approaches for constructing the set of vertices of a graph that represents X, usually define each data instance in X as a vertex of the graph being constructed, resulting in a graph with N vertices. Approaches adopting such a strategy can be found in [2, 33, 36]. In an AbDG graph the vertices represent data intervals associated with values the attributes that describe training instances can have. Thus, once attribute A_a has been divided into n_a intervals, it can be viewed as a set of disjoint intervals $A = \{I_{a,1}, \ldots, I_{a,n_1}\}$ where each interval $I_{a,i}$ stands for vertex $v_{a,i}$ in the graph.

Due to vertices being defined by data intervals, the construction of an AbDG is heavily dependent on the type of the attributes used for describing the training instances. Basically three types of attributes are commonly used for describing a given training set X, referred to as numerical (continuous-valued), categorical (whose possible values are limited and usually fixed, having no inherent order) and ordinal (whose possible values follow a particular pre-established order).

The construction of the set of vertices of an AbDG graph, representing a given data set X, starts by discretizing the values of each one of the p attributes $\{A_1, A_2, A_3, \ldots A_p\}$ that describes X. A discretization process applied to each one of the p attributes associates, to each of them, a set of disjoint intervals of attribute values. As the set of intervals associated to each attribute A_i ($i = 1, \ldots p$) depends on the type of the attribute A_i (i.e., categorical, numerical (continuous-valued) or ordinal), as well as the range of values of A_i in X, a discretization method should deal with the different types of attributes.

If an attribute is categorical, the simplest way to create its associated set of vertices is by considering each of its possible values as a degenerate interval; this is the approach used for constructing AbDGs. As an example, if the values of attribute A_5 in X (taking into account all instances in X) are 0, 1, 2, 3, 4 and 5, during the construction of the vertex set, the set of degenerate intervals: $\{[0\ 0], [1\ 1], [2\ 2], [3\ 3], [4\ 4], [5\ 5]\}$ is associated with A_5 and each interval is considered a vertex in the graph under construction.

When the attribute is numerical, the usual basic procedure a discretization method adopts is to divide the attribute range into disjoint intervals. Several discretization methods found in the literature can be used to create such set of disjoint intervals associated with a continuous-valued attribute [21].

As pointed out in [9] in relation to the problem of discretizing a continuous-valued attribute, the solution starts by finding a set of what is called *cut point candidates* in the range of the attribute, then to use a heuristic to evaluate the potentialities of the selected cut point candidates and, finally, choose the most promising subset as the actual cut points for defining the intervals. Cut point candidates are determined by sorting each attribute and then, searching for consecutive different attribute values whose corresponding instances belong to different classes, a process formalized as follows. Consider the values of attribute A and let sA represent an ordered version of the values of A. The process can be formally stated as if $sA[i] \neq sA[i + 1]$ and $class\_instance(sA[i]) \neq class\_instance(sA[i + 1])$, where $class\_instance()$ gives the class of the data instance having $A[i]$ as value for attribute A, then determine the middle point between values $sA[i]$ and $sA[i + 1]$ and assume the obtained value as a cut point. Once the vertex set has been built, edges are then established by taking into account the corresponding attribute values of patterns in X, aiming at connecting intervals (i.e. vertices) to reflect the correlations between different attributes [13]. Taking into account a p-dimensional data set, two edge structures are considered, which give rise to two different graph structures, the p-partite (Sect. 4.2.1.1) and the complete p-partite (Sect. 4.2.1.2).

4.2.1.1 The AbDG as a p-Partite Graph

Given a data set X and considering that the sets of vertices associated with each attribute have already been created, the induction of a p-partite graph, for representing X, assumes a pre-defined order among the attributes, which can be randomly established or, then, by sorting the attributes according to some criteria. So, given

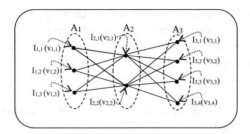

Fig. 4.1 A 3-partite AbDG structure created from a data set X with N vector-based instances, each described by 3 attributes, A_1, A_2 and A_3. The discretization process associates to each attribute A_i ($1 \leq i \leq 3$), n_i intervals (i.e., vertices), $n_1 = 3$, $n_2 = 2$ and $n_3 = 4$

the pre-defined attribute order and the sets of vertices associated to each attribute, edges can only be introduced between two consecutive (taking into account the given order) vertices. The whole graph structure is affected by the established order.

It has been empirically verified that sorting the attributes according to the descent order of their corresponding information gain values can be a convenient choice (see [9]) since it promotes and maintains connections between attributes with highest information gains. Figure 4.1 shows an example of a *3-partite* AbDG associated with a hypothetical training set X defined by three attributes A_1, A_2 and A_3 and an associated class, where the discretization process associated to each attribute A_i ($1 \leq i \leq 3$) produced, respectively, n_i intervals (i.e., vertices), namely $n_1 = 3, n_2 = 2$ and $n_3 = 4$. Figure 4.2 shows a high level pseudocode for creating a p-partite AbDG structure, given a data set with N instances described as p-dimensional vectors and an associated class, out of M possible classes.

4.2.1.2   The AbDG as a Complete p-Partite Graph

When constructing the complete p-partite structure, however, the attribute order is irrelevant, due to the intrinsic nature of complete p-partite graphs. In a complete p-partite graph all possible edges between intervals (i.e., vertices) associated with different attributes are inserted in the graph under construction. Figure 4.3 shows an example of a *3-partite* AbDG associated with a hypothetical training set X, defined by three attributes A_1, A_2 and A_3 and an associated class.

4.2.2   Assigning Weights to Vertices and Edges

This section gives the motivations for introducing labels and weights in an AbDG structure, and explains how labels associated with vertices and weights associated with edges of an AbDG are defined. Let X be a data set having N training instances from M different classes, $\{\omega_1, \omega_2, \ldots, \omega_M\}$, where each instance is described by p attributes and an associated class.

Procedure construct_p_partite_AbDG(X, λ)
Input:
X: data set with *N* instances, each described by *p* attributes {A₁, A₂, ..., Aₚ}
   and an associate class, from *M* possible classes.
λ: user defined attribute order.

% creating the set of vertices of the p-partite AbDG
Graph ← ∅
Vertices ← ∅
for $a \leftarrow 1$ to p **do**
  begin
    {I$_{a,1}$, I$_{a,2}$, ..., I$_{a,n_a}$} ←*discretize_range*(A$_a$)
    Vertices ← Vertices ∪ {I$_{a,1}$, I$_{a,2}$, ..., I$_{a,n_a}$}
  end

% creating the set of edges of the p-partite AbDG
{A₁, A₂, ..., Aₚ} ← *ordering_and_renaming_attrib*({A₁, A₂, ..., Aₚ},λ)
Edges ← ∅
for_all $x_b \in X$ **do**
  for a ← 1 to $p - 1$ **do**
    % interval I$_{a,k}$ represents vertex v$_{a,k}$ and interval I$_{a+1,q}$ vertex v$_{a+1,q}$
    if x$_{b,a}$ ∈ I$_{a,k}$ and x$_{b,a+1}$ ∈ I$_{a+1,q}$ **then**
      begin
        (v$_{a,k}$, v$_{a+1,q}$) ← *create_edge*(v$_{a,k}$, v$_{a+1,q}$)
        Edges ← Edges ∪ {(v$_{a,k}$, v$_{a+1,q}$)}
      end
Graph ← (Vertices,Edges)
Graph ← *assign_weights*(Graph)
Output: Graph {a *p*-partite AbDG}

Fig. 4.2 High-level pseudocode for constructing a *p*-partite AbDG structure

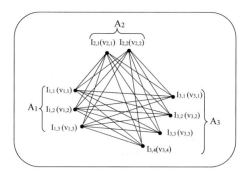

Fig. 4.3 A complete 3-partite AbDG structure created from a data set *X* with *N* vector-based instances, each described by 3 attributes, A_1, A_2 and A_3. The discretization process applied to each attribute associates to each attribute A_i ($1 \le i \le 3$), n_i intervals (i.e., vertices), namely $n_1 = 3$, $n_2 = 2$ and $n_3 = 4$. Edges are created between all vertices except those related with the same attribute

After the AbDG structure has been constructed, as a p-partite or complete p-partite, M-dimensional vectors are created and associated to each vertex and to each edge of the AbDG structure, by the two processes described next. In real domain data it is not common that instances sharing the same class are the only ones that have values of an attribute A_i $(1 \leq a \leq p)$ within one of the n_a subintervals (i.e., $I_{a,1}, \ldots, I_{a,n_a}$). Instances belonging to other classes may as well have their A_i values within that same subinterval. Aiming at evaluating the *representativeness* (i.e., the information for characterizing the class of an instance) of each particular subinterval created (i.e., each vertex of the AbDG), a M-dimensional weight vector is associated to each vertex.

4.2.2.1   Assigning Weights to Each Vertex of the AbDG

Let n_a represent the number of vertices associated with attribute A_a. The vertex $v_{a,k}$ $(1 \leq a \leq p$ and $1 \leq k \leq n_a)$ has an associated weight vector given by $\Gamma_{a,k} = \langle \gamma_1, \ldots, \gamma_j, \ldots, \gamma_M \rangle$, where γ_j relates to class ω_j, noted as $\Gamma_{a,k}(j)$. Considering that $I_{a,k}$ is the interval that defines $v_{a,k}$, $\Gamma_{a,k}(j)$ is defined by Eq. (4.1).

$$\Gamma_{a,k}(j) = P(\omega_j | I_{a,k}) = \frac{P(I_{a,k}, \omega_j)}{P(I_{a,k})} \qquad (4.1)$$

The joint probability in Eq. (4.1), $P(I_{a,k}, \omega_j)$, is the probability of a data instance having both i.e., class ω_j and its value of attribute A_a in the interval $I_{a,k}$. By rewriting the joint probability as $P(I_{a,k}, \omega_j) = P(\omega_j)P(I_{a,k}|\omega_j)$, the conditional probability $P(I_{a,k}|\omega_j)$ can be given by Eq. (4.2). $P(\omega_j)$ is the marginal probability of class ω_j, obtained by dividing the number of data instances belonging to class ω_j by the total number of data instances (i.e., N).

$$P(I_{a,k}|\omega_j) = \frac{|\{\mathbf{x}_i \in X \mid x_{i,a} \in I_{a,k} \wedge c_i = \omega_j\}|}{|\{\mathbf{x}_i \mid c_i = \omega_j\}|} \qquad (4.2)$$

In Eq. (4.1) the marginal probability, $P(I_{a,k})$, is the normalizing term, defined as the sum of the probabilities $P(I_{a,k}|\omega_j)$, for all possible classes i.e., $P(I_{a,k}) = \sum_{i=1}^{M} P(I_{a,k}|\omega_i)$.

4.2.2.2   Assigning Weights to Each Edge of the AbDG

The procedure for assigning a weight to an AbDG's edge is similar to the one for assigning a label to a vertex. Let $(v_{a,k}, v_{b,q})$ be an edge between the vertices representing the kth interval of attribute A_a and the qth interval of attribute A_b, and let this edge be weighted by the weight vector $\Delta_{k,q}^{a,b} = \langle \delta_1, \ldots, \delta_M \rangle$, where δ_j $(1 \leq j \leq M)$ is associated to class ω_j, noted as $\Delta_{k,q}^{a,b}(j)$. The edge weight δ_j $(1 \leq j \leq M)$ represents the probability of a given data instance $\mathbf{x}_i$, with attribute value $x_{i,a} \in I_{a,k}$ and $x_{i,b} \in I_{b,q}$, belonging to class ω_j, as given by Eq. (4.3).

$$\Delta_{k,q}^{a,b}(j) = P(\omega_j|I_{a,k}, I_{b,q}) = \frac{P(I_{a,k}, I_{b,q}, \omega_j)}{P(I_{a,k}, I_{b,q})} \tag{4.3}$$

Considering that $P(I_{a,k}, I_{b,q}, \omega_j) = P(\omega_j)P(I_{a,k}, I_{b,q}|\omega_j)$, then define $P(I_{a,k}, I_{b,q}|\omega_j)$ as the ratio of the number of instances belonging to class ω_j, whose values of attribute A_a lay within the kth interval and those of the attribute A_b lay within the qth interval, as in Eq. (4.4).

$$P(I_{a,k}, I_{b,q}|\omega_j) = \frac{|\{\mathbf{x}_i \in X | c_i = \omega_j \wedge x_{i,a} \in I_{a,k} \wedge x_{i,b} \in I_{b,q}\}|}{|\{\mathbf{x}_i | c_i = \omega_j\}|} \tag{4.4}$$

The probability of a data instance to have attribute values belonging to interval $I_{a,k}$ and $I_{b,q}$, regardless its class label, is the normalizing term in Eq. (4.3) and is given by the sum of Eq. (4.4), over all classes, as states Eq. (4.5).

$$P(I_{a,k}, I_{b,q}) = \sum_{j=1}^{M} P(I_{a,k}, I_{b,q}, \omega_j) \tag{4.5}$$

4.2.3 Computational Complexity for Building an AbDG

The complexity order for constructing an AbDG has been completely derived in Refs. [8, 9]. In what follows, a brief overview on the computational complexity required to build the AbDG is presented. Consider the complexity order with respect to the size of the training set, N, and to the number of attributes, p; thus building the AbDG involves:

1. **Constructing the vertex set**, which depends on the employed discretization method. As sorting is usually required as a preprocessing step to various discretization methods, building the vertex set has order of $O(pN\log N)$. If $p << N$, which is true for most domains, than building the vertex set has order of $O(N\log N)$; otherwise, if $p \approx N$ it can scale up to the order of $O(N^2\log N)$.
2. **Defining the weights of an AbDG** for vertices and edges, has complexity order of $O(N)$. The complexity order associated to the number of attributes for vertex weighting has order of $O(p)$. Edge weighting depends on the graph structure; for the p-partite structure the order is linear on the number of attributes, $O(p)$, while for the complete p-partite, the complexity order of edge weighting scales quadratically to the number of attributes, $O(p^2)$.

Therefore, building an AbDG has an order of $O(N\log N)$ with the size of the data set, when the discretization method requires sorting. What is costly about building the graph is sorting each attribute prior to apply some discretization method, as the MDLPC [18] for instance, to obtain the vertices. However, if some heuristic is employed, as dividing each attribute into subintervals of equal length, and thus not requiring sorting, building the vertex set has complexity of $O(N)$. Results regarding both ways to build the graph are reported in [9]; indeed, the equi-sized version presented comparative, or even better results than those versions which employ a discretization method.

4.3   Using the AbDG Structure for Classification Tasks

Once the construction of an AbDG for representing a given data set X of instances has finished, it can be used as the source of information on X for various tasks; among them, it can support a classifier, provided a procedure for exploring the information stored in the AbDG is defined. This section focuses on the description of such procedure. Taking into account a given AbDG, the assignment of classes to new unclassified instances (which can be modelled as a p-partite sub-graph of the given AbDG), can be conducted by checking how the subgraph defined by an instance, conforms to the existing connection patterns in the AbDG, embedded in its structure. As a consequence, the graph structure has a vital importance on the classification accuracy of AbDG-based classifiers.

Among the various possible ways to combine the information given by an AbDG, the proposal described next has shown to be a sound alternative for implementing a classification process based on a AbDG, and is based on calculating the product of vertex weights and the sum of edge weights. Consider classifying a new data instance $\mathbf{y}$. Given the AbDG and $\mathbf{y}$, two conditional probabilities, $P(\mathbf{y}|\omega_i)$ and $Q(\mathbf{y}|\omega_i)$, can be calculated. $P(\mathbf{y}|\omega_i)$ relates $\mathbf{y}$ to the vertex set of the AbDG, and $Q(\mathbf{y}|\omega_i)$ relates $\mathbf{y}$ to the edge set of the AbDG. Equations (4.6) and (4.7) describe both probabilities, respectively, for a p-partite AbDG.

$$P(\mathbf{y}|\omega_i) = \frac{PW(\mathbf{y})_{\omega_i}}{\sum_{j=1}^{M} PW(\mathbf{y})_{\omega_j}} \quad ; \quad PW(\mathbf{y})_{\omega_i} = \prod_{a=1, y_a \in I_{a,k}}^{p} \gamma_i \in \Gamma_{a,k} \tag{4.6}$$

$$Q(\mathbf{y}|\omega_i) = \frac{SW(\mathbf{y})_{\omega_i}}{\sum_{j=1}^{M} SW(\mathbf{y})_{\omega_j}} \quad ; \quad SW(\mathbf{y})_{\omega_i} = \sum_{a=1, b=a+1, y_a \in I_{a,k} \wedge y_b \in I_{b,q}}^{p-1} \delta_i \in \Delta_{k,q}^{a,b} \tag{4.7}$$

After determining both probabilities, an estimate for the class label of the $\mathbf{y}$ instance is given by Eq. (4.8). The class inferred for the new data $\mathbf{y}$, noted $\varphi(\mathbf{y})$, is the one having the greatest value for the mean of the normalized probabilities.

$$\varphi(\mathbf{y}) = \arg \max_{\{\omega_j | j=1,...,M\}} \left(\eta P(\mathbf{y}|\omega_j) + (1-\eta) Q(\mathbf{y}|\omega_j) \right) \tag{4.8}$$

In Eq. (4.8), η is a parameter of the classifier which allows to vary the emphasis between the conditional probabilities and enhances the flexibility of the model. The classification of **y** can be approached as a graph matching process. First the p attribute values of **y** help to detect the corresponding interval associated with each of the attributes i.e., the p vertices defining the subgraph representing **y**. Then edges connecting each pair of vertices in sequence are added, resulting in a subgraph structure of the AbDG. Then, the weights associated to the graph help to define the class of **y**, as the one that promotes the best matching.

An in-depth detailed analysis of the complexity orders of AbDG-related algorithms, considering both structures, the p-partite and the complete p-partite, is presented in [9]. Both structures share the same process for the vertex set definition, as well as the corresponding vertex labeling process, and differ in relation to the set of edges they have. In the referred work, both structures are approached separately when dealing with the complexity of the algorithm for constructing the edge set and the corresponding weighting process. Also, in [9], the authors present an empirical validation of the AbDG-based algorithms by conducting an experimental comparative analysis of classification results with other five well-known methods namely, the C4.5 [32], the multi-interval ID3 [18], the weighted KNN [20], the Probabilistic Neural Networks [34] and the Support Vector Machine [35]. Statistical analyses conducted by the authors show evidence that the AbDG approach has significantly better performance than four out of the five chosen algorithms.

4.4 Using an AbDG for Classification Purposes - A Case Study

This section presents a simple example illustrating the classification process based on an AbDG. Figure 4.4 shows a 3-partite AbDG structure, similar to the one depicted in Fig. 4.1, but now with the associated values for the intervals. In the figure, A_1 and A_2 are numerical attributes whose values are in the interval [0, 10] and [0, 1], respectively. A_3 is a categorical attribute having values in the set {0, 1, 2, 3}. Consider classifying the instance **y** = (5.5, 0.31, 2), whose match against the AbDG is shown

Fig. 4.4 Match of instance **y** = (5.5, 0.31, 2) to a particular 3-partite AbDG with the purpose of classification

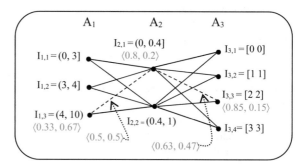

in the figure (dashed line). For the sake of visualization, only the weights associated with matching the sub-graph representing $\mathbf{y}$ against the AbDG are displayed. The task is to classify $\mathbf{y}$ into one of two possible classes $\{\omega_1, \omega_2\}$.

The classification procedure is carried out by matching the unlabeled instance against the AbDG followed by estimating the probabilities, as stated in Eqs. (4.6) and (4.7), for all classes in the problem. In the case study, given $\mathbf{y} = (5.5, 0.31, 2)$, the following matches to the AbDG are obtained: $5.5 \in I_{1,3} = (4, 10); 0.31 \in I_{2,1} = (0, 0.4]$ and $2 \in I_{3,3} = [2\ 2]$. For classification purposes, when a new value falls off the attribute range it is considered to belong to the nearest one. As a consequence, the weight vectors to be used are: $\Gamma_{1,3} = \langle 0.33, 0.67 \rangle$; $\Gamma_{2,1} = \langle 0.8, 0.2 \rangle$ and $\Gamma_{3,3} = \langle 0.85, 0.15 \rangle$ and the corresponding edges weights are $\Delta_{3,1}^{1,2} = \langle 0.5, 0.5 \rangle$ and $\Delta_{1,3}^{2,3} = \langle 0.63, 0.47 \rangle$. Next, PW and SW are calculated according to Eqs. (4.6) and (4.7).

$$PW(\mathbf{y})_{\omega_1} = 0.33 \times 0.8 \times 0.85 = 0.2244$$
$$SW(\mathbf{y})_{\omega_1} = 0.5 + 0.63 = 1.13$$
$$PW(\mathbf{y})_{\omega_2} = 0.67 \times 0.2 \times 0.15 = 0.0201$$
$$SW(\mathbf{y})_{\omega_2} = 0.5 + 0.47 = 0.97$$

Therefore the probabilities for each class are given as follows,

$$P(\mathbf{y}|\omega_1) = 0.2244/0.2445 = 0.918$$
$$Q(\mathbf{y}|\omega_1) = 1.13/2.1 = 0.538$$
$$P(\mathbf{y}|\omega_2) = 0.0201/0.2445 = 0.082$$
$$Q(\mathbf{y}|\omega_2) = 0.97/2.1 = 0.462$$

Finally, considering $\eta = 0.5$, according to Eq. (4.8) $\mathbf{y}$ is classified in class ω_1, since $0.5 \times 0.918 + 0.5 \times 0.538 > 0.5 \times 0.082 + 0.5 \times 0.462$.

4.5   Using the AbDG Structure for Imputation Tasks

Missing attribute values is a common problem present in almost every kind of real world application. The most frequent solutions to handle such problem reported in the literature are: (1) remove the instances having missing attribute values; (2) employ an imputation algorithm as a preprocessing step to the learning method and (3) adopt a learning method having internal mechanisms that enable training and classifying in the presence of missing attribute values. Regarding the previous alternatives, the first is the most used and may work well for applications having a few missing values in the training set and, also, those where ignoring a test instance with a missing value is acceptable. Clearly, this method imposes too many restrictions and it can be applied to very specific tasks. Therefore, alternatives (2) and (3) are more appealing to the majority of real world applications.

---

*Procedure imputation_from_AbDG(G, **x**)*
Input:
G: An AbDG built from a data set X with N instances, each described by p attributes $\{A_1, A_2, ..., A_p\}$ and an associate class, from M possible classes.
x: A data instance with missing value.

for $a \leftarrow 1$ to p **do**
  begin
    if $x_a = \varnothing$ **then**
    begin
      $s_a \leftarrow \{0\}_{n_a}$ % *each position in s represents an interval of A_a in G*

      for $b \leftarrow 1$ to p **do**
      begin
        if $x_b \neq \varnothing$ and $a \neq b$ **then**
        begin
          find the interval in G, into which x_b belongs to, referred to as q.
          knowing q, calculate $s_{a,k}$, for $k = 1, ..., n_a$, as in Eq. (4.9).
        end
      end
      $I_a = \text{argmax}\ s_{a,k}$, for $k = 1, ..., n_a$

      $x_a \leftarrow infer\_value(I_a)$
    end
  end
Output: Completed **x**

Fig. 4.5 High-level pseudocode for conducting an imputation process through an AbDG

The AbDG approach can be employed in both situations, either as an imputation method used to infer plausible values for the missing ones, prior to some other learning algorithm [8], or as a classification method able to handle the missing values found at the training or the classification phase [9]. The core mechanism to handle missing values through an AbDG, for both tasks, is practically the same, and is outlined in Fig. 4.5.

Let G be an AbDG and **x** be a data instance with, at least, a missing attribute value, say x_a. The imputation procedure based on AbDGs aims to find the interval of A_a in G, where the value of x_a should belong to. Thus, for each one of the n_a intervals of A_a, an estimate $s_{a,k}$ ($1 \le k \le n_a$) is calculated, taking into account the sub-graph resulted from the match between the existing values of **x** and the AbDG. Let $\Gamma_{a,k}(j)$ be the weight of the vertex associated to class ω_j which $s_{a,k}$ represents; and for each existing value in **x**, say x_b, laying in interval q of attribute A_b, let β_b, $\Gamma_{b,q}(j)$ and $\Delta_{k,q}^{a,b}(j)$ be the information gain (or some other attribute measure) of A_b, the weight of the vertex $v_{b,q}$ and the weight of the edge $(v_{a,k}, v_{b,q})$ associated to class ω_j, respectively; $s_{a,k}$, for a complete p-partite graph, is given by Eq. (4.9).

$$s_{a,k} = \Gamma_{a,k}(j) \sum_{b=1, b \neq a, \exists x_{i,b} \wedge x_{i,b} \in I_{b,q}}^{p} \beta_b \Gamma_{b,q}(j) \Delta_{k,q}^{a,b}(j) \qquad (4.9)$$

Once all the $s_{a,k}$, $1 \leq k \leq n_a$, have been calculated, the one with the highest value indicates the most plausible interval into which the missing attribute value should belong to. If the AbDG has been used as a classifier, knowing the interval is enough to proceed. However, if it has been used as an imputation method, a step to infer an actual value is necessary. In the pseudocode given in Fig. 4.5, the procedure *infer_value()* infers a single value, given an interval as argument; possible methods which could implemented are the mean, the mode, a random number within the interval, and so on. Notice that in Eq. (4.9) the used weights are all from a single class, ω_j, which is the same class as the data instance being imputed. When imputing from unlabeled data instances, the process is carried out for all possible classes and, then, the highest estimate, considering all classes, is chosen as the one which the missing value should belong to.

As commented earlier, the AbDG has been used for imputing missing data as a preprocessing step to a learning method. It has been compared against the following imputation methods: CART, MEAN, Bayesian Linear Regression, Fast Imputation, Linear Regression and Random Forests (see [8] for details). Several imputation methods were employed as a preprocessing step for the learning algorithms: Support Vector Machines, Multinomial Logistic Regression, Naive Bayes, Multilayer Perceptron, Parzen classifier, K-nearest neighbor, Classification And Regression Tree and Probabilistic Neural Networks (details can also be found in [8]).

The AbDG has showed the overall best results and the most stable performance along a varying rate of missing values. Not only has the AbDG showed its effectiveness to deal with missing value as an imputation method but, in [9], it has been tested as a classification algorithm that automatically handles missing values. When compared to the C4.5 and CART, which are two algorithms that support missing values, the AbDG had showed superior performance and has confirmed itself as an efficient alternative to cope with missing data.

4.6 Searching for Refined AbDG Structures via Genetic Algorithms

The AbDG structures reviewed in this chapter were the p-partite structure, which has subsets of vertices consecutively connected, based on a pre-defined order of attributes, and the complete p-partite. However, several other graph-based structures can be devised for the same purpose. As pointed out before, during the creation of an AbDG, the only restriction to take into account, when creating its corresponding set of edges, is not to create edges between vertices associated with the same attribute.

As far as the construction of the AbDG edge set is concerned both structures, the p-partite and the complete p-partite, have minor drawbacks, mostly related to their fixed (although dependent on the discretization process applied to all attributes) number of edges, as well as the patterns of connections they establish. Also, add to that the fact that the algorithm for inducing a p-partite AbDG expects to be given, as input, the order in which attributes should be considered (see Fig. 4.2). Depending on the number of attributes that describe a given data set X, the task of defining a convenient order is an extra task to be conducted previously to the induction of the AbDG.

Both structures have all possible connections between vertices, but still subject to the restriction above mentioned and, if a p-partite, to the given order of attributes. So, considering their vast number of edges, both structures become capable enough to represent all sorts of data. It has been empirically verified that, most times, although depending of the data set, such massive creation of edges is not necessary. A smaller subset of them would suffice for representing a given data set X.

As an strategy for searching for AbDGs having their set of edges customized to the given data set X, the work described in [4] explores the use of Genetic Algorithms (GAs) for inducing parts of an AbDG structure, specifically, a more customized set of edges, to the given set X. In the work a GA-based algorithm named *GA-AbDG* was proposed, for searching for a suitable edge set for a partially constructed AbDG, which only has its vertex set defined, aiming at finding a more refined set of edges, which could represent X better than both, *p-partite* and *complete p-partite*.

The GA starts with a population of N_P randomly generated individuals, where each individual is an AbDG classifier; individuals differ from each other only in relation to their edge set. The algorithm aims at identifying the best possible AbDG among all individuals in the population, using as criteria the value of their accuracy, by evolving their associated edge sets.

Let P represent a population of individuals such that $|P| = N_P$. At each iteration, a number of $N_{best} < N_P$ individuals from P are selected (based on their accuracy values in a given validation set) for composing the next population. The selected individuals are then used to create new individuals, which will, eventually, replace those considered not suitable enough (i.e., those with low accuracy values), when defining the new population. The new individuals are created by crossover up to restoring the population to its original size (i.e., N_P). At each generation, any individual, except for the overall best (elitism), is suitable to undergo the mutation operator. The evolutionary process is controlled by an user-defined number of iterations ($itMax$). At the end of the process, the AbDG with the highest accuracy is selected.

Before presenting the operators, a formal notation is introduced. An AbDG graph, $G = (V, E)$, can be viewed as a set of vertices (V) and a set of edges (E). If G is a p-partite graph, its set of vertices can be described as a set of disjoint vertex subsets, $V = \{V_1, \ldots, V_p\}$, where set V_a stands for the set of vertices obtained from discretizing the values associated with attribute A_a, $a = 1, \ldots, p$. Similarly, the edge set E can be written as the set of all edge sets between every pair of distinct attributes V_a and V_b, for $a = 1, \ldots, p - 1$, $b = 2, \ldots, p$ and $b > a$, as $E = \{E_{1,2}, \ldots, E_{p-1,p}\}$. Hence, resulting in $\binom{p}{2}$ subsets, where $E_{a,b}$ comprises the set of all possible edges between vertices in V_a and V_b. The description of an AbDG as a chromosome is given by the description of its edge set. Each edge set $E_{a,b}$ can be represented by a $|V_a| \times |V_b|$ matrix. In this way, an individual is represented by a set of $\binom{p}{2}$ matrices.

```
Procedure GA-AbDG(X, Y, itMax, N, Nbest, ρ, μ)
Input:
X: training labeled data set
Y: validation labeled data set
itMax: number of iterations
N: population size
Nbest: number of individuals to select at each iteration
ρ: crossover rate
μ: mutation rate

begin
    Pbest ← ∅
    build N AbDGs, each having a random edge set and compose P
    it ← 1
    while it ≤ itMax do
        begin
            P ← fitness(Y)
            Pbest ← selection(P,Nbest)
            P ← reproduction(Pbest)
            P ← crossover(P, ρ)
            P ← mutation(P,μ)
            it ← it + 1
        end
end
return: AbDG ∈ P with the highest fitness value.
```

Fig. 4.6 High-level pseudocode of the GA-AbDG procedure

As each individual in the population is an AbDG, let $G^{(i)} = (V^{(i)}, E^{(i)})$ be individual i, and straightforwardly $E^{(i)}_{a,b}$ be the set of edges between vertices from $V^{(i)}_a$ and $V^{(i)}_b$. The high level pseudocode of procedure GA-AbDG is given in Fig. 4.6. Following this notation, consider henceforward, i and j as indexes for parenting individuals, and o and m indexes for offspring individuals. In the following, each operator is described in details.

Reproduction - Reproduction is accomplished by selecting consecutive pairs of individuals ordered according to their fitness values. Each parenting pair $G^{(i)}$ and $G^{(j)}$ gives rise to two new offsprings $G^{(o)}$ and $G^{(m)}$. When obtaining each of them, each edge $(v_{a,k}, v_{b,q})$ that is common to both, $G^{(i)}$ and $G^{(j)}$, edge sets, is maintained in both offsprings i.e., $G^{(o)}$ and $G^{(m)}$. For those vertices that only belong to one of the parents, each offspring follows the configuration of one of the parents, with an associated probability of θ (whose value is a parameter to the reproduction procedure). The reproduction process implements a procedure that generates the offspring $G^{(o)}$ resembling to $G^{(i)}$; so, $G^{(o)}$ repeats the configuration of $G^{(i)}$ with probability θ and of $G^{(j)}$ with probability $1 - \theta$. Offspring $G^{(m)}$ that resembles $G^{(j)}$ is straightforward. Remember that each reproduction always generates two offspring.

Crossover - Also performed at every iteration, the crossover operator requires two parenting individuals $G^{(i)}$ and $G^{(j)}$, randomly selected from the N_{best} individuals, and also generates two offspring $G^{(o)}$ and $G^{(m)}$. Crossover is performed by randomly selecting a crossing point in the edge sets of both parents and exchanging their configuration. Crossover is considered at a rate of ρ, usually set to high values.

Mutation - At each iteration an individual has the probability μ of undergoing mutation. Mutation is implemented by randomly selecting a set of edges between two attributes e.g., $E_{a,b}$, for V_a and V_b. Then, for each possible pair in the set, with probability μ, the mutation operator is applied by adding an edge if such edge does not exist or by removing it otherwise.

The work described in [4] presents the classification results obtained with the C4.5 [32], the original AbDG, with a p-partite structure and the GA-AbDG, obtained as the result of the GA-based search process previously described, in 20 data sets from the UCI-Repository [26]. The results obtained with the GA-AbDG outperformed those obtained by both, the C4.5 and the original AbDG in 15 out of the 20 data sets used. The authors concluded that the improvements in classification performance achieved by the GA-AbDG over the original AbDG, makes the GA-based search aiming at finding a more suitable edge set worth the extra computational effort.

4.7 Conclusions

This chapter reviews a new data structure proposed in the literature as a suitable way of condensing and representing the information contained in a training set. The structure was devised to be used mainly by classification and imputation algorithms, in supervised automatic learning environments. It is named *Attribute-based Data Graph (AbDG)* and it can be described as a labeled p-partite weighted graph.

Taking into account a few other graph-based approaches for representing data, found in the literature, the main novelty introduced by AbDGs relates to the role played by the AbDG vertices. While in traditional methods vertices represent training instances, in the AbDG they represent intervals of values related to attributes that describe the training instances. The AbDG approach is a new way to build a graph from data which provides a different and more compact way of data representation for data mining tasks.

This chapter presents and discusses in detail various formal concepts and procedures related to the design, construction, and use of AbDGs, namely:

- The creation of the set of vertices of the graph, which involves the choice and use of discretization methods;
- Two different ways edges can be inserted, either constructing a *p-partite* or, then, a *complete p-partite* graph-based structure;
- Several technicalities and formal concepts involved in vertex labeling and edge weighting procedures, which play a fundamental role in adjusting the AbDG structure for representing X;

- A procedure for using a given AbDG for classification purposes, which can be approached as a method for determining how a subgraph of the AbDG, representing the new unclassified instance conforms to the AbDG structure;
- A GA-based procedure which aims at searching for a suitable set of edges of an AbDG, so to better represent a given input data set.

The chapter also briefly introduces a few other issues related to an AbDG structure, particularly its contribution for supporting imputation processes, as described in [7, 8]. Although this chapter has no focus on experiments and analyses of their results, many such results and analyses can be found in a number of works cited in this chapter. It is a fact though that most of the experimental results published can be considered evidence of the suitability of the AbDG structure for summarizing and representing data, as well as the great potential of the proposed algorithms involved in the AbDG construction and use, mainly due to their robustness, low computational costs associated with training and memory occupation.

References

1. Agarwal, S., Branson, K., Belongie, S.: Hihger order learning with graphs. In: Proceedings of the 23rd International Conference on Machine Learning (ICML 2006), pp. 17–24. ACM, New York (2006)
2. Aupetit, M.: Learning topology with the generative Gaussian graph and the em algorithm (2006). In: Weiss, Y., Schölkopf, B., Platt, J. (eds.) Advances in Neural Information Processing Systems 18, pp. 83–90. MIT Press, Cambridge (2006)
3. Belkin, M., Niyogi, P., Sindhwani, V.: Manifold regularization: a geometric framework for learning from labeled and unlabeled examples. J. Mach. Learn. Res. 1, 1–48 (2006)
4. Bertini Jr., J.R., Nicoletti, M.C.: A genetic algorithm for improving the induction of attribute-based decision graph classifiers. In: Proceedings of the 2016 IEEE Congress on Evolutionary Computation (CEC), pp. 4104–4110. IEEE Press, New York (2016)
5. Bertini Jr., J.R., Nicoletti, M.C., Zhao, L.: Attribute-based decision graphs for multiclass data classification. In: Proceedings of the IEEE Congress on Evolutionary Computation (CEC), pp. 1779–1785 (2013)
6. Bertini Jr., J.R., Nicoletti, M.C., Zhao, L.: Ensemble of complete p-partite graph classifiers for non-stationary environments. In: Proceedings of the IEEE Congress on Evolutionary Computation (CEC), pp. 1802–1809 (2013)
7. Bertini Jr., J.R., Nicoletti, M.C., Zhao, L.: Imputation of missing data supported by complete p-partite attribute-based decision graph. In: Proceedings of the International Joint Conference on Neural Networks (IJCNN), pp. 1100–1106 (2014)
8. Bertini Jr., J.R., Nicoletti, M.C., Zhao, L.: An embedded imputation method via attribute-based decision graphs. Expert Syst. Appl. 57, 159–177 (2016)
9. Bertini Jr., J.R., Nicoletti, M.C., Zhao, L.: Attribute-based decision graphs: a framework for multiclass data classification. Neural Netw. 85, 69–84 (2017)
10. Bertini Jr., J.R., Zhao, L., Motta, R., Lopes, A.A.: A nonparametric classification method based on k-associated graphs. Inf. Sci. 181, 5435–5456 (2011)
11. Bi, W., Kwok, J.: Multi-label classification on tree and dag-structured hierarchies. In: Proceedings of the 28th International Conference on Machine Learning (ICML 2011), pp. 17–24. ACM, New York (2011)
12. Chapelle, O., Zien, A., Schölkopf, B. (eds.): Semi-supervised Learning, 1st edn. MIT Press, Cambridge (2006)

13. Cheh, Z., Zhao, H.: Pattern recognition with weighted complex networks. Phys. Rev. E **78**(056107), 1–6 (2008)
14. Chen, J., Fang, H.R., Saad, Y.: Fast approximate knn graph construction for high dimensional data via recursive lanczos bisection. J. Logic Program. **10**, 1989–2012 (2009)
15. Cook, D., Holder, L.: Graph-based data mining. IEEE Intell. Syst. **15**(2), 32–41 (2000)
16. Culp, M., Michailidis, G.: Graph-based semisupervised learning. IEEE Trans. Pattern Anal. Mach. Intell. **30**(1), 174–179 (2008)
17. Eppstein, D., Paterson, M.S., Yao, F.: On nearest-neighbor graphs. Discret. Comput. Geom. **17**(3), 263–282 (1997)
18. Fayyad, U.M., Irani, K.B.: Multi-interval discretization of continuous valued attributes for classification learning. In: Proceedings of the 13th International Joint Conference on Artificial Intelligence, vol. 2, pp. 1022–1027. Morgan Kaufmann Publishers, San Francisco (1993)
19. Fortunato, S.: Community detection in graphs. Phys. Rep. **486**, 75–174 (2010)
20. Hastie, T., Tibshirani, R., Friedman, J.: The Elements of Statistical Learning: Data Mining, Inference and Prediction, 2nd edn. Springer, Canada (2009)
21. Hein, M., Audibert, J.Y., von Luxburg, U.: Discretization: an enabling technique. Data Min. Knowl. Disc. **6**, 393–423 (2002)
22. Hein, M., Audibert, J.Y., von Luxburg, U.: Graph Laplacians and their convergence on random neighborhood graphs. J. Mach. Learn. Res. **8**, 1325–1368 (2007)
23. Holder, L., Cook, D.: Graph-based relational learning: current and future directions. ACM SIGKDD Explor. **5**(1), 90–93 (2003)
24. Holder, L., Cook, D., Coble, J., Mukherjee, M.: Graph-based relational learning with application to security. Fundamenta Informaticae **66**, 1–19 (2005)
25. Jensen, D., Neville, J., Gallagher, B.: Why collective inference improves relational classification? In: Proceedings of the 10th International Conference on Knowledge Discovery and Data Mining (ACM SIGKDD'04), pp. 593–598. ACM (2004)
26. Lichman, M.: UCI machine learning repository (2013). http://archive.ics.uci.edu/ml
27. Macskassy, S., Provost, F.: A simple relational classifier. In: Proceedings of the International Conference on Knowledge Discovery and Data Mining (KDD), Workshop on Multi-Relational Data Mining, pp. 64–76 (2003)
28. Malliaros, F., Vazirgiannis, M.: Clustering and community detection in directed networks: a survey. Phys. Rep. **533**, 95–142 (2013)
29. Muggleton, S.: Inductive logic programming: issues, results and the challenge of learning language in logic. Artif. Intell. **114**, 283–296 (1999)
30. Muggleton, S., Raedt, L.D.: Inductive logic programming: theory and methods. J. Logic Progr. **19–20**, 629–679 (1994)
31. Newman, M.: The structure and function of complex networks. SIAM Rev. **45**(2), 167–256 (2003)
32. Quinlan, J.R.: C4.5 Programs for Machine Learning, 1st edn. Morgan Kaufmann Publishers, San Francisco (1993)
33. Schaeffer, S.: Graph clustering. Comput. Sci. Rev. **1**, 27–34 (2007)
34. Specht, D.F.: Probabilistic neural networks. Neural Netw. **3**, 109–118 (1990)
35. Vapnik, V.: The nature of statistical learning theory. Springer, Berlin (1999)
36. Wenga, L., Dornaikab, F., Jina, Z.: Graph construction based on data self-representativeness and Laplacian smoothness. Neurocomputing **207**, 476–487 (2016)
37. Xie, J., Kelley, S., Boleslaw, K.S.: Overlapping community detection in networks: the state-of-the-art and comparative study. ACM Comput. Surv. **45**(43), 1–35 (2013)
38. Yang, Z., Cohen, W.W., Salakhutdinov, R.: Revisiting semi-supervised learning with graph embeddings. In: Proceedings of the 33rd International Conference on Machine Learning (2016)
39. Yoshida, K., Motoda, H., Indurkhya, N.: Graph-based induction as a unified learning framework. J. Appl. Intell. **4**, 297–328 (1994)
40. Zhu, X.: Semi-supervised learning literature survey. Technical report 1530, Computer-Science, University of Wisconsin-Madison (2008)
41. Zhu, X., Lafferty, J., Ghahramani, Z.: Semi-supervised learning: from Gaussian fields to Gaussian processes. Technical report CMU-CS-03-175, Carnegie Mellon University (2003)

Chapter 5
Optimization of Decision Rules Relative to Length Based on Modified Dynamic Programming Approach

Beata Zielosko and Krzysztof Żabiński

Abstract This chapter is devoted to the modification of an extension of dynamic programming approach for optimization of decision rules relative to length. "Classical" dynamic programming approach allows one to obtain optimal rules, i.e., rules with the minimum length. This fact is important from the point of view of knowledge representation. The idea of the dynamic programming approach for optimization of decision rules is based on a partitioning of a decision table into subtables. The algorithm constructs a directed acyclic graph. Basing on the constructed graph, sets of rules with the minimum length, attached to each row of a decision table, can be described. Proposed modification is based on the idea that not the complete graph is constructed but its part. It allows one to obtain values of length of decision rules close to optimal ones, and the size of the graph is smaller than in case of "classical" dynamic programming approach. The chapter also contains results of experiments with decision tables from UCI Machine Learning Repository.

Keywords Decision rules · Dynamic programming approach · Length · Optimization

5.1 Introduction

Feature selection domain have become more and more important in recent years, especially in areas of application for which data sets contain a huge number of attributes. For example, sequence-pattern in bioinformatics, genes expression analysis, market basket analysis, stock trading, and many others. Data sets can contain either insufficient or redundant attributes used for knowledge induction from

B. Zielosko (✉) · K. Żabiński
Institute of Computer Science, University of Silesia in Katowice, 39, Będzińska St.,
41-200, Sosnowiec, Poland
e-mail: beata.zielosko@us.edu.pl

K. Żabiński
e-mail: krzysztof.kamil.zabinski@gmail.com

© Springer International Publishing AG 2018   73
U. Stańczyk et al. (eds.), *Advances in Feature Selection for Data and Pattern
Recognition*, Intelligent Systems Reference Library 138,
https://doi.org/10.1007/978-3-319-67588-6_5

these sets. The problem is how to select the relevant features that allow one to obtain knowledge stored in the data. The main objectives of variable selection are providing a better understanding of the data and improving the prediction performance of the classifiers. There are different approaches and algorithms for features selection [12–15, 18, 19, 31, 33]. They are typically divided into three groups: filter, wrapper and embedded methods, however, algorithms can be also mixed together in different variations.

Filter methods are independent from classification systems. Filters pre-process data sets without any feedback information about the classification result improvement. Their main advantage is that they tend to be faster and less resource demanding than other approaches. Their main drawback is what makes them fast and easily applicable in almost all kinds of problems, i.e., neglecting the real-time influence on the classification system.

Wrapper group of algorithms bases on the idea of examining the influence of the choice of subsets of features on the classification result. Wrapper approach can be interpreted as a system with feedback. Such an approach generally requires large computational costs as the classification step needs to be repeated many times.

The last group of algorithms is known as embedded solutions. Generally, they consist of mechanisms that are embedded directly into the learning algorithm. These mechanisms are responsible for the feature selection process at the learning stage. Their advantage is a good performance as the solutions are dedicated to the specific application. Nevertheless, they are impossible to be used without knowing the learning algorithm characteristics.

The idea presented in this chapter refers to filter methods. It is a modified extension of dynamic programming approach for optimization of rules relative to length.

Decision rules are considered as a way of knowledge representation and the aim of this study is to find the values of length of decision rules close to optimal ones, i.e., decision rules with minimum length. Shorter rules are better from the point of view of understanding and interpretation by experts. Unfortunately, the problem of construction of rules with minimum length is *NP*-hard [21, 23]. The majority of approaches, with the exception of brute-force, Boolean reasoning and dynamic programming approach (later called as "classical") cannot guarantee the construction of optimal rules.

Classical dynamic programming method of rule induction and optimization is based on the analysis of the directed acyclic graph constructed for a given decision table [2, 22]. Such graph can be huge for larger data sets. The aim of this chapter is to present a modification of the algorithm for graph construction which allows one to obtain values of length of decision rules close to optimal ones, and the size of the graph will be smaller than in case of a "classical" dynamic programming approach. This modification is based on attributes' values selection. Instead of all values of all attributes included in a data set, only one attribute with the minimum number of values and all its values are used, and for the rest of attributes – the most frequent value of each attribute is selected.

The chapter consists of six sections. Section 5.2 presents background information. Section 5.3 contains main notions connected with decision tables and decision rules.

In Sect. 5.4, modified algorithm for a directed acyclic graph construction is presented. Section 5.5 is devoted to description of a procedure of optimization of the graph relative to length. Section 5.6 contains results of experiments with decision tables from UCI Machine Learning Repository, and Sect. 5.7 – conclusions.

5.2 Background

Decision rules are a known and popular form of knowledge representation. They are used in many areas connected with data mining and knowledge discovery. A significant advantage of decision rules is their simplicity and ease in being understood and interpreted by humans.

There are many approaches to the construction of decision rules: Boolean reasoning [24, 25], brute-force approach which is applicable to tables with relatively small number of attributes, dynamic programming approach [2, 22, 36], separate-and-conquer approach (algorithms based on a sequential covering procedure) [7–11], algorithms based on decision tree construction [20, 22, 27], genetic algorithms [3, 30], ant colony optimization algorithms [16, 17], different kinds of greedy algorithms [21, 24]. Each method has different modifications. For example, in case of greedy algorithms different uncertainty measures can be used to construct decision rules. Also, there are different rule quality measures that are used for induction or classification tasks [4, 6, 29, 32, 34, 35].

Induction of decision rules can be performed from the point of view of (i) knowledge representation or (ii) classification. Since the aims are different, algorithms for construction of rules and quality measures for evaluating of such rules are also different.

In the chapter, the length is considered as a rule's evaluation measure. The choice of this measure is connected with the Minimum Description Length principle introduced by Rissanen [28]: the best hypothesis for a given set of data is the one that leads to the largest compression of data.

In this chapter, not only exact but also approximate decision rules are considered. Exact decision rules can be overfitted, i.e., dependent essentially on the noise or adjusted too much to the existing examples. If decision rules are considered as a way of knowledge representation, then instead of exact decision rules with many attributes, it is more appropriate to work with approximate ones which contain smaller number of attributes and have relatively good accuracy.

To work with approximate decision rules, an uncertainty measure $R(T)$ is used. It is the number of unordered pairs of rows with different decisions in a decision table T. A threshold β is fixed and so-called β-decision rules are considered that localize rows in subtables of T with uncertainty at most β.

The idea of the dynamic programming approach for optimization of decision rules relative to length is based on partitioning of a table T into subtables. The algorithm constructs a directed acyclic graph $\Delta_{\beta}^{*}(T)$ which nodes are subtables of the decision table T given by descriptors (pairs "attribute = value"). The algorithm finishes the partitioning of a subtable when its uncertainty is at most β. The threshold β helps one to control computational complexity and makes the algorithm applicable to solving more complex problems. For the "classical" dynamic programming approach [2] subtables of the directed acyclic graph were constructed for each value of each attribute from T. In the presented approach, subtables of the graph $\Delta_\beta^*(T)$ are constructed for all values of the one attribute from T with the minimum number of values, and for the rest of attributes - the most frequent value (value of an attribute attached to the maximum number of rows) of each attribute is chosen. So, the size of the graph $\Delta_\beta^*(T)$ (the number of nodes and edges) is smaller than the size of the graph constructed by the "classical" dynamic programming approach.

Basing on the graph $\Delta_\beta^*(T)$ sets of β-decision rules attached to rows of table T are described. Then, using a procedure of optimization of the graph $\Delta_\beta^*(T)$ relative to length, it is possible to find, for each row r of T, the shortest β-decision rule. It allows one to study how far the obtained values of length are from the optimal ones, i.e., the minimum length of rules obtained by the "classical" dynamic programming approach.

The main aim of this chapter is to study modified dynamic programming approach for optimization of decision rules relative to length, from the point of view of knowledge representation.

"Classical" dynamic programming approach, for decision rules optimization relative to length and coverage, was studied in [2, 22, 36]. Modified dynamic programming approach for optimization of decision rules relative to length was presented in [38], for optimization relative to coverage - in [37].

5.3 Main Notions

In this section, definitions of notions corresponding to decision tables and decision rules are presented. Definition of a decision table comes from Rough Sets Theory [26].

A decision table is defined as $T = (U, A \cup \{d\})$, where $U = \{r_1, \ldots, r_k\}$ is nonempty, finite set of objects (rows), $A = \{f_1, \ldots, f_n\}$ is nonempty, finite set of attributes. Elements of the set A are called conditional attributes, $d \notin A$ is a distinguishing attribute, called a decision attribute. It is assumed that decision table is consistent i.e., it does not contain rows with equal values of conditional attributes and different values of a decision attribute. An example of a decision table is presented in Table 5.1.

The number of unordered pairs of rows with different decisions is denoted by $R(T)$. This value is interpreted as an uncertainty of the table T. The table T is called *degenerate* if T is empty or all rows of T are labeled with the same decision. It is clear that in this case $R(T) = 0$.

A minimum decision value that is attached to the maximum number of rows in T is called *the most common decision for T*.

Table 5.1 Decision table T_0

$T_0 =$		f_1	f_2	f_3	
	r_1	0	0	1	1
	r_2	1	0	1	1
	r_3	1	1	1	2
	r_4	0	1	1	3
	r_5	0	0	0	3

A table obtained from T by removal of some rows is called a *subtable* of the table T. Let $f_{i_1}, \ldots, f_{i_m} \in \{f_1, \ldots, f_n\}$ be conditional attributes from T and $a_1, \ldots, a_m$ be values of such attributes. A subtable of the table T that contains only rows that have values $a_1, \ldots, a_m$ at the intersection with columns $f_{i_1}, \ldots, f_{i_m}$ is denoted by $T(f_{i_1}, a_1) \ldots (f_{i_m}, a_m)$. Such nonempty subtables (including the table T) are called *separable subtables*.

An attribute $f_i \in \{f_1, \ldots, f_n\}$ is *non-constant* on T if it has at least two different values. An attribute's value attached to the maximum number of rows in T is called *the most frequent value* of f_i.

A set of attributes from $\{f_1, \ldots, f_n\}$ which are non-constant on T is denoted by $E(T)$, and a set of attributes from $E(T)$ attached to the row r of T is denoted by $E(T, r)$.

For each attribute $f_i \in E(T)$, let us define a set $E^*(T, f_i)$ of its values. If f_i is an attribute with the minimum number of values, and it has the minimum index i among such attributes (this attribute will be called the *minimum attribute for T*), then $E^*(T, f_i)$ is the set of all values of f_i on T. Otherwise, $E^*(T, f_i)$ contains only the most frequent value of f_i on T.

The expression

$$f_{i_1} = a_1 \wedge \ldots \wedge f_{i_m} = a_m \to d \tag{5.1}$$

is called a *decision rule over T* if $f_{i_1}, \ldots, f_{i_m} \in \{f_1, \ldots, f_n\}$ are conditional attributes from T and $a_1, \ldots a_m, d$ are values of such attributes and a decision attribute, respectively. It is possible that $m = 0$. In this case (5.1) is equal to the rule

$$\to d. \tag{5.2}$$

Let $r = (b_1, \ldots, b_n)$ be a row of T. The rule (5.1) is called *realizable for r*, if $a_1 = b_{i_1}, \ldots, a_m = b_{i_m}$. If $m = 0$ then the rule (5.2) is realizable for any row from T.

Let β be a nonnegative real number. The rule (5.1) is *β-true for T* if d is the most common decision for $T' = T(f_{i_1}, a_1) \ldots (f_{i_m}, a_m)$ and $R(T') \leq \beta$. If $m = 0$ then the rule (5.2) is β-true for T if d is the most common decision for T and $R(T) \leq \beta$.

If the rule (5.1) is β-true for T and realizable for r, it is possible to say that (5.1) is a *β-decision rule for T and r*. Note that if $\beta = 0$, (5.1) is an exact decision rule for T and r.

Let τ be a decision rule over T and τ be equal to (5.1). The *length* of τ is the number of descriptors (pairs *attribute = value*) from the left-hand side of the rule, and it is denoted by $l(\tau)$.

In case of a "classical" extension of dynamic programming approach [2], so-called irredundant decision rules were considered. It was proved that by removing some descriptors from the left-hand side of each β-decision rule that is not irredundant and by changing the decision on the right-hand side of this rule it is possible to obtain an irredundant β-decision rule which length is at most the length of initial rule. It means that optimal rules (rules with the minimum length) among all β-decision rules were considered.

5.4　Modifed Algorithm for Directed Acyclic Graph Construction $\Delta_\beta^*(T)$

In this section, an algorithm which constructs a directed acyclic graph $\Delta_\beta^*(T)$ is considered (see Algorithm 5.1). Basing on this graph it is possible to describe the set of β-decision rules for T and for each row r of T.

Algorithm 5.1: Algorithm for construction of a graph $\Delta_\beta^*(T)$

Input : Decision table T with conditional attributes $f_1, \ldots, f_n$, nonnegative real number β.
Output: Graph $\Delta_\beta^*(T)$.
A graph contains a single node T which is not marked as processed;
while *all nodes of the graph are not marked as processed* **do**
    Select a node (table) Θ, which is not marked as processed;
    if $R(\Theta) \leq \beta$ **then**
    | The node is marked as processed;
    end
    if $R(\Theta) > \beta$ **then**
        for *each $f_i \in E(\Theta)$* **do**
        | draw edges from the node Θ;
        end
        Mark the node Θ as processed;
    end
end
return Graph $\Delta_\beta^*(T)$;

Nodes of the graph are separable subtables of the table T. During each step, the algorithm processes one node and marks it with the symbol *. At the first step, the algorithm constructs a graph containing a single node T which is not marked with the symbol *.

Let the algorithm have already performed p steps. Let us describe the step $(p + 1)$. If all nodes are marked with the symbol * as processed, the algorithm finishes its work and presents the resulting graph as $\Delta_\beta^*(T)$. Otherwise, choose a node (table) Θ, which has not been processed yet. If $R(\Theta) \leq \beta$ mark the considered node with symbol * and proceed to the step $(p + 2)$. If $R(\Theta) > \beta$, for each $f_i \in E(\Theta)$, draw a bundle of edges from the node Θ. Let $E^*(\Theta, f_i) = \{b_1, \ldots, b_t\}$. Then draw t edges from Θ and label these edges with pairs $(f_i, b_1), \ldots, (f_i, b_t)$ respectively. These edges enter to nodes $\Theta(f_i, b_1), \ldots, \Theta(f_i, b_t)$. If some of nodes $\Theta(f_i, b_1), \ldots, \Theta(f_i, b_t)$ are absent in the graph then add these nodes to the graph. Each row r of Θ is labeled with the set of attributes $E_{\Delta_\beta^*(T)}(\Theta, r) \subseteq E(\Theta)$. Mark the node Θ with the symbol * and proceed to the step $(p + 2)$.

The graph $\Delta_\beta^*(T)$ is a directed acyclic graph. A node of such graph will be called *terminal* if there are no edges leaving this node. Note that a node Θ of $\Delta_\beta^*(T)$ is terminal if and only if $R(\Theta) \leq \beta$.

Later, the procedure of optimization of the graph $\Delta_\beta^*(T)$ will be described. As a result a graph G is obtained, with the same sets of nodes and edges as in $\Delta_\beta^*(T)$. The only difference is that any row r of each nonterminal node Θ of G is labeled with a nonempty set of attributes $E_G(\Theta, r) \subseteq E_{\Delta_\beta^*(T)}(\Theta, r)$. It is also possible that $G = \Delta_\beta^*(T)$.

Now, for each node Θ of G and for each row r of Θ, the set of β-decision rules $Rul_G(\Theta, r)$, will be described. Let us move from terminal nodes of G to the node T.

Let Θ be a terminal node of G labeled with the most common decision d for Θ. Then

$$Rul_G(\Theta, r) = \{\to d\}.$$

Let now Θ be a nonterminal node of G such that for each child Θ' of Θ and for each row r' of Θ', the set of rules $Rul_G(\Theta', r')$ is already defined. Let $r = (b_1, \ldots, b_n)$ be a row of Θ. For any $f_i \in E_G(\Theta, r)$, the set of rules $Rul_G(\Theta, r, f_i)$ is defined as follows:

$$Rul_G(\Theta, r, f_i) = \{f_i = b_i \wedge \sigma \to k : \sigma \to k \in Rul_G(\Theta(f_i, b_i), r)\}.$$

Then

$$Rul_G(\Theta, r) = \bigcup_{f_i \in E_G(\Theta, r)} Rul_G(\Theta, r, f_i).$$

Example 5.1 To illustrate the algorithm presented in this section, a decision table T_0 depicted in Table 5.1 is considered. In the example, $\beta = 2$, so during the construction of the graph $\Delta_2^*(T_0)$ the partitioning of a subtable Θ is stopped when $R(\Theta) \leq 2$.

The set of non-constant attributes, for table T_0, is the following: $E_G(T_0) = \{f_1, f_2, f_3\}$. Each of attributes contains two values, $E^*(T_0, f_i) = \{0, 1\}$, $i = 1, 2, 3$, so the minimum attribute for T_0 is f_1. For this attribute all its values will be considered and separable subtable $\Theta_1 = T_0(f_1, 0)$ and $\Theta_2 = T_0(f_1, 1)$ will be created during the construction of the first level of the graph. For attributes f_2 and f_3 the most frequent values are 0 and 1 respectively, so separable subtables $\Theta_3 = T_0(f_2, 0)$ and $\Theta_4 = T_0(f_3, 1)$ are created. The first level of the graph $\Delta_2^*(T_0)$ is depicted in Fig. 5.1. In case of "classical" dynamic programming approach all values of each attribute from $E(T_0)$ are considered during directed acyclic graph construction, so for studied table T_0, the upper bound of the number of edges incoming to corresponding separable subtables at the first level of the graph is 6, each attribute has two values.

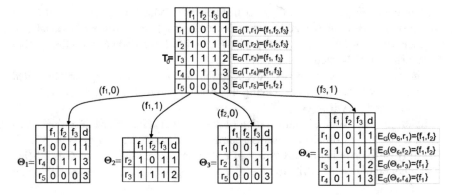

Fig. 5.1 The first level of the directed acyclic graph $G = \Delta_2^*(T_0)$

The condition for stop partitionig table into seprable subtables is $R(\Theta) \leq 2$, so it holds for Θ_1, Θ_2 and Θ_3. For subtable Θ_4, $R(\Theta_4) > 2$, so this table should be divided into subtables. $E_G(\Theta_4) = \{f_1, f_2\}$. Attributes f_1 and f_2 contains two values, so the minimum attribute for Θ_4 is f_1. For this attribute all its values are considered during the construction of the second level of the graph. Separable subtable $\Theta_4(f_1, 1)$ contains the same rows as Θ_2 which exists in the graph, so corresponding edge is guided from Θ_4 to Θ_2. Subtable $\Theta_5 = \Theta_4(f_1, 0)$ does not exist in the graph so it is created at the second level of the graph. For the attribute f_2 the most frequent value is 0, so corresponding subtable $\Theta_6 = \Theta_4(f_2, 0)$ is created.

The second level of the obtained graph is depicted in Fig. 5.2. The stopping condition for partitioning table into separable subtables holds for Θ_5 and Θ_6, so this graph is a directed acyclic graph for table T_0, and it is denoted by $G = \Delta_2^*(T_0)$. Green color denotes connections between parent and child tables containing row r_1 from table T_0. Labels of edges (descriptors) are used during description of rules for this row. As a result set $Rul_G(T, r_1)$ was obtained (see description below).

Now, for each node Θ of the graph G and for each row r of Θ the set $Rul_G(\Theta, r)$ is described. Let us move from terminal nodes of G to the node T. Terminal nodes of the graph G are Θ_1, Θ_2, Θ_3, Θ_5, and Θ_6. For these nodes,

$Rul_G(\Theta_1, r_1) = Rul_G(\Theta_1, r_4) = Rul_G(\Theta_1, r_5) = \{\to 3\}$;
$Rul_G(\Theta_2, r_2) = Rul_G(\Theta_2, r_3) = \{\to 1\}$;
$Rul_G(\Theta_3, r_1) = Rul_G(\Theta_3, r_2) = Rul_G(\Theta_3, r_5) = \{\to 1\}$;
$Rul_G(\Theta_5, r_1) = Rul_G(\Theta_5, r_4) = \{\to 1\}$;
$Rul_G(\Theta_6, r_1) = Rul_G(\Theta_6, r_2) = \{\to 1\}$.

Now, the sets of rules attached to rows of nonterminal node Θ_4 can be described. Children of this subtable (subtables Θ_2, Θ_5 and Θ_6) have already been treated.

$Rul_G(\Theta_4, r_1) = \{f_1 = 0 \to 1, f_2 = 0 \to 1\}$;
$Rul_G(\Theta_4, r_2) = \{f_1 = 1 \to 1, f_2 = 0 \to 1\}$;
$Rul_G(\Theta_4, r_3) = \{f_1 = 1 \to 1\}$;
$Rul_G(\Theta_4, r_4) = \{f_1 = 0 \to 1\}$.

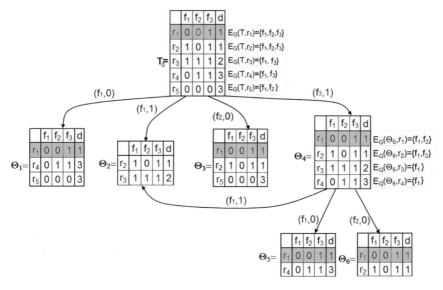

Fig. 5.2 Directed acyclic graph $G = \Delta_2^*(T_0)$ and description of rules for row r_1 from T

Finally, the sets of rules attached to rows of T_0, can be described:

$Rul_G(T_0, r_1) = \{f_1 = 0 \rightarrow 3, f_2 = 0 \rightarrow 1, f_3 = 1 \wedge f_1 = 0 \rightarrow 1, f_3 = 1 \wedge f_2 = 0 \rightarrow 1\}$;

$Rul_G(T_0, r_2) = \{f_1 = 1 \rightarrow 1, f_2 = 0 \rightarrow 1, f_3 = 1 \wedge f_2 = 0 \rightarrow 1, f_3 = 1 \wedge f_1 = 1 \rightarrow 1\}$;

$Rul_G(T_0, r_3) = \{f_1 = 1 \rightarrow 1, f_3 = 1 \wedge f_1 = 1 \rightarrow 1\}$;

$Rul_G(T_0, r_4) = \{f_1 = 0 \rightarrow 3, f_3 = 1 \wedge f_1 = 0 \rightarrow 1\}$;

$Rul_G(T_0, r_5) = \{f_1 = 0 \rightarrow 3, f_2 = 0 \rightarrow 1\}$.

5.5 Procedure of Optimization Relative to Length

In this section, a procedure of optimization of the graph G relative to the length l, is presented. Let $G = \Delta_\beta^*(T)$.

For each node Θ in the graph G, this procedure assigns to each row r of Θ the set $Rul_G^l(\Theta, r)$ of decision rules with the minimum length from $Rul_G(\Theta, r)$ and the number $Opt_G^l(\Theta, r)$ – the minimum length of a decision rule from $Rul_G(\Theta, r)$. The set $E_G(\Theta, r)$ attached to the row r in the nonterminal node Θ of G is changed. The obtained graph is denoted by G^l.

Let us move from the terminal nodes of the graph G to the node T. Let Θ be a terminal node of G and d be the most common decision for Θ. Then each row of Θ has been assigned the number

$$Opt_G^l(\Theta, r) = 0.$$

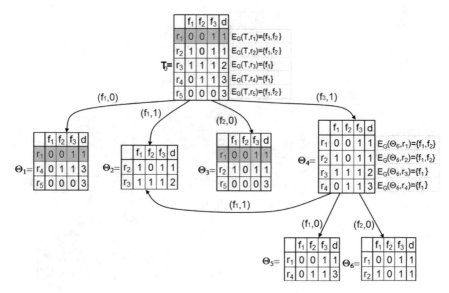

Fig. 5.3 Graph G^l

Let Θ be a nonterminal node of G and all children of Θ have already been treated. Let $r = (b_1, \ldots, b_n)$ be a row of Θ. The number

$$Opt^l_G(\Theta, r) = \min\{Opt^l_G(\Theta(f_i, b_i), r) + 1 : f_i \in E_G(\Theta, r)\}$$

is assigned to the row r in the table Θ and we set

$$E_{G^l}(\Theta, r) = \{f_i : f_i \in E_G(\Theta, r), Opt^l_G(\Theta(f_i, b_i), r) + 1 = Opt^l_G(\Theta, r)\}.$$

Example 5.2 The directed acyclic graph G^l obtained from the graph G (see Fig. 5.2) by the procedure of optimization relative to the length is presented in Fig. 5.3. Sets of nodes and edges are the same as in the Fig. 5.2. The difference is that any row r of each nonterminal node Θ of G is labeled with a nonempty set of attributes $E_{G^l}(\Theta, r) \subseteq E_{\Delta^*_\beta(T)}(\Theta, r)$.

Basing on the graph G^l it is possible to describe, for each row $r_i, i = 1, \ldots, 5$, of the table T_0, the set $Rul^l_G(T_0, r_i)$ of decision rules for T_0 and r_i with the minimum length from $Rul_G(T_0, r_i)$. Additionally, the value $Opt^l_G(T_0, r_i)$ was obtained during the procedure of optimization of the graph G relative to length. Green color presents connections between parent and child tables containing row r_1 from the table T_0 which allow us to describe the shortest rules for this row from $Rul_G(T_0, r_1)$. It is easy to see that only two edges (which labels correspond to descriptors) are considered. As a result set $Rul^l_G(T_0, r_1)$ was obtained. The minimum length of a decision rule assigned to each row of T_0 after the procedure of optimization relative to length, is equal to 1.

$Rul_G(T_0, r_1) = \{f_1 = 0 \to 3, f_2 = 0 \to 1\}, Opt^l_G(T_0, r_1) = 1;$

$Rul_G(T_0, r_2) = \{f_1 = 1 \rightarrow 1, f_2 = 0 \rightarrow 1\}, Opt^l_G(T_0, r_2) = 1;$
$Rul_G(T_0, r_3) = \{f_1 = 1 \rightarrow 1\}, Opt^l_G(T_0, r_3) = 1;$
$Rul_G(T_0, r_4) = \{f_1 = 0 \rightarrow 3\}, Opt^l_G(T_0, r_4) = 1;$
$Rul_G(T_0, r_5) = \{f_1 = 0 \rightarrow 3, f_2 = 0 \rightarrow 1\}, Opt^l_G(T_0, r_5) = 1.$

5.6 Experimental Results

Experiments were made using decision tables from UCI Machine Learning Repository [5] and Dagger software system [1]. When for some of the decision tables there were attributes taking unique value for each row, such attributes were removed. When some of the decision tables contained missing values, each of these values was replaced with the most common value of the corresponding attribute. When, in some of the decision tables, there were equal values of conditional attributes but different decisions, then each group of identical rows (identical conditional attributes values) was replaced with a single row from the group with the most common decision for this group.

Let T be one of these decision tables. Values of β from the set $B(T) = \{R(T) \times 0.0, R(T) \times 0.01, R(T) \times 0.1, R(T) \times 0.2, R(T) \times 0.3\}$, are studied for table T. To remind, $R(T)$ denotes the number of unordered pairs of rows with different decisions in a decision table T. Value $R(T)$ is different for different data sets, so values of $\beta \in B(T)$, for each decision table, are considered. They are used as thresholds for stop of partitioning of a decision table into subtables and descritption of β-decision rules.

In this section, experiments connected with size of the directed acyclic graph, length of β-decision rules and accuracy of rule based classifiers, are presented.

5.6.1 Attributes' Values Selection and Size of the Graph

In this subsecion, results of attributes' values selection as well as their influence on the size of the directed acyclic graph constructed by modified and "classical" algorithms, are considered.

Table 5.2 presents attributes' values considered during construction of the first level of a directed acyclic graph by modified algorithm and "classical" algorithm. Column *Attribute* contains name of the attribute, column *values* contains chosen values of attributes. Data set Cars is considered as an exemplary decision table. It contains six conditional attributes and 1728 rows. This data set directly relates car evaluation to the attributes [5]: buying (buying price), maint (price of the maintenance), doors (number of doors), persons (capacity in terms of persons to carry), lug_boot (the size of luggage boot), safety (estimated safety of the car).

Table 5.2 Values of attributes chosen during construction of the first level of the graph by modified and "classical" algorithms, for decision table Cars

Attribute	Modified algorithm	"Classical" algorithm
Buying	v-high	v-high, high, med, low
Maint	v-high	v-high, high, med, low
Doors	2	2, 3, 4, 5-more
Persons	2, 4, more	2, 4, more
Lug_boot	Small	Small, med, big
Safety	Low	Low, med, high

Table 5.3 An upper bound of the number of edges for the first level of the graph constructed by modified and "classical" algorithms

Decision table	Rows	Attr	Modified algorithm	"Classical" algorithm
Adult-stretch	16	4	5	8
Balance-scale	625	4	8	20
Breast-cancer	266	9	10	41
Cars	1728	6	8	21
Hayes-roth-data	69	4	6	15
House-votes	279	16	18	48
Lymphography	148	18	19	59
Soybean-small	47	35	35	72
Teeth	23	8	9	28
Tic-tac-toe	958	9	11	27

Table 5.3 presents an upper bound of the number of edges for the first level of the directed acyclic graph constructed by modified algorithm and "classical" algorithm. Label of each edge corresponds to the descriptor considered during partition of a decision table T into separable subtables. In case of "classical" algorithm for graph construction, the upper bound of the number of edges for the first level of the graph is equal to the number of all values of all non-constant attributes from T. The difference regarding to the number of edges, for modified and "classical" algorithms, is noticable for all decision tables. It influences directly the size of a directed acyclic graph.

Table 5.4 presents the number of nodes and edges of the directed acyclic graph constructed by the modified algorithm (columns *nd* and *edg*) and "classical" algorithm (columns *nd-dp* and *edg-dp*), for exact decision rules. Columns *nd-diff* and *edg-diff* present difference, i.e., values of these columns are equal to the number of nodes/edges in the directed acyclic graph constructed by the "classical" algorithm divided by the number of nodes/edges in the directed acyclic graph constructed by the proposed algorithm. Values in bold denote difference greater than three. In particular, for a decision table Cars, the difference referring to the number of nodes is more than eight, and referring to the number of edges is more than seventeen.

Table 5.4 Comparison of the size of graph for exact decision rules

Decision table	Rows	Attr	Modified algorithm		"Classical" algorithm		Difference	
			nd	edg	nd-dp	edg-dp	nd-diff	edg-diff
Adult-stretch	16	4	36	37	72	108	2.00	2.92
Balance-scale	625	4	654	808	1212	3420	1.85	**4.23**
Breast-cancer	266	9	2483	9218	6001	60387	2.42	**6.55**
Cars	1728	6	799	1133	7007	19886	**8.77**	**17.55**
Hayes-roth-data	69	4	140	215	236	572	1.69	2.66
House-votes	279	16	123372	744034	176651	1981608	1.43	2.66
Lymphography	148	18	26844	209196	40928	814815	1.52	**3.89**
Soybean-small	47	35	3023	38489	3592	103520	1.19	2.69
Teeth	23	8	118	446	135	1075	1.14	2.41
Tic-tac-toe	958	9	14480	41214	42532	294771	2.94	**7.15**

The number of nodes and edges of the directed acyclic graph constructed by the proposed algorithm and "classical" algorithm, for $\beta \in B(T)$, were obtained. Table 5.5 presents comparison of the size of the graph, for $\beta \in \{R(T) \times 0.01, R(T) \times 0.1, R(T) \times 0.2, R(T) \times 0.3\}$. Columns *nd-diff* and *edg-diff* contain, respectively, the number of nodes/edges in the directed acyclic graph constructed by the "classical" algorithm divided by the number of nodes/edges in the directed acyclic graph constructed by the proposed algorithm.

Presented results show that the size of the directed acyclic graph constructed by the proposed algorithm is smaller than the size of the directed acyclic graph constructed by the "classical" algorithm. Values in bold denote difference greater than three.

The size of the directed acyclic graph is related in some way with the properties of a given data set, e.g., number of attributes, distribution of attribute values, number of rows. To understand these relationships, the structure of the graph, for example, the number of nodes in each layer of the graph, will be considered "deeply" in the future.

5.6.2 Comparison of Length of β-Decision Rules

Modified as well as "classical" dynamic programming approach was applied for optimization of decision rules relative to length.

For each of the considered decision tables T and for each row r of the given table T, the minimum length of a decision rule for T and r was obtained. After that, for rows of T, the minimum (column *min*), average (column *avg*), and maximum (column *max*) values of length of rules with the minimum length were obtained.

Table 5.5 Comparison of the size of graph for β-decision rules

Decision table	Rows	Attr	$\beta = R(T) \times 0.01$		$\beta = R(T) \times 0.1$		$\beta = R(T) \times 0.2$		$\beta = R(T) \times 0.3$	
			nd-diff	edg-diff	nd-diff	edg-diff	nd-diff	edg-diff	nd-diff	edg-diff
Adult-stretch	16	4	2.00	2.92	2.00	2.59	2.00	2.59	3.33	4.00
Balance-scale	625	4	3.23	5.00	2.33	2.50	2.33	2.50	2.33	2.50
Breast-cancer	266	9	5.24	6.90	5.07	5.18	4.50	4.91	3.50	4.06
Cars	1728	6	6.48	8.00	5.13	5.86	2.44	2.63	2.44	2.63
Hayes-roth-data	69	4	2.27	2.62	2.92	3.50	2.29	2.50	2.29	2.50
House-votes	279	16	2.21	2.25	2.30	2.13	2.20	1.98	2.05	1.87
Lymphography	148	18	2.56	3.95	2.69	3.25	2.46	3.13	2.41	3.14
Soybean-small	47	35	1.32	2.82	2.06	2.61	2.20	2.62	2.20	2.58
Teeth	23	8	1.22	2.42	1.83	2.59	1.97	2.80	1.93	2.88
Tic-tac-toe	958	9	4.82	5.37	3.47	4.14	2.33	2.45	2.33	2.45

Table 5.6 Minimum, average and maximum length of exact decision rules and number of rules optimized by modified and "classical" dynamic programming approach

Decision table	Rows	Attr	Modified algorithm				"Classical" algorithm			
			min	avg	max	nr	nr-dp	min-dp	avg-dp	max-dp
Adult-stretch	16	4	1	1.75	4	22	24	1	1.25	2
Balance-scale	625	4	3	3.48	4	1475	8352	3	3.20	4
Breast-cancer	266	9	3	4.88	8	9979	15905	1	2.67	6
Cars	1728	6	1	2.72	6	3928	138876	1	2.43	6
Hayes-roth-data	69	5	2	3.1	4	138	450	1	2.15	4
Hause-votes-84	279	16	2	3.13	8	9787	16246	2	2.54	5
Lymphography	148	18	2	3.14	7	2611	3334	1	1.99	4
Soybean-small	47	35	1	1.64	2	148	141	1	1.00	1
Teeth	23	8	2	3.35	4	123	310	1	2.26	4
Tic-tac-toe	958	9	3	3.54	6	10746	10776	3	3.02	4

Table 5.6 presents the minimum, average and maximum length of exact decision rules ($\beta = R(T) \times 0.0$) obtained by modified dynamic programming approach and "classical" dynamic programming approach. The number of rules after the procedure of optimization relative to length (respectively, columns *nr* and *nr-dp*) is also presented. If the same rules exist for different rows of T, they are counted each time. Column *Attr* contains the number of conditional attributes, column *Rows* - the number of rows. To make comparison with the optimal values obtained by the "classical" dynamic programming approach, some experiments were performed and the results are presented. Columns *min-dp*, *avg-dp*, and *max-dp* present, respectively, minimum, average, and maximum values of length of optimal rules.

The number of rules constructed by the modified dynamic programming approach is smaller than the number of rules constructed by the "classical" dynamic programming approach. Presented results show that the number of rules obtained by modified dynamic programming approach after optimization relative to length (column *nr*) is smaller than the number of rules obtained by the "classical" dynamic programming approach after optimization relative to length (column *nr-dp*), for almost all decision tables. Only for Soybean-small, the number of rules after optimization relative to length for modified dynamic programming approach is equal to 148, in case of "classical" dynamic programming approach–it is 141. Note, that in case of "classical algorithm" (see Table 5.6) the minimum, average and maximum length of rules, for Soybean-small, is equal to 1.

Table 5.7 presents comparison of length of exact decision rules. Values in columns *min-diff*, *avg-diff*, *max-diff* are equal to values of the minimum, average and maximum length of β-decision rules obtained by the modified dynamic programming approach divided by the corresponding values obtained by the "classical" dynamic programming approach. Values in bold denote that length of rules obtained by the

Table 5.7 Comparison of length of exact decision rules

Decision table	Rows	Attr	min-diff	avg-diff	max-diff
Adult-stretch	16	4	**1.00**	1.40	2.00
Balance-scale	625	4	**1.00**	1.09	**1.00**
Breast-cancer	266	9	3.00	1.83	1.33
Cars	1728	6	**1.00**	1.12	**1.00**
Hayes-roth-data	69	5	2.00	1.45	**1.00**
Hause-votes-84	279	16	**1.00**	1.23	1.60
Lymphography	148	18	2.00	1.58	1.75
Soybean-small	47	35	**1.00**	1.64	2.00
Teeth	23	8	2.00	1.48	**1.00**
Tic-tac-toe	958	9	**1.00**	1.17	1.50

proposed algorithm is equal to these for optimal ones. The difference of average length of rules, for each decision table, is less than two.

Presented results show that values of average length of exact decision rules, usually are not far from optimal ones. In particular, for data set Cars the difference regarding the size of the graph is big (see Table 5.4), however, values of minimum and maximum length of exact decision rules are equal to optimal ones, and difference referring to the average length is very small. The number of rules is significantly reduced (see Table 5.6).

Table 5.8 presents the minimum, average and maximum length of β-decision rules obtained by modified dynamic programming approach, $\beta \in \{R(T) \times 0.01, R(T) \times 0.1, R(T) \times 0.2, R(T) \times 0.3\}$.

Results presented in Tables 5.6 and 5.8 show that the length of β-decision rules is non-increasing when the value of β is increasing.

Table 5.9 presents comparison of the average length of β-decision rules, $\beta \in \{R(T) \times 0.01, R(T) \times 0.1, R(T) \times 0.2, R(T) \times 0.3\}$. Optimal values were obtained by "classical" dynamic programming approach. Cells in columns of this table $(R(T) \times 0.01, R(T) \times 0.1, R(T) \times 0.2, R(T) \times 0.3)$ are equal to the average length of β-decision rules obtained by the modified dynamic programming approach divided by the corresponding values obtained by the "classical" dynamic programming approach.

Values in bold show that the average length of β-decision rules optimized by the proposed algorithm is equal to optimal one. For all decision tables, with the exception of Breast-cancer and $\beta = R(T) \times 0.01$, the difference is less than two, and usually, it is non-increasing when β grows. In case of the Balance-scale decision table, the average length of β-decision rules is equal to optimal one, the difference regarding the size of the graph is noticeable (see Table 5.5).

Table 5.8 Minimum, average and maximum length of β-decision rules optimized by modified dynamic programming approach

Decision table	Rows	Attr	$\beta = R(T) \times 0.01$			$\beta = R(T) \times 0.1$			$\beta = R(T) \times 0.2$			$\beta = R(T) \times 0.3$		
			min	avg	max	min	avg	max	min	avg	max	min	avg	max
Adult-stretch	16	4	1	1.75	4	1	1.25	2	1	1.25	2	1	1.00	1
Balance-scale	625	4	2	2.00	2	1	1.00	1	1	1.00	1	1	1.00	1
Breast-cancer	266	9	2	2.95	6	1	1.84	3	1	1.26	3	1	1.11	3
Cars	1728	6	1	1.67	3	1	1.15	2	1	1.00	1	1	1.00	1
Hayes-roth-data	69	4	2	2.45	3	1	1.51	2	1	1.00	1	1	1.00	1
House-votes	279	16	2	2.47	4	1	1.39	3	1	1.34	2	1	1.31	2
Lymphography	148	18	2	2.52	4	1	1.60	2	1	1.16	2	1	1.00	1
Soybean-small	47	35	1	1.64	2	1	1.43	2	1	1.00	1	1	1.00	1
Teeth	23	8	2	2.74	3	1	1.74	2	1	1.44	2	1	1.13	2
Tic-tac-toe	958	9	2	2.93	3	1	1.93	2	1	1.00	1	1	1.00	1

Table 5.9 Comparison of the average length of β-decision rules

Decision table	$\beta = R(T) \times 0.01$	$\beta = R(T) \times 0.1$	$\beta = R(T) \times 0.2$	$\beta = R(T) \times 0.3$
Adult-stretch	1.40	**1.00**	**1.00**	**1.00**
Balance-scale	**1.00**	**1.00**	**1.00**	**1.00**
Breast-cancer	2.11	1.84	1.26	1.11
Cars	1.16	1.15	**1.00**	**1.00**
Hayes-roth-data	1.57	1.51	**1.00**	**1.00**
House-votes	1.22	1.38	1.34	1.31
Lymphography	1.69	1.60	1.16	**1.00**
Soybean-small	1.64	1.43	**1.00**	**1.00**
Teeth	1.57	1.74	1.44	1.13
Tic-tac-toe	1.14	1.21	**1.00**	**1.00**

Table 5.10 Average test error

Decision table	Modified algorithm		"Classical" algorithm	
	Test error	Std	Test error	Std
Balance-scale	0.28	0.04	0.31	0.04
Breast-cancer	0.32	0.05	0.28	0.04
Cars	0.29	0.03	0.19	0.03
House-votes	0.09	0.06	0.08	0.09
Lymphography	0.27	0.04	0.23	0.05
Soybean-small	0.10	0.15	0.17	0.08
Tic-tac-toe	0.20	0.03	0.19	0.03
Average	0.22		0.21	

5.6.3 Classifier Based on Rules Optimized Relative to Length

Experiments connected with accuracy of classifiers basing on approximate decision rules optimized relative to length for modified dynamic programming approach and "classical" dynamic programming approach, were performed.

Table 5.10 presents an average test error for two-fold cross validation method (experiments were repeated for each decision table 30 times). Each data set was randomly divided into three parts: train–30%, validation–20%, and test–50%. Classifier was constructed on the train part, then pruned taking into account the minimum error on the validation set. Exact decision rules (0-rules) were constructed on the train part of a data set. Then, these rules were pruned and, for increasing value of β, β-decision rules were obtained. The set of rules, having value of β which gives the minimum error on validation set was chosen. This model was then used on a test part of the decision table as a classifier. Test error is a result of classification. It is the number of improperly classified rows from the test part of a decision table divided by the number of all rows in the test part of the decision table. Columns *test error* and *std* denote, respectively, average test error and standard deviation. The last row in Table 5.10 presents the average test error for all decision tables. It shows that accuracy of constructed classifiers for modified dynamic programming approach and "classical" dynamic programming approach, is comparable.

5.7 Conclusions

In the paper, a modification of the dynamic programming approach for optimization of β-decision rules relative to length, was presented.

"Classical" dynamic programming approach for optimization of decision rules allows one to obtain optimal rules, i.e., rules with minimum length. This fact is important from the point of view of knowledge representation. However, the size of the directed acyclic graph based on which decision rules are described can be huge for larger data sets.

Proposed modification is based on attributes' values selection. Instead of all values of all attributes, during construction of the graph only one attribute with the minimum number of values is selected and for the rest of attributes - the most frequent value of each attribute is chosen.

Experimental results show that the size of the directed acyclic graph constructed by the proposed algorithm is smaller than the size of the directed acyclic graph constructed by the "classical" algorithm. The biggest difference regarding the number of nodes is more than eight, and regarding the number of edges – more than seventeen (decision table Cars and $\beta = 0$).

Values of length of decision rules optimized by proposed algorithm, usually, are not far from the optimal ones. In general, the difference regarding the average length of β-decision rules, $\beta \in B(T)$, is less than two for all decision tables (with the exception of Breast-cancer and $\beta = R(T) \times 0.01$). Such differences, usually, are decreasing when β is increasing.

Accuracy of rule based classifiers constructed by modified dynamic programming approach and "classical" dynamic programming approach, is comparable.

In the future study, another possibilities for decreasing the size of a graph and selection of attributes will be considered.

References

1. Alkhalid, A., Amin, T., Chikalov, I., Hussain, S., Moshkov, M., Zielosko, B.: Dagger: a tool for analysis and optimization of decision trees and rules. In: Ficarra, F.V.C. (ed.) Computational Informatics, Social Factors and New Information Technologies: Hypermedia Perspectives and Avant-Garde Experiences in the Era of Communicability Expansion, pp. 29–39. Blue Herons, Bergamo, Italy (2011)
2. Amin, T., Chikalov, I., Moshkov, M., Zielosko, B.: Dynamic programming approach to optimization of approximate decision rules. Inf. Sci. **119**, 403–418 (2013)

3. Ang, J., Tan, K., Mamun, A.: An evolutionary memetic algorithm for rule extraction. Expert Syst. Appl. **37**(2), 1302–1315 (2010)
4. An, A., Cercone, N.: Rule quality measures improve the accuracy of rule induction: An experimental approach. In: Raś, Z.W., Ohsuga, S. (eds.) ISMIS. Lecture Notes in Computer Science, vol. 1932, pp. 119–129. Springer (2000)
5. Asuncion, A., Newman, D.J.: UCI Machine Learning Repository (2007). http://www.ics.uci.edu/~mlearn/
6. Błaszczyński, J., Słowiński, R., Susmaga, R.: Rule-based estimation of attribute relevance. In: Yao, J., Ramanna, S., Wang, G., Suraj, Z. (eds.) RSKT 2011. LNCS, vol. 6954, pp. 36–44. Springer (2011)
7. Błaszczyński, J., Słowiński, R., Szeląg, M.: Sequential covering rule induction algorithm for variable consistency rough set approaches. Inf. Sci. **181**(5), 987–1002 (2011)
8. Clark, P., Niblett, T.: The cn2 induction algorithm. Mach. Learn. **3**(4), 261–283 (1989)
9. Dembczyński, K., Kotłowski, W., Słowiński, R.: Ender: a statistical framework for boosting decision rules. Data Min. Knowl. Discov. **21**(1), 52–90 (2010)
10. Fürnkranz, J.: Separate-and-conquer rule learning. Artif. Intell. Rev. **13**(1), 3–54 (1999)
11. Grzymała-Busse, J.W.: Lers – a system for learning from examples based on rough sets. In: Słowiński, R. (ed.) Intelligent Decision Support. Handbook of Applications and Advances of the Rough Sets Theory, pp. 3–18. Kluwer Academic Publishers (1992)
12. Guyon, I., Elisseeff, A.: An introduction to variable and feature selection. J. Mach. Learn. Res. **3**, 1157–1182 (2003)
13. Guyon, I., Gunn, S., Nikravesh, M., Zadeh, L. (eds.): Feature Extraction: Foundations and Applications, Studies in Fuzziness and Soft Computing, vol. 207. Physica, Springer (2006)
14. Janusz, A., Ślęzak, D.: Rough set methods for attribute clustering and selection. Appl. Artif. Intell. **28**(3), 220–242 (2014)
15. Jensen, R., Shen, Q.: Computational Intelligence and Feature Selection: Rough and Fuzzy Approaches. IEEE Press Series on Computational Intelligence. Wiley, New York (2008)
16. Kozak, J., Juszczuk, P.: Association ACDT as a tool for discovering the financial data rules. In: Jedrzejowicz, P., Yildirim, T., Czarnowski, I. (eds.) IEEE International Conference on INnovations in Intelligent SysTems and Applications, INISTA 2017, Gdynia, Poland, July 3–5 2017, pp. 241–246. IEEE (2017)
17. Liu, B., Abbass, H.A., McKay, B.: Classification rule discovery with ant colony optimization. In: IAT 2003, pp. 83–88. IEEE Computer Society (2003)
18. Liu, H., Motoda, H.: Guest editors' introduction: feature transformation and subset selection. IEEE Intell. Syst. **13**(2), 26–28 (1998)
19. Liu, H., Motoda, H.: Computational Methods of Feature Selection (Chapman & Hall/Crc Data Mining and Knowledge Discovery Series). Chapman & Hall/CRC, Boca Raton (2007)
20. Michalski, S., Pietrzykowski, J.: iAQ: A program that discovers rules. AAAI-07 AI Video Competition (2007). http://videolectures.net/aaai07_michalski_iaq/
21. Moshkov, M., Piliszczuk, M., Zielosko, B.: Partial Covers, Reducts and Decision Rules in Rough Sets - Theory and Applications, Studies in Computational Intelligence, vol. 145. Springer, Heidelberg (2008)
22. Moshkov, M., Zielosko, B.: Combinatorial Machine Learning - A Rough Set Approach, Studies in Computational Intelligence, vol. 360. Springer, Heidelberg (2011)
23. Nguyen, H.S., Ślęzak, D.: Approximate reducts and association rules - correspondence and complexity results. In: RSFDGrC 1999, LNCS, vol. 1711, pp. 137–145. Springer (1999)
24. Nguyen, H.S.: Approximate boolean reasoning: foundations and applications in data mining. In: Peters, J.F., Skowron, A. (eds.) T. Rough Sets. LNCS, vol. 4100, pp. 334–506. Springer (2006)
25. Pawlak, Z., Skowron, A.: Rough sets and boolean reasoning. Inf. Sci. **177**(1), 41–73 (2007)
26. Pawlak, Z., Skowron, A.: Rudiments of rough sets. Inf. Sci. **177**(1), 3–27 (2007)
27. Quinlan, J.R.: C4.5: Programs for Machine Learning. Morgan Kaufmann Publishers Inc, Massachusetts (1993)
28. Rissanen, J.: Modeling by shortest data description. Automatica **14**(5), 465–471 (1978)

29. Sikora, M., Wróbel, Ł.: Data-driven adaptive selection of rule quality measures for improving rule induction and filtration algorithms. Int. J. General Syst. **42**(6), 594–613 (2013)
30. Ślęzak, D., Wróblewski, J.: Order based genetic algorithms for the search of approximate entropy reducts. In: Wang, G., Liu, Q., Yao, Y., Skowron, A. (eds.) RSFDGrC 2003. LNCS, vol. 2639, pp. 308–311. Springer (2003)
31. Stańczyk, U.: Feature evaluation by filter, wrapper, and embedded approaches. In: Stańczyk, U., Jain, L.C. (eds.) Feature Selection for Data and Pattern Recognition. Studies in Computational Intelligence, vol. 584, pp. 29–44. Springer (2015)
32. Stańczyk, U., Zielosko, B.: On combining discretisation parameters and attribute ranking for selection of decision rules. In: Polkowski, L., Yao, Y., Artiemjew, P., Ciucci, D., Liu, D., Ślęzak, D., Zielosko, B. (eds.) Proceedings of Rough Sets - International Joint Conference, IJCRS 2017, Olsztyn, Poland, 3–7 July 2017, Part I. Lecture Notes in Computer Science, vol. 10313, pp. 329–349. Springer (2017)
33. Stańczyk, U., Jain, L.C. (eds.): Feature Selection for Data and Pattern Recognition. Studies in Computational Intelligence, vol. 584. Springer, Berlin (2015)
34. Stefanowski, J., Vanderpooten, D.: Induction of decision rules in classification and discovery-oriented perspectives. Int. J. Intell. Syst. **16**(1), 13–27 (2001)
35. Wieczorek, A., Słowiński, R.: Generating a set of association and decision rules with statistically representative support and anti-support. Inf. Sci. **277**, 56–70 (2014)
36. Zielosko, B., Chikalov, I., Moshkov, M., Amin, T.: Optimization of decision rules based on dynamic programming approach. In: Faucher, C., Jain, L.C. (eds.) Innovations in Intelligent Machines-4 - Recent Advances in Knowledge Engineering. Studies in Computational Intelligence, vol. 514, pp. 369–392. Springer (2014)
37. Zielosko, B.: Optimization of approximate decision rules relative to coverage. In: Kozielski, S., Mrózek, D., Kasprowski, P., Małysiak-Mrózek, B., Kostrzewa, D. (eds.) Proceedings of Beyond Databases, Architectures, and Structures - 10th International Conference, BDAS 2014, Ustron, Poland, 27–30 May 2014. Communications in Computer and Information Science, vol. 424, pp. 170–179. Springer (2014)
38. Zielosko, B.: Optimization of exact decision rules relative to length. In: Czarnowski, I., Howlett, R.J., Jain, L.C. (eds.) Intelligent Decision Technologies 2017: Proceedings of the 9th KES International Conference on Intelligent Decision Technologies (KES-IDT 2017) – Part I, pp. 149–158. Springer (2018)

Part II
Ranking and Exploration of Features

Chapter 6
Generational Feature Elimination and Some Other Ranking Feature Selection Methods

Wiesław Paja, Krzysztof Pancerz and Piotr Grochowalski

Abstract Feature selection methods are effective in reducing dimensionality, removing irrelevant data, increasing learning accuracy, and improving result comprehensibility. However, the recent increase of dimensionality of data poses a severe challenge to many existing feature selection methods with respect to efficiency and effectiveness. In this chapter, both an overview of reasons for using ranking feature selection methods and the main general classes of this kind of algorithms are described. Moreover, some background of ranking method issues is defined. Next, we are focused on selected algorithms based on random forests and rough sets. Additionally, a newly implemented method, called Generational Feature Elimination (GFE), based on decision tree models, is introduced. This method is based on feature occurrences at given levels inside decision trees created in subsequent generations. Detailed information, about its particular properties and results of performance with comparison to other presented methods, is also included. Experiments are performed on real-life data sets as well as on an artificial benchmark data set.

Keywords Ranking feature selection · Recursive feature elimination
Generational feature elimination · Boruta algorithm · Random forest

W. Paja (✉) · K. Pancerz · P. Grochowalski
Faculty of Mathematics and Natural Sciences, Department of Computer Science,
University of Rzeszów, Pigonia Str. 1, 35-310 Rzeszów, Poland
e-mail: wpaja@ur.edu.pl

K. Pancerz
e-mail: kpancerz@ur.edu.pl

P. Grochowalski
e-mail: piotrg@ur.edu.pl

© Springer International Publishing AG 2018   97
U. Stańczyk et al. (eds.), *Advances in Feature Selection for Data and Pattern
Recognition*, Intelligent Systems Reference Library 138,
https://doi.org/10.1007/978-3-319-67588-6_6

6.1   Introduction

The main task of the feature selection (FS) process is to determine which predictors should be included in a model to make the best prediction results. It is one of the most critical questions as data are becoming increasingly highly-dimensional. Over the last decade, a huge number of highly efficient systems has produced big data sets with extraordinary numbers of descriptive variables. For this reason, effective feature selection methods have become exceptionally important in a wide range of disciplines [3, 13, 16]. These areas include business research, where FS is used to find important relationships between customers, products, and transactions, in pharmaceutical research, where it is applied to define relationships between structures of molecules and their activities, in a wide area of biological applications for the analysis of different aspects of genetic data to find relationships for diseases, and in scientific model and phenomenon simulations [13, 17, 19, 20, 23, 28]. Basically, three general schemes for feature selection are identified depending on how they combine the selection algorithm and the model building: filter methods, wrapper methods and embedded methods. The filter methods take place before the induction step. This kind of methods is independent of the induction algorithm and rely on intrinsic properties of the data (Fig. 6.1). Two steps are identified: ranking of features and subset selection. The subset selection has to be done before applying the learning algorithm and the performance testing process. Filter methods receive no feedback from the classifier, i.e., these methods select only the most interesting attributes that will be a part of a classification model [1]. However, due to the lack of the feedback, some redundant, but relevant features could not be recognized.

The wrapper methods (Fig. 6.2) are classifier-dependent approaches. It means that they evaluate the *goodness* of a selected feature subset directly based on a fixed classifier performance. Wrapper methods allow to detect the possible interactions

Fig. 6.1   Filter method scheme

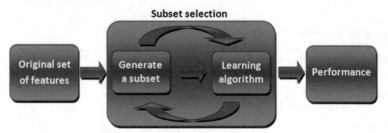

Fig. 6.2   Wrapper method scheme

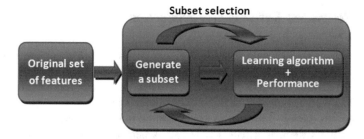

Fig. 6.3 Embedded method scheme

between variables [12, 19, 25, 31] and they use the learning algorithm as a subroutine to evaluate the feature subsets. Intuitively, this kind of methods should yield better performance, but it involves the high computational complexity.

The third type of feature selection scheme (Fig. 6.3) consists of embedded methods. They are similar to wrapper approaches. Features are specifically selected for a certain inducer. However, in embedded methods, the inducer selects the features in the process of learning (explicitly or implicitly), i.e., the classification model evolves simultaneously with the selected subset of features.

In this study, we focus on three selected ranking methods. The first method is known in the domain of rough set theory, and is a proper example of application of fuzzy indiscernibility relation to find superreducts based on the investigated data set [5, 9]. The second algorithm is based on random forest formalism and information about each attribute importance is gathered from the Braiman's forest [2]. The last method is a newly developed algorithm which applies information from a decision tree model developed in subsequent generations. Summarizing, the goal of proposed generational feature elimination is to simplify and improve feature selection process by relevant feature elimination during recursive generation of decision tree or other learning model. The hypothesis is that by removing subsets of relevant features in each step gradually all-relevant feature subset could be discovered. The details of each formalism are presented in the next section. The paper is divided into three main sections besides Introduction. In the second section, selected methods and algorithms are described in details. Next, utilized experiments are briefly characterized and the last section contains gathered results of experiments and some conclusions.

6.2 Selected Methods and Algorithms

In this section, a background for investigated algorithms is presented. The main idea of each of them is explained.

6.2.1 Rough Set Based Feature Selection

In this section, we briefly describe an idea of feature selection using methods of rough set theory. At the beginning, we recall a series of fundamental notions of rough set theory (cf. [22]). Rough sets proposed by Pawlak [21] are a mathematical tool to analyze data with vagueness, uncertainty or imprecision.

In rough set theory, information systems are a tool to represent data. Formally, an information system is a pair $S = (U, A)$, where U is the nonempty finite set of objects known as the universe of discourse and A is the nonempty finite set of attributes (features). Each attribute $a \in A$ is a function $a : U \rightarrow V_a$, where V_a is the set of values of a. Sometimes, we distinguish special attributes, in the set of attributes, called decision attributes, determining classes of objects in S. An information system with distinguished decision attributes is called a decision system. More formally, a decision system is a pair $S = (U, A \cup D)$, where $A \cap D = \emptyset$. Attributes from A are called condition attributes, whereas attributes from D are called decision attributes. Further, a special case of a decision system will be considered, when only one decision attribute is distinguished, i.e., $S = (U, A \cup \{d\})$.

In case of discrete values of condition attributes in S, we define a binary relation, called an indiscernibility relation, for each subset $B \subseteq A$:

$$Ind_B = \{(u, v) \in U \times U : \underset{a \in B}{\forall} \, a(u) = a(v)\}. \tag{6.1}$$

The indiscernibility relation Ind_B is an equivalence relation. The equivalence class of $u \in U$ with respect to Ind_B is denoted by $[u]_B$.

A family $\xi = \{X_1, X_2, \ldots, X_k\}$ of sets, where $X_1, X_2, \ldots, X_k \in U$, is called classification of U in S if and only if $X_i \cap X_j = \emptyset$ for $i, j = 1, 2, \ldots, k$, $i \neq j$, and $\bigcup_{i=1}^{k} X_i = U$. In many cases, classification of U is determined by values of the decision attribute d. For example, each value of d forms one class, called a decision class.

Let $X \subseteq U$ and $B \subseteq A$. The B-lower approximation of X in S is the set $\underline{B}(X) = \{u \in U : [u]_B \subseteq X\}$. The B-upper approximation of X in S is the set $\overline{B}(X) = \{u \in U : [u]_B \cap X \neq \emptyset\}$. The B-positive region of classification ξ in S is the set $Pos_B(\xi) = \bigcup_{X \in \xi} \underline{B}(X)$.

Let ξ be classification determined by the decision attribute d. The degree of dependency of the decision attribute d on the set $B \subseteq A$ of condition attributes is defined as

$$\gamma_B(\xi) = \frac{card(Pos_B(\xi))}{card(U)}, \tag{6.2}$$

where $card$ denotes the cardinality of a set.

In rough set theory, feature selection refers to finding the so-called decision reducts in a decision system $S = (U, A \cap \{d\})$. Let ξ be classification determined by the decision attribute d. A decision reduct R in S is a minimal (with respect to set inclusion) set $R \subseteq A$ such that $\gamma_R(\xi) = \gamma_A(\xi)$, i.e., for every $R' \subset R$, $\gamma'_R < \gamma_R$.

Various rough set methods were proposed to calculate decision reducts in decision systems. In general, we can divide them into three groups:

- methods producing a decision superreduct that is not necessarily a decision reduct (i.e., it is a subset of condition attributes that may be not minimal),
- methods producing a single decision reduct,
- methods producing all decision reducts.

Calculation of all decision reducts in a given decision system is the *NP*-hard problem (see [27]). Therefore, there is a need to use more efficient methods for real-life data. The *QUICKREDUCT* algorithm proposed in [26] is an example of the method producing a decision superreduct. This algorithm is an example of ranking methods of feature selection. In consecutive steps, we add, to the current quasi-reduct R, the attribute a that causes the highest increase of the degree of dependency $\gamma_{R \cup \{a\}}(\xi)$.

Algorithm 6.1: *QUICKREDUCT* feature selection algorithm

Input : $S = (U, A \cap \{d\})$ - a decision system; ξ - classification determined by d.
Output: R - a superreduct, $R \subseteq A$.

$R \leftarrow \emptyset$
repeat
   $T \leftarrow R$
   for *each* $a \in (A - R)$ **do**
      if $\gamma_{R \cup \{a\}}(\xi) > \gamma_T(\xi)$ **then**
        $T \leftarrow R \cup \{a\}$
   $R \leftarrow T$
until $\gamma_R(\xi) = \gamma_A(\xi)$
return T

In case of continuous values of condition attributes in a given decision system, we can use, in rough set based feature selection, some other relations between attribute values (instead of an indiscernibility relation Ind_B), for example:

- a dominance relation, cf. [6],
- a modified indiscernibility relation, cf. [4],
- a fuzzy indiscernibility relation, cf. [24].

In experiments, we have used the *QUICKREDUCT* algorithm based on a fuzzy indiscernibility relation $FInd_B$. This relation is defined as

$$FInd_B(u, v) = \underset{a \in B}{\mathscr{T}} \; FTol_a(u, v) \tag{6.3}$$

for any $u, v \in U$, where $\mathcal{T}$ is a T-norm operator [11] or, in general, an aggregation operator, and $FTol_a$ is a fuzzy tolerance relation defined for an attribute $a \in A$. Several fuzzy tolerance relations are defined in [9]. One of them is as follows:

$$FTol_a(u, v) = \frac{|a(u) - a(v)|}{a_{max} - a_{min}}, \tag{6.4}$$

where $a_{max} = \underset{u \in U}{max} \, a(u)$ and $a_{min} = \underset{u \in U}{min} \, a(u)$.

The degree of dependency of the decision attribute d on the set $B \subseteq A$ of condition attributes is defined as

$$\gamma_B = \frac{\sum\limits_{x \in U} Pos_B(x)}{card(U)}, \tag{6.5}$$

where $card$ denotes the cardinality of a set and $Pos_B(x)$ is the fuzzy B-positive region of $y \in X$. The task of calculating the fuzzy B-positive region is more complicated. We refer the readers to [24].

6.2.2 Random Forest Based Feature Selection

The next technique, well known in the domain of ranking feature selection, is the Boruta algorithm [14, 15] which uses random forest models for feature relevance estimation. The random forest is an algorithm based on ensemble of decision trees [2]. Each tree is developed using a random sample of the original dataset. In this way, the correlation between basic learners is removed. Additionally, each split inside the tree is also created using only a random subset of attributes. Their number influences the balance between bias and variance for the training set. The default value for classification tasks is the square root of the total number of attributes, and it is usually very powerful selection. The random forest is very popular and simple for application in the domain of different classification and regression tasks. During application of the random forest we can estimate classification quality but additional advantage of this model is the ability to estimate the importance of features. The importance computation is possible by measures of accuracy decreasing when information about attributes in a node is removed from the system.

The Boruta algorithm is based on the same idea, namely, by adding randomness to the system and collecting results from the ensemble of randomized samples one can reduce the misleading impact of random fluctuations and correlations. In this algorithm (see Algorithm 6.2), the original dataset is extended by adding copies of original attributes (*shadowAttr*) but their values are randomly permuted among the learning cases to remove their correlations with a decision attribute. The pseudo-code version of the Boruta algorithm created for this paper is based on [14, 15] and it is defined as Algorithm 6.2. Boruta is a kind of ranking algorithm. The random forest classifier is run on the extended set of data (*extendedData*) and in this way

---

Algorithm 6.2: Boruta algorithm

Input   : *originalData* - input dataset; *RFruns* - the number of random forest runs.
Output: *finalSet* that contains relevant and irrelevant features.

confirmedSet = ∅
rejectedSet = ∅
for *each RFruns* **do**
    │ *originalPredictors* ← *originalData(predictors)*
    │ *shadowAttr* ← *permute(originalPredictors)*
    │ *extendedPredictors* ← *cbind(originalPredictors, shadowAttr)*
    │ *extendedData* ← *cbind(extendedPredictors, originalData(decisions))*
    │ *zScoreSet* ← *randomForest(extendedData)*
    │ *MZSA* ← *max(zScoreSet(shadowAttr))*
    │ **for** *each a ∈ originalPredictors* **do**
    │   │ **if** *zScoreSet(a) > MZSA* **then**
    │   │ └ *hit(a) + +*

for *each a ∈ originalPredictors* **do**
    │ *significance(a)* ← *twoSidedEqualityTest(a)*
    │ **if** *significance(a) ≫ MZSA* **then**
    │ └ *confirmedSet* ← *finalSet ∪ a*
    │ **else if** *significance(a) ≪ MZSA* **then**
    │ └ *rejectedSet* ← *rejectedSet ∪ a*
return *finalSet* ← *rejectedSet ∪ confirmedSet*

Z score is calculated for each attribute (*zScoreSet*). Then, the maximum Z score among shadow attributes (*MZSA*) is identified and a hit is assigned to every attribute that is scored better than *MZSA*. The two-sided equality test with *MZSA* is applied. The attributes which have importance significantly lower than *MZSA* are treated as irrelevant (*rejectedSet*), in turn, the attributes which have importance significantly higher than *MZSA* are treated as relevant (*confirmedSet*). The details of this test are clearly described in [15]. The procedure is repeated until all attributes have importance estimated or if the algorithm reaches the limit of the random forest runs, previously set.

6.2.3   Generational Feature Elimination Algorithm

The *DTLevelImp* algorithm [18] is used to define ranking values for each investigated feature. These rank values are used to define the importance of each feature in the General Feature Elimination (GFE) algorithm. However, also other ranking methods could be applied inside the GFE algorithm for the feature importance estimation. A proposed measure is focused on the occurrence $\omega(a)$ of a given feature a inside the decision tree model which is extracted from a dataset. In this way, three different ranking measures could be used. The first one, called *Level-Weight based*

Importance (LWI), is defined by a weighting factor w_j assigned to a given level j of the tree model in which the feature a occurs (see Eq. 6.6).

$$LWI_a = \sum_{j=1}^{l} \sum_{node=1}^{x} w_j \cdot \omega(a) \qquad (6.6)$$

where l is the number of levels inside the decision tree model *Tree*, x is the number of nodes at a given level j of *Tree* and $\omega(a)$ describes the occurrence of the attribute a, usually 1 (attribute occurred) or 0 (attribute did not occur). We assume that $j = 1$ for a root of *Tree*. The weighting factor w_j of the level j is computed as follows (Eq. 6.7):

$$w_j = \begin{cases} 1 & \text{if } j = 1, \\ \frac{w_{j-1}}{2} & \text{if } 1 < j \le l. \end{cases} \qquad (6.7)$$

The second measure, called **Level-Percentage based Importance** (LPI), is defined by the percentage $\pi(node)$ of cases that appear in a given tree *node* at each level j, where the investigated attribute a is tested (Eq. 6.8).

$$LPI_a = \sum_{j=1}^{l} \sum_{node=1}^{x} \pi(node) \cdot \omega(a). \qquad (6.8)$$

In turn, the third measure, called **Level-Instance based Importance** (LII), is defined by the number $v(node)$ of cases that appear in a given tree *node* at each level j, where the investigated attribute a is tested (Eq. 6.9).

$$LII_a = \sum_{j=1}^{l} \sum_{node=1}^{x} w_j \cdot v(node) \cdot \omega(a). \qquad (6.9)$$

Additionally, the sum of *LPI* and *LWI* measures for each examined attribute could also be defined and investigated. Here, we use only *LII* measure as an example in the process of ranking and selection of the important feature set.

The proposed Generational Feature Elimination approach could be treated as an all-relevant feature selection method [25]. It means that during selection not only the most important features are selected but also that with the lowest level of relevance (greater than random shadow). These kind of methods are called all-relevant feature selection. Some motivation for GFE methodology was the Recursive Feature Elimination (*RFE*) algorithm in which, by application of external estimator, specific weight values are assigned to features [8, 10]. This process is repeated recursively, and in each step, attributes whose weights are the smallest ones are removed from the current set. It works until the desired set of features to select from is eventually reached. In the RFE approach, a number of features to select from should be initially defined.

On the contrary, in this research, the *contrast variable* concept [25] has been used to distinguish between relevant and irrelevant features. These variables do not carry information on the decision variable by design. They are added to the system in order to discern relevant and irrelevant variables. Here, they are obtained from the real variables by random permutation of values between objects of the analyzed dataset. The use of contrast variables was first proposed by Stoppiglia [29] and next by Tuv [30]. The goal of the proposed methodology is to simplify and improve feature selection process by relevant feature elimination during recursive generation of a decision tree. The hypothesis is that by removing subsets of relevant features in each step gradually all-relevant feature subset could be discovered.

The algorithm was initially called Generational Feature Elimination and it is presented as Algorithm 6.3. Generally, this algorithm is able to apply four importance measures (*impMeasure*) during the feature ranking process. The original data (*originalData*) set is extended by adding contrast variables (*contrastData*) which are the result of permutation of original predictors. While $x = 0$, the algorithm recursively develops a classification tree model (*treeModel*) from the current data set (*currentSet*). Next, based on the applied importance measure, the set of important features (*impDataSet*) is defined from the features which have a rank value (*featureRankValue*) greater than the maximal rank value of a contrast variable (*maxCFRankValue*). Then, the selected feature set is removed from current data. These iterations are executed until no original feature has the ranking measure value greater than the contrast variable, i.e., *impDataSet* is the empty set. Finally, the selected relevant feature subset *finalSet* for each importance measure may be defined.

It is worth noting that the GFE algorithm could be treated as a heuristic procedure designed to extract all relevant attributes, including weakly relevant attributes which are important when we can find a subset of attributes among which the attribute is not redundant. The heuristic used in GFE algorithm implies that the attributes which are significantly correlated with the decision variables are relevant, and the significance here means that the correlation is higher than that of the randomly generated attributes.

6.3 Feature Selection Experiments

We conducted experiments on artificial and real-world datasets to test the GFE algorithm's capabilities. Having done the first experiment, we present detailed functionality of the GFE algorithm in the domain of feature selection. The GFE algorithm was developed and tested on the basis of the Madelon data set. It is an artificial data set, which was one of the Neural Information Processing Systems challenge problems in 2003 (called NIPS2003) [7]. The data set contains 2600 objects (2000 of training cases + 600 of validation cases) corresponding to points located in 32 vertices of a 5-dimensional hypercube. Each vertex is randomly assigned to one of the two classes: -1 or $+1$, and the decision for each object is a class of its vertex. Each object is described by 500 features which were constructed in the following

---

Algorithm 6.3: Generational Feature Elimination

Input   : *originalData* - the analyzed dataset; *ncross* is a number of cross validation steps;
              impMeasure $\in \{LWI, LPI, LII, LWI + LPI\}$ is an importance measure used to rank
              variables.
Output: *finalSet* that contains all relevant features for each *impMeasure*.

$contrastData \leftarrow permute(originalData(predictors))$
$mergedPredictors \leftarrow cbind(originalData(predictors), contrastData)$
$mergedData \leftarrow cbind(mergedPredictors, originalData(decisions))$
for *each impMeasure* $\in \{LWI, LPI, LII, LWI + LPI\}$ **do**
  $finalSet = \varnothing$
  for *each* $n = 1, 2, .., ncross$ **do**
    $maxCFRankValue = 0$
    $featureRankValue = 0$
    $cvDataSet \leftarrow selectCVData(mergedData - n)$
    $currentSet \leftarrow cvDataSet$
    $x = 0$
    while $x = 0$ **do**
      $treeModel \leftarrow treeGeneration(currentSet)$
      $featureRank \leftarrow DTLevelImp(treeModel, impMeasure)$
      $maxCFRankValue \leftarrow maxValue(featureRank(contrastData))$
      $impDataSet = \varnothing$
      for *each featureRankValue* $\in featureRank(originalData)$ **do**
        if *featureRankValue* > *maxCFRankValue* **then**
          $impDataSet \leftarrow impDataSet \cup feature(featureRankValue)$

      $currentSet \leftarrow currentSet - impDataSet$
      if $impDataSet = \varnothing$ **then**
        $x{+}{+}$

    $resultImpDataSet[n] \leftarrow cvDataSet - currentSet$
  for *each feature* $\in resultImpDataSet$ **do**
    $z \leftarrow countIfImp(feature)$
    if $z \geq ncross/2$ **then**
      $finalSet_{impMeasure} \leftarrow finalSet_{impMeasure} \cup feature$
  return *finalSet*

---

way: 5 of them are randomly jittered coordinates of points, other 15 attributes are random linear combinations of the first five attributes. The rest of the data set is a uniform random noise. The goal is to select 20 important attributes from the system without false attribute selection. In our second experiment, 10 real-life data sets from the UCI Machine Learning Repository were also investigated. In this experiment, we applied the 10-fold cross validation procedure to prepare training and testing subsets. Then, for the training data set, all three presented ranking feature selection algorithms were applied. We obtained selected subsets of important features for each applied algorithm. In this way, we compared the results gathered by means of these approaches. The detailed discussion of the results of our experiments is presented in the next section.

6.4   Results and Conclusions

To illustrate the proposed Generational Feature Elimination algorithm, only experimental results for the second sample fold of the Madelon data set using the *LII* measure are presented (Table 6.1). Here, four iterations of this algorithm are applied. During the first iteration, the classification tree (the first generation of a tree) is built on the basis of the whole input data set. By applying the *LII* importance measure values, the subset of 10 features could be marked as important (grey cells marked), according to the decreased value of *LII* computed from the gathered tree model. The eleventh feature, *f335_1*, which is a contrast variable, defines a threshold for selection between relevant and irrelevant subsets. Next, the subset of selected features is removed from a data set. Then, in the second iteration of the algorithm, the next subset of six probably important features is selected using the *LII* parameter calculated from the tree built on the reduced dataset. The seventh feature, *f115_1*,

Table 6.1 The example of results gathered in the second fold using the *LII* measure. Bold names denote truly relevant features, other names denote irrelevant features. Additionally, names with *_1* index denote contrast variables. The grey colored cells contain the feature set found as important and removed from the data for the next iteration

1st iteration		2nd iteration		3rd iteration		4th iteration	
feature name	LII value	feature name	LII value	feature nam e	LII value	feature name	LII value
f476	2 368.05	**f242**	2 353.78	**f337**	2 359.33	f187_1	2 340.66
f339	625.03	**f129**	764.19	**f454**	1 104.63	f5	1 148.00
f154	603.13	**f319**	552.59	**f452**	752.50	f231_1	456.00
f379	455.50	**f473**	206.94	f256	92.50	f223_1	150.88
f29	171.78	**f282**	150.00	**f65**	54.06	f291_1	118.00
f443	83.54	**f434**	128.80	f112	23.13	f41 4	77.13
f106	82.94	f115_1	124.25	f291_1	23.13	f86_1	51.19
f85	39.38	**f65**	83.53	f258_1	23.00	f267_1	49.25
f494	23.40	f409_1	60.63	f365_1	14.50	f11	32.75
f49	22.75	f45 2	39.22	f119	12.71	f178_1	28.44
f335_1	19.50	f200_1	23.69	f385_1	12.59	f264_1	23.88
f454	13.94	f42 3	18.81	f16 4	12.06	f411_1	22.00
f337	13.03	f16 3	16.19	f468_1	12.00	f17	20.38
f319	11.74	f45 4	14.98	f157_1	10.69	f45 9	14.25
f74	11.69	f44 2	11.50	f368_1	10.14	f13 5	12.50
f322_1	10.38	f473_1	7.25	f307_1	9.72	f51_1	11.47
f176_1	6.06	f293	6.63	f40 6	8.69	f56	9.86
f33_1	5.47	f491	6.53	f20	8.13	f30 7	9.35
f432	4.97	f214	4.69	f375_1	5.97	f10 9	8.00
...	...	...	...	...	...	...	...

Table 6.2 The number of features in the original and reduced data sets during the 10-fold cross-validation procedure and using the GFE algorithm with the *LII* measure applied

Dataset	# Original features	Fold number										AVG
		1	2	3	4	5	6	7	8	9	10	
Madelon	500	23	22	17	22	22	26	20	21	15	20	20.8 ± 2.9

which is a contrast variable, defines a threshold for selection of the next important subset. This subset is also removed from a data set. Later, in the third iteration of the algorithm, the next subset of six probably important features is selected using the *LII* measure calculated from the tree built on the subsequently reduced dataset. The seventh feature, *f291_1*, defines a threshold for selection of the next important subset. This subset is therefore removed from a data set. Finally, in the fourth iteration, the subset of important features is empty, because the highest value of the *LII* measure is reached by a contrast variable *f187_1*. In this way, the algorithm ends the execution, and the subset of 22 features is defined as an important one. The truly important features in the Madelon data set are written in bold. It could be found that three non-informative attributes: *f85*, *f256* and *f112*, are also observed in the discovered subset. However, their importance values are very random and unique what is presented in Table 6.3, where these features are selected only once. They reach threshold ≥ 0.8 of probability estimator of attribute removal (see Eq. 6.10) during the 10-fold cross-validation procedure of the investigated Madelon data set. The proposed removing probability estimator P_{rm} of a given attribute a is defined as follows:

$$P_{rm}(a) = \frac{ncross - nsel(a)}{ncross}, \tag{6.10}$$

where *ncross* means a number of folds during cross validation, *nsel* means a number of selections of a given descriptive attribute a.

During the *10*-fold procedure, different subsets of important attributes were defined (see Table 6.2). However, an average number of the selected features was about 20.9 ± 2.9, i.e., it is similar to the defined source subset in the Madelon data. To find a strict set of important attributes, we should study how many times each feature was found to be important during the 10-fold experiment. These statistics are presented in Table 6.3.

According to our assumptions, that if a feature reaches the removing probability value of *0.5* then it could not be treated as the relevant one and thus it should be removed from the final reduced data set. Based on this calculations, our final subset (Table 6.3, grey colored cells) contains all 20 truly informative features from the Madelon data set (Table 6.3, bold faced names). For comparison, experiments with 10 real-life data sets were conducted. We focused on measuring the number of finally

Table 6.3 The number of selections of features used during the 10-fold cross-validation procedure applying the *LII* measure. Bold names denote 20 truly informative features, other names denote non-informative features. The grey colored cells contain a feature set found as important. *1—relevant feature name, 2—number of selections, 3—removing probability*

1	2	3	1	2	3	1	2	3	1	2	3
f29	10	0.0	**f379**	10	0.0	f74	1	0.9	f148	1	0.9
f49	10	0.0	**f434**	8	0.2	f177	1	0.9	f472	1	0.9
f65	7	0.3	**f443**	10	0.0	f359	1	0.9	f203	1	0.9
f106	10	0.0	**f452**	10	0.0	f414	1	0.9	f211	1	0.9
f129	10	0.0	**f454**	6	0.4	f85	1	0.9	f304	1	0.9
f154	10	0.0	**f456**	5	0.5	f256	1	0.9	f7	1	0.9
f242	10	0.0	**f473**	10	0.0	f112	1	0.9	f440	1	0.9
f282	10	0.0	**f476**	10	0.0	f286	2	0.8	f323	1	0.9
f319	10	0.0	**f494**	6	0.4	f216	1	0.9			
f337	10	0.0	f292	2	0.8	f5	4	0.6			
f339	10	0.0	f343	1	0.9	f245	1	0.9			

selected features (Table 6.4) and the accuracy of classification using the original and reduced data sets (Table 6.5).

The results gathered in Table 6.5 show that all accuracy scores gathered using the Boruta, Rough Set, and GFE algorithms are very similar, both before and after selection. The difference is only a few percent more or less, not so significant. However, comparison of quantitative feature selection in Table 6.4 shows the greater variance. Mostly, subsets selected using the GFE algorithm were similar to that selected using the Boruta algorithm. The reason is that both methods applied a similar model for importance estimation, but in a different way. During experiments, the GFE algorithm seems to be rather faster than the Boruta algorithm, and also than the Rough

Table 6.4 The number of features in the original and reduced data sets using three FS algorithms

Dataset	# Original features	# Selected feature		
		Boruta	RoughSet	GFE
Climate	18	4.5 ± 0.5	5.6 ± 0.5	6.5 ± 0.9
Diabetes	8	7.2 ± 0.4	8.0 ± 0.0	7.8 ± 0.6
Glass	9	8.9 ± 0.3	9.0 ± 0.0	9.0 ± 0.0
Ionosphere	34	33.0 ± 0.0	7.2 ± 0.6	33.0 ± 0.0
Wine	13	13.0 ± 0.0	5.0 ± 0.0	13.0 ± 0.0
Zoo	16	13.5 ± 0.5	7.1 ± 1.2	15.0 ± 0.5
German	20	11.6 ± 0.9	11.0 ± 0.0	15.4 ± 1.2
Seismic	18	10.8 ± 0.6	13.0 ± 0.0	10.5 ± 1.4
Sonar	60	27.7 ± 1.4	5.1 ± 0.3	49.2 ± 8.8
Spect	22	9.3 ± 1.0	2.1 ± 0.3	20.0 ± 0.8

Table 6.5 The classification accuracy results using the original and reduced data sets

Dataset	Boruta		RoughSet		GFE	
	Original	Selected	Original	Selected	Original	Selected
Climate	0.92 ± 0.02	0.92 ± 0.02	0.92 ± 0.02	0.91 ± 0.03	0.92 ± 0.04	0.92 ± 0.05
Diabetes	0.75 ± 0.06	0.75 ± 0.05	0.75 ± 0.06	0.75 ± 0.06	0.72 ± 0.09	0.72 ± 0.10
Glass	0.70 ± 0.13	0.70 ± 0.13	0.70 ± 0.13	0.70 ± 0.13	0.71 ± 0.16	0.71 ± 0.16
Ionosphere	0.91 ± 0.04	0.91 ± 0.04	0.91 ± 0.04	0.92 ± 0.04	0.87 ± 0.06	0.87 ± 0.06
Wine	0.92 ± 0.09	0.92 ± 0.09	0.92 ± 0.09	0.95 ± 0.08	0.94 ± 0.07	0.94 ± 0.07
Zoo	0.92 ± 0.08	0.92 ± 0.08	0.92 ± 0.08	0.93 ± 0.06	0.90 ± 0.08	0.90 ± 0.08
German	0.73 ± 0.04	0.71 ± 0.05	0.73 ± 0.04	0.72 ± 0.04	0.71 ± 0.04	0.71 ± 0.04
Seismic	0.93 ± 0.01	0.93 ± 0.01	0.93 ± 0.01	0.93 ± 0.01	0.89 ± 0.01	0.90 ± 0.02
Sonar	0.70 ± 0.06	0.75 ± 0.11	0.70 ± 0.06	0.74 ± 0.06	0.77 ± 0.08	0.77 ± 0.08
Spect	0.80 ± 0.09	0.80 ± 0.09	0.80 ± 0.09	0.79 ± 0.03	0.78 ± 0.08	0.79 ± 0.09

Set algorithm. In GFE, the decision tree was built only in few generations. For example, in case of the second fold of the Madelon data set (see Table 6.1), only four generations (iterations) are required to eliminate and to extract the subset of relevant features. A default number of trees in the Boruta algorithm is equal to 500, and for the Madelon data set, the runtime of the experiment could be rather long. On the other hand, the Rough Set feature selection algorithm is also a time consuming solution and in case of the Madelon data set, the computation time could be even longer. The presented results are promising and the proposed GFE algorithm may be applied in the domain of FS. However, an important question is how to define the threshold used to separate a truly informative feature from the other non-informative ones. For example, in case of the Madelon data set, feature $f5$, which is the random noise, was found 4 times during the 10-fold cross-validation procedure (see Table 6.3), thus its probability estimator for pruning is 0.6. If we define non-informative features using threshold ≥ 0.8, then $f5$ may be treated as the important one. Therefore, a proper definition of the threshold is needed.The introduced algorithm of Generational Feature Elimination seems to be robust and effective and it is able to find weakly relevant important attributes due to sequential elimination of strongly relevant attributes.

Acknowledgements This work was supported by the Center for Innovation and Transfer of Natural Sciences and Engineering Knowledge at the University of Rzeszów.

References

1. Bermingham, M., Pong-Wong, R., Spiliopoulou, A., Hayward, C., Rudan, I., Campbell, H., Wright, A., Wilson, J., Agakov, F., Navarro, P., Haley, C.: Application of high-dimensional feature selection: evaluation for genomic prediction in man. Sci. Rep. **5** (2015). https://doi.org/10.1038/srep10312

2. Breiman, L.: Random forests. Mach. Learn. **45**(1), 5–32 (2001). https://doi.org/10.1023/A:1010933404324
3. Cheng, X., Cai, H., Zhang, Y., Xu, B., Su, W.: Optimal combination of feature selection and classification via local hyperplane based learning strategy. BMC Bioinform. **16**, 219 (2015). https://doi.org/10.1186/s12859-015-0629-6
4. Cyran, K.A.: Modified indiscernibility relation in the theory of rough sets with real-valued attributes: application to recognition of Fraunhofer diffraction patterns. In: Peters, J.F., Skowron, A., Rybiński, H. (eds.) Transactions on Rough Sets IX, pp. 14–34. Springer, Berlin (2008). https://doi.org/10.1007/978-3-540-89876-4_2
5. Dubois, D., Prade, H.: Rough fuzzy sets and fuzzy rough sets. Int. J. Gen. Syst. **17**(2–3), 191–209 (1990). https://doi.org/10.1080/03081079008935107
6. Greco, S., Matarazzo, B., Slowinski, R.: Rough sets theory for multicriteria decision analysis. Eur. J. Oper. Res. **129**(1), 1–47 (2001)
7. Guyon, I., Gunn, S., Hur, A.B., Dror, G.: Result analysis of the NIPS 2003 feature selection challenge. In: Proceedings of the 17th International Conference on Neural Information Processing Systems, pp. 545–552 (2004)
8. Guyon, I., Weston, J., Barnhill, S., Vapnik, V.: Gene selection for cancer classification using support vector machines. Mach. Learn. **46**(1), 389–422 (2002). https://doi.org/10.1023/A:1012487302797
9. Jensen, R., Shen, Q.: New approaches to fuzzy-rough feature selection. IEEE Trans. Fuzzy Syst. **17**(4), 824–838 (2009). https://doi.org/10.1109/TFUZZ.2008.924209
10. Johannes, M., Brase, J., Frohlich, H., Gade, S., Gehrmann, M., Falth, M., Sultmann, H., Beissbarth, T.: Integration of pathway knowledge into a reweighted recursive feature elimination approach for risk stratification of cancer patients. Bioinformatics **26**(17), 2136–2144 (2010). https://doi.org/10.1093/bioinformatics/btq345
11. Klement, E.P., Mesiar, R., Pap, E.: Triangular Norms. Kluwer Academic Publishers, Dordrecht (2000)
12. Kohavi, R., John, G.H.: Wrappers for feature subset selection. Artif. Intell. **97**(1), 273–324 (1997). https://doi.org/10.1016/S0004-3702(97)00043-X
13. Kuhn, M., Johnson, K.: Applied Predictive Modeling. Springer, New York, NY (2013)
14. Kursa, M., Rudnicki, W.: Feature selection with the Boruta package. J. Stat. Softw. **36**(1), 1–13 (2010)
15. Kursa, M.B., Jankowski, A., Rudnicki, W.R.: Boruta—a system for feature selection. Fundam. Inf. **101**(4), 271–285 (2010). https://doi.org/10.3233/FI-2010-288
16. Li, J., Cheng, K., Wang, S., Morstatter, F., Trevino, R.P., Tang, J., Liu, H.: Feature selection: A data perspective. CoRR. arXiv:1601.07996 (2016)
17. Nilsson, R., Peña, J.M., Björkegren, J., Tegnér, J.: Detecting multivariate differentially expressed genes. BMC Bioinform. **8**(1), 150 (2007). https://doi.org/10.1186/1471-2105-8-150
18. Paja, W.: Feature selection methods based on decision rule and tree models. In: Czarnowski, I., Caballero, A.M., Howlett, R.J., Jain, L.C. (eds.) Intelligent Decision Technologies 2016: Proceedings of the 8th KES International Conference on Intelligent Decision Technologies (KES-IDT 2016)—Part II, pp. 63–70. Springer International Publishing, Cham (2016). https://doi.org/10.1007/978-3-319-39627-9_6
19. Paja, W., Wrzesien, M., Niemiec, R., Rudnicki, W.R.: Application of all-relevant feature selection for the failure analysis of parameter-induced simulation crashes in climate models. Geosci. Model Dev. **9**(3), 1065–1072 (2016). https://doi.org/10.5194/gmd-9-1065-2016
20. Pancerz, K., Paja, W., Gomuła, J.: Random forest feature selection for data coming from evaluation sheets of subjects with ASDs. In: Ganzha, M., Maciaszek, L., Paprzycki, M. (eds.) Proceedings of the 2016 Federated Conference on Computer Science and Information Systems (FedCSIS), pp. 299–302. Gdańsk, Poland (2016)
21. Pawlak, Z.: Rough sets. Int. J. Comput. Inf. Sci. **11**(5), 341–356 (1982)
22. Pawlak, Z.: Rough Sets: Theoretical Aspects of Reasoning about Data. Kluwer Academic Publishers, Dordrecht (1991)

23. Phuong, T.M., Lin, Z., Altman, R.B.: Choosing SNPs using feature selection. In: Proceedings of the 2005 IEEE Computational Systems Bioinformatics Conference (CSB'05), pp. 301–309 (2005). https://doi.org/10.1109/CSB.2005.22

24. Radzikowska, A.M., Kerre, E.E.: A comparative study of fuzzy rough sets. Fuzzy Sets Syst. 126(2), 137–155 (2002). https://doi.org/10.1016/S0165-0114(01)00032-X

25. Rudnicki, W.R., Wrzesień, M., Paja, W.: All relevant feature selection methods and applications. In: Stańczyk, U., Jain, L.C. (eds.) Feature Selection for Data and Pattern Recognition, pp. 11–28. Springer, Berlin (2015). https://doi.org/10.1007/978-3-662-45620-0_2

26. Shen, Q., Chouchoulas, A.: A modular approach to generating fuzzy rules with reduced attributes for the monitoring of complex systems. Eng. Appl. Artif. Intell. 13(3), 263–278 (2000). https://doi.org/10.1016/S0952-1976(00)00010-5

27. Skowron, A., Rauszer, C.: The discernibility matrices and functions in information systems. In: Słowiński, R. (ed.) Intelligent Decision Support: Handbook of Applications and Advances of the Rough Sets Theory, pp. 331–362. Kluwer Academic Publishers, Dordrecht (1992). https://doi.org/10.1007/978-94-015-7975-9_21

28. Stoean, C., Stoean, R., Lupsor, M., Stefanescu, H., Badea, R.: Feature selection for a cooperative coevolutionary classifier in liver fibrosis diagnosis. Comput. Biol. Med. 41(4), 238–246 (2011). https://doi.org/10.1016/j.compbiomed.2011.02.006

29. Stoppiglia, H., Dreyfus, G., Dubois, R., Oussar, Y.: Ranking a random feature for variable and feature selection. J. Mach. Learn. Res. 3, 1399–1414 (2003)

30. Tuv, E., Borisov, A., Torkkola, K.: Feature selection using ensemble based ranking against artificial contrasts. In: Proceedings of the 2006 IEEE International Joint Conference on Neural Network, pp. 2181–2186 (2006). https://doi.org/10.1109/IJCNN.2006.246991

31. Zhu, Z., Ong, Y.S., Dash, M.: Wrapper-filter feature selection algorithm using a memetic framework. IEEE Trans. Syst. Man Cybern. Part B (Cybernetics) 37(1), 70–76 (2007). https://doi.org/10.1109/TSMCB.2006.883267

Chapter 7
Ranking-Based Rule Classifier Optimisation

Urszula Stańczyk

Abstract Ranking is a strategy widely used for estimating relevance or importance of available characteristic features. Depending on the applied methodology, variables are assessed individually or as subsets, by some statistics referring to information theory, machine learning algorithms, or specialised procedures that execute systematic search through the feature space. The information about importance of attributes can be used in the pre-processing step of initial data preparation, to remove irrelevant or superfluous elements. It can also be employed in post-processing, for optimisation of already constructed classifiers. The chapter describes research on the latter approach, involving filtering inferred decision rules while exploiting ranking positions and scores of features. The optimised rule classifiers were applied in the domain of stylometric analysis of texts for the task of binary authorship attribution.

Keywords Attribute · Ranking · Rule classifier · DRSA · Stylometry · Authorship attribution

7.1 Introduction

Considerations on relevance of the characteristic features can constitute a part of the initial phase of input data preparation for some inducer [23]. When early rejection of some elements forms the set of available attributes, it influences the execution of the data mining processes and obtained solutions. A study of importance of variables does not end with construction of a classifier, as it is then evaluated and obtained performance can be exploited as an indicator of significance of roles played by individual attributes or their groups. These observations in turn can be used to modify a system, and optimise it for a certain recognition task. Such post-processing methodologies enable closer tailoring of the inducer to particular properties, giving improved

U. Stańczyk (✉)
Institute of Informatics, Silesian University of Technology, Akademicka 16,
44-100 Gliwice, Poland
e-mail: urszula.stanczyk@polsl.pl

© Springer International Publishing AG 2018
U. Stańczyk et al. (eds.), *Advances in Feature Selection for Data and Pattern Recognition*, Intelligent Systems Reference Library 138,
https://doi.org/10.1007/978-3-319-67588-6_7

performance or reduced structures, but typically they are wrapped around the specific classifier, and either they are not applicable to other systems or their application does not bring comparable gains.

Many data analysis approaches perceive characteristic features through their ordering, which reflects their estimated importance [18]. The scores assigned to attributes result in their ranking, with the highest weights typically given to the highest ranking positions, and the lowest weights to the lowest ranking. Ordering of variables can be obtained by calculation of some statistical measures that base on concepts from information theory [7], such as entropy, or by algorithms like Relief, which computes differences between instances, or OneR, which relies on generalisation and descriptive properties possessed by short rules [17]. These methods work as filters, as they are independent on an inducer used for recognition, contrasting with wrapper approaches that condition the observation of attribute importance on the performance of the employed classifier [36].

The chapter describes research on optimisation of rule classifiers inferred within Dominance-Based Rough Set Approach (DRSA) [14]. It is a rough set processing methodology which by observation of orderings and preferences in value sets for attributes enables not only nominal but also ordinal classification. DRSA is capable of operating on both continuous and nominal features and is employed in multi-criteria decision making problems [32].

In post-processing the rule classifiers were optimised by filtering constituent decision rules governed by attribute rankings [35]. The considered rule subsets were retrieved from the sets of all available elements by referring to their condition attributes and by evaluation of the proposed rule quality measure based on weights assigned to features, corresponding to ranking positions.

The domain of application for the presented research framework was stylometric analysis of texts, a branch of science that focuses on studying writing styles and their characteristics [2], and descriptive stylometric features of quantitative type [8]. Textual descriptors referring to linguistic characteristics of styles allow to treat the most prominent task of authorship attribution as classification, where class labels correspond to recognised authors [33].

The text of the chapter is organised as follows. Section 7.2 provides the theoretical background. In Sect. 7.3 details of the framework for the performed experiments are presented. Section 7.4 shows tests results. Concluding remarks and comments on future research are listed in Sect. 7.5.

7.2 Background

In the research described in this chapter there were combined elements of feature selection approaches, in particular attribute rankings, and filtering rules as a way of classifier optimisation, briefly presented in the following sections.

7.2.1 Attribute Ranking

Attribute rankings belong with feature selection approaches dedicated to estimation
of relevance or importance of variables [16]. Through some measures referring to
statistics or algorithms to available features there are assigned some scores which lead
to their ordering [20]. Some approaches are capable of detecting irrelevant elements
while others for all return nonzero scores. Still other methodologies simply organise
all available features in an order reflecting descending importance of elements. Thus
ranked elements may have assigned weights calculated by the ranking algorithm or
just ranking positions are specified.

Statistical measures typically use concepts from information theory for calculations, in particular entropy, and probabilities of occurrences for features and classes.
Here, the χ^2 coefficient, employed in the research, estimates relations between random variables [7], by the following equation:

$$\chi^2(x_f, Cl) = \sum_{i,j} \frac{\left(P(x_f = x_j, Cl_i) - P(x_f = x_j)P(Cl_i)\right)^2}{P(x_f = x_j)P(Cl_i)}, \qquad (7.1)$$

where $P(x_f)$ is the probability of occurrence for the feature x_f, and $P(x_f, Cl_i)$ the
probability of the joint occurrence of x_f and Cl_i, for the feature x_f and the class Cl_i.

Another way of obtaining scores for attributes is to calculate them with the help
of some algorithm such as Relief or OneR [17]. The latter, used in the performed
experiments, with the pseudo-code listed as Algorithm 7.1, relies on the observation
that short and simple rules often perform well due to their good descriptive and
generalisation properties.

Algorithm 7.1: Pseudo-code for OneR classifier

Input: set of learning instances X;
           set A of all attributes;
Output: $1\text{-}r_B$ rule;
begin
      $CandidateRules \leftarrow \emptyset$
      for each attribute $a \in A$ **do**
           for each value v_a of attribute a **do**
               count how often each class appears
               find the most frequent class Cl_F
               construct a rule *IF* $a = v_a$ *THEN* Cl_F
           endfor
           calculate classification accuracy for all constructed rules
           choose the best performing rule r_B
           $CandidateRules \leftarrow r_B$
      endfor
      choose as $1\text{-}r_B$ rule the best one from $CandidateRules$
end {algorithm}

In the first stage of algorithm execution for all attributes, their values and class labels these values most frequently classify to, there is constructed the set of candidate rules. From this set for each feature there is selected a single best rule, evaluated by establishing the highest classification accuracy obtained for the training samples. Then in the second stage from all best rules the one with the lowest error is chosen as $1\text{-}r_B$ rule. Calculated errors are used as weighting scores for the attributes leading to their ranking.

Rankings are typically a part of initial pre-processing of input data, obtained independently on classification systems used later for recognition. Hence they belong with the filtering approaches to feature selection [9]. Embedded methods exploit inherent mechanisms of data mining techniques and are inseparable from them [30, 34], while wrappers condition the process of assessing attribute importance on the performance of the particular inducer [36]. In the research as a representative from this group the wrapper based on Artificial Neural Networks (ANN) was employed. ANN had a topology of widely used Multi-Layer Perceptron (MLP), a feed-forward network with backpropagation algorithm for its learning rule [11]. Training the neural network involves changing weights assigned to interconnections in order to minimise the error on the output, calculated as the difference between received and expected outcome, for all outputs and all training facts. However, since this wrapper was used for the sole purpose of ranking features, in the described research it also worked as a filter.

7.2.2 Rule Classifiers

Any inducer used in data mining somehow stores information discovered during training and applies it to previously unknown samples in order to classify them. One of the greatest advantages of rule classifiers is their explicit form of storing learned knowledge by listing particular and detailed conditions on attributes that correspond to detected patterns specific to the recognised classes. It enhances understanding of described concepts and makes the workings of the classifier transparent [40].

Dominance-Based Rough Set Approach uses elements of rough set theory, invented by Zdzisław Pawlak [26, 27], to solve tasks from the domain of multi-criteria decision making [12]. Through the original Classical Rough Set Approach (CRSA) the Universe becomes granular, with granules corresponding to equivalence classes of objects that cannot be discerned while taking into account values of their attributes. These values need to be nominal and only nominal classification is possible [10]. In DRSA indiscernibility relation is replaced by dominance, thus granules turn into dominance cones, dominated and dominating sets used to approximate upward and downward unions of decision classes $Cl = \{Cl_t\}$, with $t = 1, \ldots, n$,

$$Cl_t^{\geq} = \bigcup_{s \geq t} Cl_s \qquad Cl_t^{\leq} = \bigcup_{s \leq t} Cl_s. \tag{7.2}$$

For all value sets preference orders are either discovered or arbitrarily assigned, which allows for nominal and numerical attributes, in particular continuous valued as well, and both nominal and ordinal classification [13].

The structure of a rule classifier depends on rules that form a decision algorithm, their number and properties. The numbers of rules can greatly vary as there are many rule induction algorithms [31]. Minimal cover decision algorithms are relatively quickly generated as they include only so many elements that are sufficient to provide a cover for the learning samples. But such short algorithms often exhibit unsatisfactory performance as coverage does not equal high quality of rules. On the other end there are placed all rules on examples algorithms that include all rules that can be inferred from the training instances. Such algorithms can be unfeasibly long, causing the necessity of additional processing in order to avoid ambiguous decisions and arrive at some acceptable performance. In between there are included all algorithms with neither minimum nor maximal numbers of constituent rules, inferred taking into account some specific requirements or guidelines, possibly based on some rule parameters [6].

Constructed rules are characterised by their included conditions on attributes, but also by length, support, coverage, strength. A rule length equals the number of conditions included in its premise, while support indicates the number of learning instances on which the rule is based, which support the rule. The number of samples that match all listed conditions of a rule, regardless of classes they belong to, specifies the coverage. A rule strength or confidence shows the ratio between the number of instances with matching conditions on attributes to the total number of samples included in the same decision class. Both support and coverage can also be given as percentage in relation to the total number of available samples.

Rule classifiers can be optimised by pre-processing, which is in fact some transformation of input data, at the induction stage when only interesting rules are inferred, or in post-processing, by analysis of induced rule sets and their filtering. This last approach was used in the described experiments.

7.2.3 Filtering Rules

Regardless of the original cardinality of an inferred rule set, particular methodology employed, algorithm used for induction, or performance of the constructed classifier, once the set of rules becomes available, it can be analysed with respect to possible optimisation. As typical optimisation criteria there are used either classification accuracy or the length of the algorithm corresponding to the number of constituent rules, which reflects on the required storage and processing time.

The simplest rule selection approaches take into account explicit properties of rules, length, support, particular conditions [15], or coverage [42, 43]. Shorter rules usually possess better descriptive and generalisation properties than long, since the latter express some discovered patterns with such minutiae that run into the risk of overfitting [17]. High support values indicate that the pattern captured by a rule matches many training instances, which makes it more probable to be also present in test or previously unknown samples.

Another way of processing relies on condition attributes included in rules, which somehow indicate elements from the rule sets to be selected or discarded. Information leading to such procedures can be based on domain expert knowledge specifying preferences of characteristic features, but also on other methodologies or algorithms allowing for estimation of relevance of individual variables or their groups, such as attribute rankings [37].

Yet another approach to rule selection employs measures defined for rules, which are used to evaluate their quality or interestingness [41]. Values of quality measures calculated for all rules result in their ordering or grouping, depending on types of assigned scores, which can be used as rule weights [25, 39]. In such case imposing some thresholds allows to filter only these rules that satisfy requirements.

7.3 Research Framework

The framework of described research consisted of several steps:

- preparation of input datasets,
- calculation of selected rankings,
- induction of decision rules,
- filtering rules by two approaches: directly by condition attributes and by values of rule quality measures, and
- analysis of results,

which are presented in detail in the following sections.

7.3.1 Preparation of the Input Datasets

As the application domain for described procedures of rule filtering there was chosen stylometric analysis of texts [3]. It relies on quantitative, as opposed to qualitative, definitions of writing styles [29], expressed by their authorial invariants — groups of features characteristic and unique for each writer, regardless of the text genre or specific topic. These features allow to describe linguistic characteristics of styles, compare styles of various writers while looking for shared and differentiating elements, and recognise authorship in the most popular stylometric task of authorship attribution [8]. Through defining styles by their features this task is transformed into classification, where class labels correspond to the recognised authors. In stylometric analysis typically there are used statistic-based calculations [21, 28], or machine learning techniques [1, 19].

To be reliable descriptors of writing styles, used linguistic markers need to be based on satisfactory number of text samples of sufficient length. Very short texts force the usage of other features than longer ones. To provide variety of samples for the training phase the longer works are usually divided into smaller parts, ensuring both more detailed information about variations of style which would be otherwise hidden in some general and averaged characteristics, and higher cardinalities of sets with available text samples.

As writing styles of female authors tend to exhibit noticeably different characteristics than male [22, 34], for experiments two pairs of authors were chosen, two female, and two male, namely Edith Wharton and Jane Austen, and Henry James and Thomas Hardy. The authors selected are famous for their long works.[1] The novels were split into two sets, to provide a base for separate learning and test samples. The literary works were next divided into smaller samples of comparable size. With such approach to construction of training datasets their balance was ensured [38] and classification binary.

For thus prepared text samples there were calculated frequencies of usage for selected 17 function words and 8 punctuation marks, which gave the total set of 25 characteristic features of continuous type, reflecting lexical and syntactic properties of texts [4, 24]. The function words chosen were based on the list of the most commonly used words in English language.

Text samples based on the same longer work share some characteristics, thus with the structure of the training sets as described above, evaluation of the classification systems by cross-validation tends to return falsely high recognition [5]. To ensure more reliable results in the experiments performed the evaluation was executed with the independent test sets.

7.3.2 Rankings of Features

In the experiments conducted within the research described, three rankings of features were calculated for both datasets, as listed in Table 7.1, χ^2 as a representative of statistics-based approaches, OneR from algorithmic, and a wrapper basing on Artificial Neural Networks with backward sequential selection of considered features. For all three approaches the results were obtained with 10-fold cross-validation on the training sets, with averaged merit and rank from all 10 runs. Only in this part of experiments cross-validation was used, while in all other cases the evaluation of performance was executed with test sets.

Comparative analysis of ranking positions taken by the considered features shows some expected similarities, especially between χ^2 and OneR, but also differences as ranking mechanisms based on different elements. It is worth to notice that for female writer dataset for all three rankings the two highest ranking attributes are the same. For male writer dataset the same holds true but only for the highest ranking feature.

---

[1]The works are available for on-line reading and download in various e-book formats thanks to Project Gutenberg (see http://www.gutenberg.org).

Table 7.1 Obtained rankings of attributes for both datasets

Ranking position	Female writer dataset			Male writer dataset		
	χ^2	OneR	Wrp-ANN	χ^2	OneR	Wrp-ANN
1	not	not	not	and	and	and
2	:	:	:	but	but	:
3	;	,	?	by	from	not
4	,	—	(	that	by	from
5	—	;	from	what	what	;
6	on	on	to	from	that	,
7	?	?	,	for	for	but
8	but	that	but	?	if	by
9	(	as	at	—	with	to
10	that	.	in	if	not	of
11	as	this	with	with	—	.
12	by	by	if	not	?	that
13	for	to	of	at	:	at
14	to	(	by	:	this	in
15	at	from	for	to	!	if
16	.	for	.	in	to	what
17	and	with	and	as	(	on
18	!	but	this	(	as	—
19	with	at	that	;	at	!
20	in	and	what	!	in	?
21	this	of	!	.	on	(
22	of	!	as	on	.	this
23	what	if	—	,	;	with
24	if	in	on	this	,	as
25	from	what	;	of	of	for

7.3.3 DRSA Decision Rules

DRSA is capable of working both for nominal and continuous values of condition attributes, the only requirement being the definition of preference orders for all value sets, which can be arbitrary or a result of some processing. These preferences are either of *gain*, the higher value the higher class, or *cost* type, the lower value the higher class, and classes are always ordered increasingly. With intended stylistic characteristic features these preferences were assigned arbitrarily for both datasets and then the decision rules could be induced. The performance of rule classifiers in all stages of experiments was evaluated by using test sets.

In the first step only minimal cover algorithms were inferred, but their performance was unacceptably low. Thus all rules on training examples were generated, returning

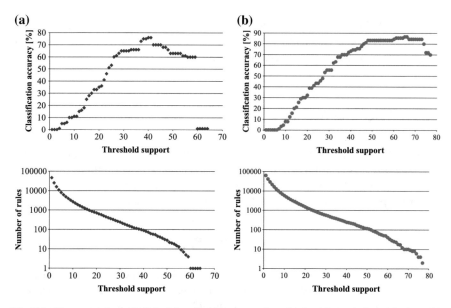

Fig. 7.1 Characteristics of full decision algorithms, number of rules (shown in logarithmic scale) and classification accuracy with respect to imposed threshold value on required rule support for: **a** male writer dataset, **b** female writer dataset

Table 7.2 Parameters of reference rule classifiers for both datasets

Dataset	Number of rules	Minimal support	Classification accuracy [%]	Algorithm
Female writer	17	66	86.67	*F-BAlg17*
Male writer	80	41	76.67	*M-BAlg80*

46,191 rules for male writers and 62,383 rules for female writers, which will be from this point on referred to in the chapter as *M-FAlg* and *F-FAlg* respectively. When performance of these sets of rules was tested without any additional processing (such as filtering rules or some kind of voting to solve conflicts), no sample was correctly recognised due to the very high numbers of matching rules with conflicting decisions. Such ambiguities were always treated as wrong decisions. These characteristics of algorithms are shown in Fig. 7.1.

In order to find the highest classification accuracy the two full algorithms were next processed by imposing hard constraints on minimal supports required of rules to be included in classification. Rules with supports lower than the required thresholds were gradually rejected until the maximal prediction ratio was detected, with the summary for both datasets presented in Table 7.2.

For male writers with the shortened algorithm consisting of 80 rules with supports equal at least 41 the prediction was at the level of 76.67%, which was denoted as *M-BAlg80*. For female writers the best performance was the correct recognition of

86.67% with 17 rules supported by not fewer than 66 learning samples, denoted as *F-BAlg17*. These two shortened algorithms were used as reference points for comparison in research on rule classifiers optimisation.

7.3.4 Weighting Rules

As attribute rankings were obtained with executing cross-validation on the training sets, the results were given as averaged merits and ranks, which were not suitable to use as weights of positions. Therefore, to all ranking positions a set of weights was assigned, spanned over the range of $(0,1]$, with the equation as follows,

$$\forall_{i \in \{1, \dots, N\}} WR_i = \frac{1}{i}, \tag{7.3}$$

where N is the number of considered attributes, and i an attribute ranking position. Thus the weight of the highest ranking attribute was 1, for the second in ranking it was $1/2$, and so on, to the lowest ranking variable with the weight equal $1/N$. It can be observed that with such formulation the distances between weights of subsequent ranking positions while following down the ranking were gradually decreasing.

Basing on these weights a quality measure *QMWR* for r rule was defined as a product of weights assigned to attributes included in the premise part of the rule,

$$QMWR(r) = \prod_{i=1}^{NrCond} Weight(a_i). \tag{7.4}$$

This definition led to obtaining the value of 1 only for rules with single conditions being the highest ranking variables, while for all other cases fractions were calculated. In fact the longer the rule the lower value became, thus typical characteristics of rules (that is dependence of quality on rule length) were preserved. Multiplication of the measure by the rule support results in

$$QMWRS(r) = QMWR(r) \cdot Support(r). \tag{7.5}$$

Such additional operation helped to reject rules with low support values at the earlier stage of considerations.

Since three distinct rankings were considered, for each rule three values of the measure were obtained, denoted respectively as $QMWR\text{-}\chi^2(r)$, $QMWR\text{-}OneR(r)$, and $QMWR\text{-}Wrp\text{-}ANN(r)$. These values were used to weight rules by putting them in descending order and selecting gradually increasing subsets of rules with the highest weights from the entire sets of available rules. The process of rule filtering was stopped once all rules from the best performing short algorithms *F-BAlg17* and *M-BAlg80* were recalled.

7.4 Experimental Results

In the experiments performed the rules were filtered by two distinct approaches: directly by included condition attributes, and by weighting decision rules, the results of which are described in this section.

7.4.1 Filtering Rules by Attributes

When selection of rules is driven by attribute ranking, the process is executed in steps, the first of which corresponds to taking into consideration a single, the highest ranking feature, and in the following steps to this initial subset gradually one by one other variables are added. The process can be stopped when some criteria are met, or it can continue till all attributes from the entire available set are considered. In each step from the set of all inferred rules there are recalled these that include conditions only on features present in the currently considered subset of variables selected from the ranking. If even one condition of a rule refers to an attribute not included in the considered set, that rule is disregarded.

Characteristics of decision algorithms constructed by rule selection governed by attribute rankings for the female writer dataset are listed in Table 7.3. Each retrieved subset of rules was treated as a decision algorithm by assuming the classification strategy that all cases of no rules matching or conflicts were treated as incorrect decisions. Each decision algorithm was further filtered by imposing hard constraints on minimal rule support leading to obtaining the maximal classification accuracy.

In the first step of processing a single feature was selected from each ranking, as it happened, it was the same for all three rankings, corresponding to the frequency of usage for "not" word, and from the set of all rules only these with single conditions on this particular attribute were recalled. There were 10 such rules, giving the correct classification of 61.11% of test samples. With the imposed limit on rule support to be at least 55, the number of rules was reduced to just 4, and the maximal classification accuracy obtained was the same as without any additional processing. In the second step, two highest ranking features were selected, "not" and semicolon, and rules with conditions on either one of these two or both of them were retrieved, returning the subset of rules with 27 elements. These rules correctly classified 80.00% of test samples, but selecting 13 rules with supports equal or higher than 55 led to the classification accuracy of 81.11%. The process continued in this manner till all 25 features were considered, but starting with the 3rd highest ranking position attributes were no longer the same for all three rankings studied.

In the initial steps, when just few features were considered, it can be observed that the numbers of recalled rules were rather low and the classification accuracy fell below the expected level of the reference algorithm *F-BAlg17*. This level was reached the soonest while following OneR ranking, in the 3rd step, and then in the 4th for χ^2 ranking, but this performance was obtained with lower support values than in the reference algorithm, thus the numbers of included rules were higher.

Table 7.3 Characteristics of decision algorithms for female writer dataset with pruning rules by condition attributes while following rankings: N indicates the number of considered attributes, (a) number of recalled rules, (b) classification accuracy without any constraints on rules [%], (c) constraints on rule support, (d) number of rules satisfying condition on support, and (e) maximal classification accuracy [%]

N	χ^2					OneR					Wrp-ANN				
	(a)	(b)	(c)	(d)	(e)	(a)	(b)	(c)	(d)	(e)	(a)	(b)	(c)	(d)	(e)
1	10	61.11	55	4	61.11	10	61.11	55	4	61.11	10	61.11	55	4	61.11
2	27	80.00	55	13	81.11	27	80.00	55	13	81.11	27	80.00	55	13	81.11
3	36	80.00	55	13	81.11	68	84.44	55	27	86.67	42	78.89	55	13	81.11
4	79	84.44	55	27	86.67	78	84.44	55	27	86.67	57	72.22	46	18	83.33
5	91	83.33	55	27	86.67	91	83.33	55	27	86.67	115	70.00	46	26	85.56
6	128	78.89	55	27	86.67	128	78.89	55	27	86.67	228	67.78	46	33	85.56
7	167	75.56	55	27	86.67	167	75.56	55	27	86.67	355	67.78	62	16	86.67
8	272	70.00	55	27	86.67	266	73.33	55	27	86.67	547	58.89	62	16	86.67
9	337	64.44	55	27	86.67	429	63.33	66	11	**86.67**	741	53.33	62	16	86.67
10	513	57.78	55	27	86.67	585	57.78	66	11	86.67	1098	41.11	62	16	86.67
11	832	51.11	66	11	**86.67**	893	53.33	66	13	86.67	1549	32.22	62	16	86.67
12	1415	37.78	66	12	86.67	1574	35.56	66	14	86.67	2166	25.56	62	16	86.67
13	2201	35.56	66	14	86.67	2242	31.11	66	14	86.67	3268	14.44	62	16	86.67
14	3137	30.00	66	14	86.67	2825	25.56	66	14	86.67	4692	14.44	62	17	86.67
15	4215	25.56	66	14	86.67	3868	11.11	66	14	86.67	6861	12.22	66	13	**86.67**
16	5473	18.89	66	14	86.67	5651	8.89	66	16	86.67	9617	3.33	66	13	86.67
17	7901	6.67	66	14	86.67	7536	2.22	66	16	86.67	12648	0.00	66	13	86.67
18	10024	6.67	66	15	86.67	9833	2.22	66	16	86.67	16818	0.00	66	15	86.67
19	13084	4.44	66	15	86.67	12921	1.11	66	16	86.67	21153	0.00	66	15	86.67
20	17770	2.22	66	15	86.67	17905	0.00	66	16	86.67	27407	0.00	66	15	86.67
21	23408	1.11	66	17	86.67	24095	0.00	66	16	86.67	35416	0.00	66	16	86.67
22	31050	0.00	66	17	86.67	30674	0.00	66	17	86.67	44945	0.00	66	17	86.67
23	39235	0.00	66	17	86.67	38363	0.00	66	17	86.67	52587	0.00	66	17	86.67
24	48538	0.00	66	17	86.67	50356	0.00	66	17	86.67	58097	0.00	66	17	86.67
25						62383	0.00	66	17	86.67					

Again for OneR ranking in the 9th step there were recalled 429 rules, which constrained by support equal or higher than 66 were limited to 11 that gave the shortest reduced version of *F-BAlg17* algorithm maintaining its accuracy. For χ^2 ranking the same algorithm was found in the 11th step, whereas for Wrp-ANN ranking the shortest reduced algorithm with the same power, discovered in the 15th step, included 13 rules. Thus from these three rankings for female writers the best results of filtering were obtained for OneR ranking and the worst for Wrp-ANN. All rules from *F-BAlg17* algorithm were retrieved in the 21st step for χ^2 ranking, and in the 22nd step for the other two.

For the male writer dataset the results of filtering are given in Table 7.4. Generally for this dataset the recognition was more difficult and required higher numbers of attributes to be taken into considerations to reach the classification accuracy of the reference algorithm *M-BAlg80*. Also in this case the results were the best for OneR

Table 7.4 Characteristics of decision algorithms for male writer dataset with pruning rules by condition attributes while following rankings: N is the number of considered attributes, (a) number of recalled rules, (b) classification accuracy without any constraints on rules [%], (c) constraints on rule support, (d) number of rules satisfying condition on support, and (e) maximal classification accuracy [%]

N	χ^2					OneR					Wrp-ANN				
	(a)	(b)	(c)	(d)	(e)	(a)	(b)	(c)	(d)	(e)	(a)	(b)	(c)	(d)	(e)
1	6	13.33	14	4	13.33	6	13.33	14	4	13.33	6	13.33	14	4	13.33
2	20	36.67	14	10	36.67	20	36.67	14	10	36.67	15	31.67	14	8	31.67
3	52	61.67	10	34	61.67	44	48.33	13	30	50.00	49	38.33	14	28	38.33
4	73	61.67	10	41	61.67	111	71.67	21	61	75.00	117	56.67	27	29	60.00
5	100	65.00	12	43	61.67	150	73.33	21	64	75.00	191	51.67	21	63	66.67
6	198	68.33	21	65	75.00	198	68.33	21	65	75.00	313	43.33	21	66	66.67
7	239	66.67	26	46	75.00	239	66.67	26	46	75.00	360	43.33	21	72	66.67
8	337	65.00	26	50	75.00	311	60.00	26	54	75.00	635	38.33	30	65	73.33
9	422	65.00	21	79	75.00	389	58.33	26	61	75.00	885	33.33	30	67	73.33
10	531	58.33	21	89	75.00	637	41.67	25	99	75.00	1312	30.00	30	78	71.67
11	659	53.33	21	98	75.00	812	38.33	25	103	75.00	1960	15.00	30	83	71.67
12	1063	33.33	26	100	73.33	1063	33.33	26	100	73.33	2285	11.67	30	84	71.67
13	1395	28.33	32	65	75.00	1352	30.00	26	104	73.33	2838	6.67	30	89	70.00
14	1763	25.00	32	67	75.00	1796	28.33	34	66	73.33	4083	5.00	41	39	73.33
15	2469	16.67	32	67	75.00	2415	26.67	34	72	76.67	5040	3.33	41	42	73.33
16	3744	15.00	41	42	73.33	3327	23.33	34	72	76.67	5869	1.67	41	42	73.33
17	4651	15.00	41	43	73.33	3936	16.67	34	89	75.00	8080	1.67	41	43	73.33
18	5352	13.33	41	57	75.00	5168	13.33	34	90	75.00	10152	1.67	41	43	73.33
19	7402	10.00	41	60	75.00	6809	8.33	34	92	75.00	13265	1.67	41	46	73.33
20	9819	8.33	41	63	75.00	9593	5.00	41	69	**76.67**	15927	0.00	41	49	73.33
21	13106	5.00	41	63	75.00	14515	3.33	41	70	76.67	18794	0.00	41	66	75.00
22	18590	5.00	41	64	75.00	18882	1.67	41	70	76.67	25544	0.00	41	75	**76.67**
23	24404	3.33	41	65	75.00	24960	0.00	41	73	76.67	31420	0.00	41	79	76.67
24	32880	0.00	41	76	**76.67**	32880	0.00	41	76	76.67	38741	0.00	41	80	76.67
25						46191	0.00	41	80	76.67					

ranking, with reduced reference algorithm recalled in the 20th step, including 69 constituent rules. Wrp-ANN seemed to be the second best, as reduction to 75 rules was obtained in the 22nd step, whereas for χ^2 just 4 rules were discarded in the 24th step. For both χ^2 and OneR rankings all rules from *M-BAlg80* algorithm were retrieved only when all available attributes were taken into considerations, and for Wrp-ANN in the 24th step.

Comparison of results of rule filtering for both datasets brings the conclusion that the best reduction was obtained for OneR ranking, and it was found sooner, for fewer considered condition attributes, for fewer recalled rules, which can be explained by the fact that DRSA classifier is also a rule classifier as OneR, therefore they are bound to share some characteristics. Thus OneR ranking can be considered as showing some preferential bias towards other rule classifiers.

7.4.2 Filtering Rules by Weights

Weighting rules constitutes another approach to rule filtering, within which each rule is assigned some score [39]. This value is then used to impose an order on rules similar to a ranking. All available rules are put in descending order and then from this ordered set gradually increasing subsets of rules are retrieved. Depending on particular formulation of the weighting measure its values for different rules can be distinct or the same, thus in the process of selection either single elements or their groups can be recalled.

In case of sets of available rules with high cardinalities retrieving subsets increasing with each step by single included elements would make the filtering process unfeasibly long. Thus the more reasonable attitude is to impose some thresholds on considered values and perform some arbitrarily set numbers of steps. In the experiments 10 such steps were executed for both datasets and the processing was stopped when all rules from the reference algorithms F-$BAlg17$ and M-$BAlg80$ were included in the recalled subsets of rules.

Detailed parameters of decision algorithms constructed by weighting rules for the female writers are listed in Table 7.5, the upper part dedicated to $QMWR$ measures, and the bottom part to $QMWRS$. It can be observed that in this case not only reduction of the reference algorithm was possible, but also some cases of increased classification accuracy were detected.

$QMWR$-Wrp-ANN measure brought no noticeable reduction of the F-$BAlg17$ algorithm, but enabled increased performance, at the level of 90.00%, in the best case (that is for the shortest algorithm) for 38 rules with supports equal at least 52. For the other two measures generally the results were similar, the increase in recognition was rather slight, but they both led to obtaining the noticeably reduced length of the reference algorithm for the same level of accuracy, with the smallest number of rules equal 11, the same as in case of filtering by attributes, but discovered in the subsets of retrieved rules with smaller cardinalities than before. For $QMWR$ measures all rules from the F-$BAlg17$ algorithm were recalled for all three rankings in the subsets with cardinalities below 500 elements, and for $QMWRS$ measures below 200, in both cases the fewest rules were for χ^2.

For the male writer dataset the characteristics of constructed rule classifiers are listed in Table 7.6. For this dataset no increase in recognition was detected, but more rules could be rejected from the reference algorithm of M-$BAlg80$ without undermining its power than it happened for filtering by condition attributes.

For $QMWR$-χ^2 measure the shortest such algorithm contained 67 rules, while for $QMWR$-$OneR$ the smallest length was 63, and for $QMWR$-Wrp-ANN it was 66. Versions of $QMWRS$ did not bring any further reduction, on the contrary, but allowed for recalling about three times fewer rules from the entire available set as rules with low support values were immediately transferred to lower positions in the rule rankings. For male writers OneR ranking gave the best results in rule filtering by measures, while for other two rankings the gains were similar.

Table 7.5 Characteristics of decision algorithms for female writer dataset with pruning rules by measures based on ranked condition attributes: (a) number of recalled rules, (b) classification accuracy without any constraints on rules [%], (c) constraints on rule support, (d) number of rules satisfying condition on support, and (e) maximal classification accuracy [%]

Step	$QMWR\text{-}\chi^2$					$QMWR\text{-}OneR$					$QMWR\text{-}Wrp\text{-}ANN$				
	(a)	(b)	(c)	(d)	(e)	(a)	(b)	(c)	(d)	(e)	(a)	(b)	(c)	(d)	(e)
1	10	61.11	55	4	61.11	10	61.11	55	4	61.11	21	64.44	46	9	65.56
2	27	80.00	55	13	81.11	27	80.00	55	13	81.11	52	82.22	46	24	85.56
3	46	85.56	52	25	**87.78**	44	85.56	52	25	**87.78**	122	77.78	46	42	**90.00**
4	107	82.22	55	27	86.67	84	84.44	55	27	86.67	190	71.11	52	38	**90.00**
5	140	80.00	66	11	**86.67**	114	81.11	66	11	**86.67**	232	70.00	52	43	90.00
6	179	78.89	66	12	86.67	153	80.00	66	13	86.67	283	65.56	52	52	90.00
7	194	78.89	66	14	86.67	195	78.89	66	14	86.67	325	62.22	52	57	90.00
8	229	76.67	66	14	86.67	251	74.44	66	14	86.67	363	60.00	52	60	90.00
9	297	74.44	66	15	86.67	307	68.89	66	16	86.67	427	55.56	52	63	88.89
10	373	66.67	66	17	86.67	447	62.22	66	17	86.67	466	55.56	66	17	**86.67**

Step	$QMWRS\text{-}\chi^2$					$QMWRS\text{-}OneR$					$QMWRS\text{-}Wrp\text{-}ANN$				
	(a)	(b)	(c)	(d)	(e)	(a)	(b)	(c)	(d)	(e)	(a)	(b)	(c)	(d)	(e)
1	15	81.11	55	12	81.11	7	61.11	55	4	61.11	7	61.11	55	4	61.11
2	19	81.11	55	13	81.11	15	81.11	55	12	81.11	17	82.22	55	13	82.22
3	36	87.78	52	25	**87.78**	23	86.67	55	17	86.67	25	88.89	46	24	**88.89**
4	40	85.56	55	18	86.67	31	87.78	51	25	**87.78**	72	86.67	46	43	**90.00**
5	45	85.56	55	20	86.67	43	85.56	55	22	86.67	77	86.67	52	39	**90.00**
6	51	85.56	55	26	86.67	55	85.56	66	11	**86.67**	85	85.56	52	43	90.00
7	55	85.56	66	11	**86.67**	65	84.44	66	13	86.67	96	85.56	52	47	90.00
8	67	85.56	66	14	86.67	82	82.22	66	14	86.67	102	85.56	52	50	90.00
9	94	81.11	66	14	86.67	107	82.22	66	16	86.67	166	78.89	66	16	**87.78**
10	130	81.11	66	17	86.67	141	82.22	66	17	86.67	170	78.89	66	17	**86.67**

7.4.3 Summary of Results

Comparison of the results for rule selection driven by condition attributes and by defined measures for female writer dataset brought several conclusions, a summary of which is given by Table 7.7. Obtaining both increased classification accuracy and shortened decision algorithm at the same time was not possible, but separately these two popular aims of filtering were achieved.

The improved prediction was detected only for filtering while using measures based on Wrp-ANN ranking, but for this ranking the smallest reduction of the reference *F-BAlg17* algorithm was obtained, by 23.53%. For the other two rankings only small improvement in prediction accuracy was detected while filtering by measures, but they allowed to shorten the reference algorithm by 35.29%.

Table 7.6 Characteristics of decision algorithms for male writer dataset with pruning rules by measures based on ranked condition attributes: (a) number of recalled rules, (b) classification accuracy without any constraints on rules [%], (c) constraints on rule support, (d) number of rules satisfying condition on support, and (e) maximal classification accuracy [%]

Step	$QMWR$-χ^2					$QMWR$-$OneR$					$QMWR$-Wrp-ANN				
	(a)	(b)	(c)	(d)	(e)	(a)	(b)	(c)	(d)	(e)	(a)	(b)	(c)	(d)	(e)
1	103	65.00	10	62	66.67	121	61.67	10	71	66.67	239	58.33	27	62	63.33
2	237	61.67	12	124	70.00	220	63.33	26	78	73.33	402	50.00	21	136	70.00
3	284	56.67	26	88	73.33	441	53.33	21	159	75.00	601	41.67	26	132	73.33
4	820	50.00	26	136	75.00	835	48.33	41	56	75.00	1222	36.67	41	53	73.33
5	1411	40.00	41	58	75.00	1372	43.33	41	59	75.00	1355	31.67	41	56	75.00
6	2715	23.33	41	67	**76.67**	1744	35.00	41	63	**76.67**	3826	15.00	41	62	75.00
7	4560	18.33	41	70	76.67	2710	28.33	41	68	76.67	4466	13.33	41	66	**76.67**
8	6515	13.33	41	74	76.67	4549	18.33	41	74	76.67	5837	10.00	41	71	76.67
9	7936	8.33	41	76	76.67	7833	8.33	41	77	76.67	7568	6.67	41	76	76.67
10	9520	5.00	41	80	76.67	9819	3.33	41	80	76.67	10268	3.33	41	80	76.67

Step	$QMWRS$-χ^2					$QMWRS$-$OneR$					$QMWRS$-Wrp-ANN				
	(a)	(b)	(c)	(d)	(e)	(a)	(b)	(c)	(d)	(e)	(a)	(b)	(c)	(d)	(e)
1	128	70.00	26	89	70.00	58	60.00	27	41	60.00	57	58.33	27	38	58.33
2	230	71.67	12	191	75.00	135	70.00	26	96	73.33	254	56.67	30	99	71.67
3	599	53.33	41	60	75.00	351	68.33	41	58	75.00	383	55.00	41	55	75.00
4	889	50.00	41	68	**76.67**	650	51.67	41	64	**76.67**	1320	30.00	41	64	75.00
5	1273	38.33	41	70	76.67	930	48.33	41	68	76.67	1595	28.33	41	68	**76.67**
6	1747	35.00	41	73	76.67	1337	38.33	41	69	76.67	2094	23.33	41	73	76.67
7	2021	30.00	41	75	76.67	1478	38.33	41	74	76.67	2303	23.33	41	75	76.67
8	2382	26.67	41	77	76.67	1606	36.67	41	76	76.67	2551	23.33	41	77	76.67
9	2086	25.00	41	79	76.67	2728	25.00	41	78	76.67	2868	20.00	41	78	76.67
10	3240	23.33	41	80	76.67	3210	25.00	41	80	76.67	3408	18.33	41	80	76.67

Table 7.7 Gains of rule filtering driven by condition attributes and measures for female writer dataset: (a) length reduction of decision algorithms [%], (b) obtained classification accuracy [%]

Ranking	Filtering decision rules by				
	Attributes	$QMWR$		$QMWRS$	
	(a)	(a)	(b)	(a)	(b)
χ^2	35.29	35.29	87.78	35.29	87.78
OneR	35.29	35.29	87.78	35.29	87.78
Wrp-ANN	23.53		90.00		90.00

As can be seen in Table 7.8, for male writer dataset no increase in classification accuracy was detected in rule filtering, but the reference algorithm *M-BAlg80* was reduced to the highest extent thanks to the usage of *QMWR* measure, and the best results were always obtained for OneR ranking. The length of the algorithm was shortened at maximum by 21.25%.

Table 7.8 Length reduction of the reference decision algorithm obtained in rule filtering driven by condition attributes and measures for male writer dataset [%]

Ranking	Filtering decision rules by		
	Attributes	*QMWR*	*QMWRS*
χ^2	5.00	16.25	15.00
OneR	13.75	21.25	20.00
Wrp-ANN	6.25	17.50	15.00

Both filtering by condition attributes and by evaluation of the defined quality measure for rules and weighting them by this measure resulted in construction of decision algorithms with reduced numbers of rules (with respect to the reference algorithms), and also some cases of improved performance of rule classifiers. Hence both these approaches to rule selection show some merit and can be considered as optimisation possibilities.

7.5 Conclusions

The chapter describes research on rule filtering governed by attribute rankings, applied in the task of binary authorship recognition with balanced classes for two datasets. For available stylistic characteristic features three distinct rankings were obtained, one referring to statistics by the usage of χ^2 coefficient, the second algorithmic employing OneR, and the third the wrapper based on artificial neural network with sequential backward reduction of features. Next the exhaustive decision algorithms were induced with Dominance-Based Rough Set Approach.

The sets of inferred rules were then filtered in two different ways, firstly directly by included condition attributes, while following a chosen ranking, then by weighting all rules through the defined quality measure based on weights assigned arbitrarily to attribute ranking positions. Rule weighting led to their ordering and then recalling rules with highest values of the calculated measures. In all tested approaches it was possible to obtain some reduction of the reference best performing algorithms, and in some cases also improved classification accuracy was achieved.

In the future research other attribute rankings and other quality measures will be tested on rules induced by other methodologies, not only for continuous valued characteristic features but also for discretised.

Acknowledgements In the research there was used WEKA workbench [40]. 4eMka Software exploited for DRSA processing [32] was developed at the Laboratory of Intelligent Decision Support Systems, Poznań, Poland. The research was performed at the Silesian University of Technology, Gliwice, within the project BK/RAu2/2017.

References

1. Ahonen, H., Heinonen, O., Klemettinen, M., Verkamo, A.: Applying data mining techniques in text analysis. Technical report C-1997-23, Department of Computer Science, University of Helsinki, Finland (1997)
2. Argamon, S., Burns, K., Dubnov, S. (eds.): The Structure of Style: Algorithmic Approaches to Understanding Manner and Meaning. Springer, Berlin (2010)
3. Argamon, S., Karlgren, J., Shanahan, J.: Stylistic analysis of text for information access. In: Proceedings of the 28th International ACM Conference on Research and Development in Information Retrieval, Brazil (2005)
4. Baayen, H., van Haltern, H., Tweedie, F.: Outside the cave of shadows: using syntactic annotation to enhance authorship attribution. Lit. Linguist. Comput. **11**(3), 121–132 (1996)
5. Baron, G.: Comparison of cross-validation and test sets approaches to evaluation of classifiers in authorship attribution domain. In: Czachórski, T., Gelenbe, E., Grochla, K., Lent, R. (eds.) Proceedings of the 31st International Symposium on Computer and Information Sciences, Communications in Computer and Information Science, vol. 659, pp. 81–89. Springer, Cracow (2016)
6. Bayardo Jr., R., Agrawal, R.: Mining the most interesting rules. In: Proceedings of the 5th ACM SIGKDD International Conference on Knowledge Discovery and Data Mining, pp. 145–154 (1999)
7. Biesiada, J., Duch, W., Kachel, A., Pałucha, S.: Feature ranking methods based on information entropy with Parzen windows. In: Proceedings of International Conference on Research in Electrotechnology and Applied Informatics, pp. 109–119, Katowice (2005)
8. Craig, H.: Stylistic analysis and authorship studies. In: Schreibman, S., Siemens, R., Unsworth, J. (eds.) A Companion to Digital Humanities. Blackwell, Oxford (2004)
9. Dash, M., Liu, H.: Feature selection for classification. Intelligent Data Analysis **1**, 131–156 (1997)
10. Deuntsch, I., Gediga, G.: Rough Set Data Analysis: A Road to Noninvasive Knowledge Discovery. Mathoδos Publishers, Bangor (2000)
11. Fiesler, E., Beale, R.: Handbook of Neural Computation. Oxford University Press, Oxford (1997)
12. Greco, S., Matarazzo, B., Słowiński, R.: The use of rough sets and fuzzy sets in multi criteria decision making. In: Gal, T., Hanne, T., Stewart, T. (eds.) Advances in Multiple Criteria Decision Making, Chap. 14, pp. 14.1–14.59. Kluwer Academic Publishers, Dordrecht (1999)
13. Greco, S., Matarazzo, B., Słowiński, R.: Rough set theory for multicriteria decision analysis. Eur. J. Oper. Res. **129**(1), 1–47 (2001)
14. Greco, S., Matarazzo, B., Słowiński, R.: Dominance-based rough set approach as a proper way of handling graduality in rough set theory. Trans. Rough Sets VII **4400**, 36–52 (2007)
15. Greco, S., Słowiński, R., Stefanowski, J.: Evaluating importance of conditions in the set of discovered rules. Lect. Notes Artif. Intell. **4482**, 314–321 (2007)
16. Guyon, I., Elisseeff, A.: An introduction to variable and feature selection. J. Mach. Learn. Res. **3**, 1157–1182 (2003)
17. Holte, R.: Very simple classification rules perform well on most commonly used datasets. Mach. Learn. **11**, 63–91 (1993)
18. Jensen, R., Shen, Q.: Computational Intelligence and Feature Selection. Wiley, Hoboken (2008)
19. Jockers, M., Witten, D.: A comparative study of machine learning methods for authorship attribution. Lit. Linguist. Comput. **25**(2), 215–223 (2010)
20. John, G., Kohavi, R., Pfleger, K.: Irrelevant features and the subset selection problem. In: Cohen, W., Hirsh, H. (eds.) Machine Learning: Proceedings of the 11th International Conference, pp. 121–129. Morgan Kaufmann Publishers (1994)
21. Khmelev, D., Tweedie, F.: Using Markov chains for identification of writers. Lit. Linguist. Comput. **16**(4), 299–307 (2001)
22. Koppel, M., Argamon, S., Shimoni, A.: Automatically categorizing written texts by author gender. Lit. Linguist. Comput. **17**(4), 401–412 (2002)

23. Liu, H., Motoda, H.: Computational Methods of Feature Selection. Chapman & Hall/CRC, Boca Raton (2008)
24. Lynam, T., Clarke, C., Cormack, G.: Information extraction with term frequencies. In: Proceedings of the Human Language Technology Conference, pp. 1–4. San Diego (2001)
25. Moshkov, M., Piliszczuk, M., Zielosko, B.: On partial covers, reducts and decision rules with weights. Trans. Rough Sets VI **4374**, 211–246 (2006)
26. Pawlak, Z.: Computing, artificial intelligence and information technology: rough sets, decision algorithms and Bayes' theorem. Eur. J. Oper. Res. **136**, 181–189 (2002)
27. Pawlak, Z.: Rough sets and intelligent data analysis. Inf. Sci. **147**, 1–12 (2002)
28. Peng, R.: Statistical aspects of literary style. Bachelor's thesis, Yale University (1999)
29. Peng, R., Hengartner, H.: Quantitative analysis of literary styles. Am. Stat. **56**(3), 15–38 (2002)
30. Shen, Q.: Rough feature selection for intelligent classifiers. Trans. Rough Sets VII **4400**, 244–255 (2006)
31. Sikora, M.: Rule quality measures in creation and reduction of data rule models. In: Greco, S., Hata, Y., Hirano, S., Inuiguchi, M., Miyamoto, S., Nguyen, H., Słowiński, R. (eds.) Rough Sets and Current Trends in Computing, Lecture Notes in Computer Science, vol. 4259, pp. 716–725. Springer, Berlin (2006)
32. Słowiński, R., Greco, S., Matarazzo, B.: Dominance-based rough set approach to reasoning about ordinal data. Lect. Notes Comput. Sci. (Lect. Notes Artif. Intell.) **4585**, 5–11 (2007)
33. Stamatatos, E.: A survey of modern authorship attribution methods. J. Am. Soc. Inf. Sci. Technol. **60**(3), 538–556 (2009)
34. Stańczyk, U.: Weighting of attributes in an embedded rough approach. In: Gruca, A., Czachórski, T., Kozielski, S. (eds.) Man-Machine Interactions 3, Advances in Intelligent and Soft Computing, vol. 242, pp. 475–483. Springer, Berlin (2013)
35. Stańczyk, U.: Attribute ranking driven filtering of decision rules. In: Kryszkiewicz, M., Cornelis, C., Ciucci, D., Medina-Moreno, J., Motoda, H., Raś, Z. (eds.) Rough Sets and Intelligent Systems Paradigms. Lecture Notes in Computer Science, vol. 8537, pp. 217–224. Springer, Berlin (2014)
36. Stańczyk, U.: Feature evaluation by filter, wrapper and embedded approaches. In: Stańczyk, U., Jain, L. (eds.) Feature Selection for Data and Pattern Recognition. Studies in Computational Intelligence, vol. 584, pp. 29–44. Springer, Berlin (2015)
37. Stańczyk, U.: Selection of decision rules based on attribute ranking. J. Intell. Fuzzy Syst. **29**(2), 899–915 (2015)
38. Stańczyk, U.: The class imbalance problem in construction of training datasets for authorship attribution. In: Gruca, A., Brachman, A., Kozielski, S., Czachórski, T. (eds.) Man-Mach. Interact. 4. Advances in Intelligent and Soft Computing, vol. 391, pp. 535–547. Springer, Berlin (2016)
39. Stańczyk, U.: Weighting and pruning of decision rules by attributes and attribute rankings. In: Czachórski, T., Gelenbe, E., Grochla, K., Lent, R. (eds.) Proceedings of the 31st International Symposium on Computer and Information Sciences, Communications in Computer and Information Science, vol. 659, pp. 106–114. Springer, Cracow (2016)
40. Witten, I., Frank, E., Hall, M.: Data Mining: Practical Machine Learning Tools and Techniques, 3rd edn. Morgan Kaufmann, San Francisco (2011)
41. Wróbel, L., Sikora, M., Michalak, M.: Rule quality measures settings in classification, regression and survival rule induction – an empirical approach. Fundamenta Informaticae **149**, 419–449 (2016)
42. Zielosko, B.: Application of dynamic programming approach to optimization of association rules relative to coverage and length. Fundamenta Informaticae **148**(1–2), 87–105 (2016)
43. Zielosko, B.: Optimization of decision rules relative to coverage–comparison of greedy and modified dynamic programming approaches. In: Gruca, A., Brachman, A., Kozielski, S., Czachórski, T. (eds.) Man-Machine Interactions 4. Advances in Intelligent and Soft Computing, vol. 391, pp. 639–650. Springer, Berlin (2016)

Chapter 8
Attribute Selection in a Dispersed Decision-Making System

Małgorzata Przybyła-Kasperek

Abstract In this chapter, the use of a method for attribute selection in a dispersed decision-making system is discussed. Dispersed knowledge is understood to be the knowledge that is stored in the form of several decision tables. Different methods for solving the problem of classification based on dispersed knowledge are considered. In the first method, a static structure of the system is used. In more advanced techniques, a dynamic structure is applied. Different types of dynamic structures are analyzed: a dynamic structure with disjoint clusters, a dynamic structure with inseparable clusters and a dynamic structure with negotiations. A method for attribute selection, which is based on the rough set theory, is used in all of the methods described here. The results obtained for five data sets from the UCI Repository are compared and some conclusions are drawn.

Keywords Dispersed knowledge · Attribute selection · Rough set theory

8.1 Introduction

Nowadays, a classification that is made based on single data set may not be sufficient. The possibility of classification on the basis of dispersed knowledge is becoming increasingly important. We understand dispersed knowledge to be the knowledge that is stored in the form of several decision tables. Many times knowledge is stored in a dispersed form for various reasons. For example, when knowledge in the same field is accumulated by separate units (hospitals, medical centers, banks). Another reason for knowledge dispersion is when knowledge is not available at the same time but only at certain intervals. Knowledge can be stored in a dispersed form when the data is too large to store and process in a single decision table.

M. Przybyła-Kasperek (✉)
Institute of Computer Science, University of Silesia in Katowice, Będzińska 39,
41-200 Sosnowiec, Poland
e-mail: malgorzata.przybyla-kasperek@us.edu.pl

© Springer International Publishing AG 2018
U. Stańczyk et al. (eds.), *Advances in Feature Selection for Data and Pattern Recognition*, Intelligent Systems Reference Library 138,
https://doi.org/10.1007/978-3-319-67588-6_8

When knowledge is accumulated by separate units, it is difficult to require that the local decision tables have the same sets of attributes or the same sets of objects. It is also difficult to require that these sets are disjoint. Rather, a more general approach is needed. The problem that is considered here concerns the use of dispersed knowledge in the classification process. The knowledge that is used is set in advance. Local knowledge bases have sets of attributes that do not need to be equal or disjoint. The same applies to sets of objects. Thus, the form of these sets may be very different, some attributes or objects may be common for several local bases, while others may be unique and specific to a single knowledge base.

In multi-agent systems [10], the concept of distributed knowledge can be found. A group has distributed knowledge if by putting all their information together, the members of the group can solve a problem. In this issue, much attention is paid to consensus protocols and sensors for data gathering. The problem discussed here may seem similar, but is quite different because the knowledge that is being used has already been collected; the process of its creation is complete. This knowledge was stored separately and was not intended to be merged or combine together. Therefore, dispersed knowledge will be discussed here.

Research on using the method of attribute selection, which is based on the rough set theory, in dispersed systems is presented. The issues related to decision-making that are based on dispersed knowledge have been considered in the literature. We will focus on methods that solve these problems by analyzing conflicts and by the creating coalitions. The first method that will be considered assumes that the system has a static structure, i.e. coalitions were created only once [16]. In the following methods, a dynamic structure has been applied. Different types of coalitions in a dynamic structure were obtained. Firstly, disjoint clusters were created [19], then inseparable clusters were used [18]. However, the most extensive process of conflict analysis was applied in the method with negotiations [17]. In this chapter, all of the systems that were used are discussed.

The chapter is organized as follows. Section 8.2 discusses related works. Section 8.3 summarizes basics of the rough set theory and the method of attribute selection that was used. A brief overview of dispersed decision-making systems is presented in Sect. 8.4. Section 8.5 shows the methodology of experiments on some data sets from the UCI Repository. The results are discussed in Sect. 8.6. Conclusions are drawn in the final section.

8.2 Related Works

The issue of combining classifiers is a very important aspect in the literature [1, 4, 7, 9]. Various terms are used for this concept: combination of multiple classifiers [20], classifier fusion [29] or classifier ensembles [8, 23]. In the technique, the results of the prediction of the base classifiers are combined in order to improve the quality of the classification. There are two basic approaches to this topic: classifier selection and classifier fusion. In the classifier selection method, each classifier is an expert in some local area. If the object from the classifier area appears, the classifier is responsible for assigning the class label to the object [5, 6]. In the classifier fusion approach, it is assumed that all classifiers are trained over the whole feature space, and that they are competitive rather than complementary [11].

In this chapter, the classifier fusion method is considered. This issue is discussed, for example, in the multiple model approach [7, 15] or in the distributed decision-making [21, 22]. In these techniques, one data set is divided into smaller sets. One of the methods for decomposition is to use the domain knowledge to decompose the nature of the decisions into a hierarchy of layers [12]. In [13, 26, 32], an ensemble of feature subsets is considered. Ślęzak and Widz investigated the correspondence between the mathematical criteria for feature selection and the mathematical criteria for voting between the resulting decision models [27]. In this chapter, it is assumed that the dispersed knowledge is given in advance and collected by separate and independent units. Therefore, the form of local knowledge bases can be very diverse in terms of sets of attributes and sets of objects.

The problem of reducing dimensionality is also an important and widely discussed issue. The two main approaches that are discussed in the literature are feature extraction and feature selection. The feature extraction method involves mapping the original set of attributes in the new space with lower dimensionality. This method has many applications [2, 3] such as signal processing, image processing or visualization. In this chapter, a method based on the feature selection is being considered. In this approach, attributes are selected from the original set of attributes, which provide the most information, and irrelevant or redundant attributes are removed. The existence of redundant attributes in the data set increases the execution time of algorithms and reduces the accuracy of classification. There are many methods of attribute selection [31, 33], for example, information gain, t-statistic or correlation. In this chapter, a method that is based on the rough set theory was used.

The rough sets was proposed by Pawlak [14]. The concept is widely used in various fields [24, 28]. One of the important applications of rough sets is attribute selection, which is realized by examining certain dependencies in the data. In the concept, some equivalent (indiscernibility) classes are defined in a data set. Based on these equivalent classes, unnecessary attributes that can be removed from the data set without losing the ability to classify are determined. This operation often results in improving the quality of classification, because these unnecessary data, which are removed from the data set, often deform the result, especially when a similarity measure or distance measure is used during classification.

8.3 Basics of the Rough Set Theory

Let $D = (U, A, D)$ be a decision table, where U is the universe; A is a set of conditional attributes, V^a is a set of attribute a values; D is a set of decision attributes. For any set $B \subseteq A$, an indiscernibility relation is defined

$$IND(B) = \{(x, y) \in U \times U : \forall_{a \in B} \, a(x) = a(y)\}. \tag{8.1}$$

If $B \subseteq A$ and $X \subseteq U$ then the lower and upper approximations of X, with respect to B, can be defined

$$\underline{B}X = \{x \in U : [x]_{IND(B)} \subseteq X\}, \tag{8.2}$$

$$\overline{B}X = \{x \in U : [x]_{IND(B)} \cap X \neq \emptyset\}, \tag{8.3}$$

where

$$[x]_{IND(B)} = \{y \in U : \forall_{a \in B} \, a(x) = a(y)\} \tag{8.4}$$

is the equivalence class of x in $U/IND(B)$.

A B-positive region of D is a set of all objects from the universe U that can be classified with certainty to one class of $U/IND(D)$ employing attributes from B,

$$POS_B(D) = \bigcup_{X \in U/IND(D)} \underline{B}X. \tag{8.5}$$

An attribute $a \in A$ is dispensable in B if $POS_B(D) = POS_{B \setminus \{a\}}(D)$, otherwise a is an indispensable attribute in B with respect to D. A set $B \subseteq A$ is called independent if all of its attributes are indispensable.

A reduct of set of attributes A can be defined as follows, a set of attributes $B \subseteq A$ is called the reduct of A, if B is independent and $POS_B(D) = POS_A(D)$.

The aim of this study was to apply the method of attribute selection that is based on the rough set theory to dispersed knowledge that is stored in a set of decision tables. This was realized in the following way. For each local decision table a set of reducts was generated. For this purpose, a program Rough Set Exploration System (RSES [25]) was used. The program was developed at the University of Warsaw. For each local decision table, one reduct, which contained the smallest number of attributes was selected. The conditional attributes that did not occur in the reduct were removed from the local decision table. Based on the modified local decision tables, decisions were taken using a dispersed system.

8.4 An Overview of Dispersed Systems

The system for making decisions based on dispersed knowledge was considered for the first time in [30]. It was assumed that the dispersed knowledge is stored in separate sets that were collected by different, independent units. As was mentioned earlier, in this situation, the system has very general assumptions. The separability or equality of sets of attributes or sets of objects of different decision tables are not fulfilled. The main assumptions that were adopted in the proposed model are given in the following sections.

8.4.1 Basic Definitions

Knowledge is stored in several decision tables. There is a set of resource agents; one agent has access to one decision table. ag in Ag is called a resource agent if it has access to the resources that are represented by decision table $D_{ag} := (U_{ag}, A_{ag}, d_{ag})$, where U_{ag} is "the universe"; A_{ag} is a set of conditional attributes and V_{ag}^{a} is a set of attribute a values that contain the special signs * and ?. The equation $a(x) = *$ for some $x \in U_{ag}$ means that for an object x, the value of attribute a has no influence on the value of the decision attribute, while the equation $a(x) = ?$ means that the value of attribute a for object x is unknown; d_{ag} is referred to as a decision attribute. The only condition that must be satisfied by the decision tables of agents is the occurrence of the same decision attributes in all of the decision tables of the agents.

The resource agents, which are similar in some specified sense, are combined into a group that is called a cluster. In the process of creating groups, elements of conflict analysis and negotiation are used. Different approaches to creating a system's structure have been proposed from a very simple solution to a more complex method of creating groups of agents. In the following subsections, a brief overview of the proposed methods is presented. However, now we will focus on the common concepts that are used in a dispersed system.

A system has a hierarchical structure. For each group of agents, a superordinate agent is defined – a synthesis agent. The synthesis agent as has access to knowledge that is the result of the process of inference that is carried out by the resource agents that belong to its subordinate group. For each synthesis agent, aggregated knowledge is defined.

A significant problem that must be solved when making decisions based on dispersed knowledge is that inconsistencies in the knowledge may occur within the clusters. This problem stems from the fact that there are no assumptions about the relation between the sets of the conditional attributes of different local decision tables in the system. An inconsistency in the knowledge is understood to be a situation in which conflicting decisions are made on the basis of two different decision tables that have common conditional attributes and for the same values for the common attributes using logical implications.

The process of generating the common knowledge (an aggregated decision table) for each cluster was proposed. This process was termed the process for the elimination of inconsistencies in the knowledge. The method consists in constructing new objects that are based on the relevant objects from the decision tables of the resource agents that belong to one cluster (m_2 objects from each decision class of the decision tables of the agents that carry the greatest similarity to the test object are selected). The aggregated objects are created by combining only those relevant objects for which the values of the decision attribute and common conditional attributes are equal. A formal definition of the aggregated tables of synthesis agents is given below.

Definition 8.1 Let $\{ag_{i_1}, \ldots, ag_{i_k}\}$ be a cluster in the multi-agent decision-making system WSD_{Ag}^{dyn}, and as be a synthesis agent of this cluster. Then the resources of the as agent can be recorded in decision table

$$D_{as} = (U_{as}, A_{as}, \overline{d}_{as}), \tag{8.6}$$

where $A_{as} = \{\overline{b} : U_{as} \to V^b, b \in \bigcup_{j=1}^{k} A_{ag_{i_j}}\}$ and $\overline{d}_{as}$ is a function $\overline{d}_{as} : U_{as} \to V^d$. Moreover, for each maximal, due to inclusion, set $G = \{ag_{i_{j_1}}, \ldots, ag_{i_{j_k}}\} \subseteq \{ag_{i_1}, \ldots, ag_{i_k}\}$ and each set of objects $\{x_{i_{j_1}}, \ldots, x_{i_{j_k}}\}$, $x_{i_m} \in U_{ag_{i_m}}^{rel}$, $j_1 \leq m \leq j_k$, meeting the two following conditions:

$$\forall_{b \in Wsp_G} \exists_{v_b \in V^b} \forall_{j_1 \leq m \leq j_k} \left[b \in A_{ag_{i_m}} \Longrightarrow b(x_{i_m}) = v_b\right], \tag{8.7}$$

$$d_{ag_{i_{j_1}}} (x_{i_{j_1}}) = \ldots = d_{ag_{i_{j_k}}} (x_{i_{j_k}}), \tag{8.8}$$

where $Wsp_G = \{b : b \in A_{ag_i} \cap A_{ag_j} \text{ and } ag_i, ag_j \in G\}$, an object $x \in U_{as}$ is defined as follows:

$$\forall_{ag_j \in G} \forall_{b \in A_{ag_j}} \left[\overline{b}(x) = b(x_j) \text{ and } \overline{d}_{as}(x) = d_{ag_j}(x_j)\right], \tag{8.9}$$

$$\forall_{b \in \bigcup_{j=1}^{k} A_{ag_{i_j}} \setminus \bigcup_{m=j_1}^{j_k} A_{ag_{i_m}}} \overline{b}(x) = ?. \tag{8.10}$$

A detailed discussion of the approximated method of the aggregation of decision tables can be found in [18, 30].

Based on the aggregated decision tables, the values of decisions with the highest support of the synthesis agents are selected. At first, a c-dimensional vector of values $[\mu_{j,1}(x), \ldots, \mu_{j,c}(x)]$ is generated for each jth aggregated decision table, where c is the number of all of the decision classes. The value $\mu_{j,i}(x)$ determines the level of certainty with which the decision v_i is taken by agents from the jth cluster for a given test object x. The value $\mu_{j,i}(x)$ is equal to the maximum value of the similarity measure of the objects from the decision class v_i of the decision table of synthesis agent as_j to the test object x.

Based on local decisions taken by synthesis agents, global decisions are generated using certain fusion methods and methods of conflict analysis. The sum of the vectors is calculated and the DBSCAN algorithm is used to select a set of global decisions. The generated set will contain not only the value of the decisions that have the greatest support among all of the agents, but also those for which the support is relatively high. This is accomplished in the following way. At first, the decision with the greatest support is selected. Then, decisions with a relatively high support are determined using the DBSCAN algorithm. A description of the method can be found in [17–19].

8.4.2 Static Structure

In [16, 30], the first method for creating groups of agents was considered. In these articles, definitions of resource agents and synthesis agents are given and the hierarchical structure of the system was established.

In this first approach, it is assumed that the resource agents taking decisions on the basis of common conditional attributes form a group of agents called a cluster. For each cluster that contains at least two resource agents, a superordinate agent is defined, which is called a synthesis agent as. A dispersed system is one in which

$$WSD_{Ag} = \langle Ag, \{D_{ag} : ag \in Ag\}, As, \delta \rangle, \tag{8.11}$$

where Ag is a finite set of resource agents; $\{D_{ag} : ag \in Ag\}$ is a set of decision tables of resource agents; As is a finite set of synthesis agents, $\delta : As \rightarrow 2^{Ag}$ is an injective function which each synthesis agent assign a cluster.

8.4.3 Dynamic Structure with Disjoint Clusters

In [19], the second approach is considered. In the previous method, a dispersed system had a static structure (created once, for all test objects) and this time a dynamic structure is used (created separately for each test object). The aim of this method is to identify homogeneous groups of resource agents. The agents who agree on the classification into the decision classes for a test object will be combined in a group.

The definitions of the relations of friendship and conflict as well as a method for determining the intensity of the conflict between agents are used. Relations between agents are defined by their views for the classification of the test object to the decision classes.

In the first step of the process of creating clusters for each resource agent ag_i, a vector of probabilities that reflects the classification of the test object is generated. This vector will be defined on the basis of certain relevant objects. That is, m_1 objects from each decision class of the decision tables of the agents that carry the greatest similarity to the test object. The value of the parameter m_1 is selected experimentally. The value of the ith coordinate is equal to the average value of the similarity measure of the relevant objects from the ith decision class. Then, based on this vector, the vector of ranks $[r_{i,1}(x), \ldots, r_{i,c}(x)]$, where c – is the number of all of the decision classes, is generated. The function $\phi_{v_j}^x$ for the test object x and each value of the decision attribute v_j is defined as: $\phi_{v_j}^x : Ag \times Ag \rightarrow \{0, 1\}$

$$\phi_{v_j}^x(ag_i, ag_k) = \begin{cases} 0 & \text{if } r_{i,j}(x) = r_{k,j}(x) \\ 1 & \text{if } r_{i,j}(x) \neq r_{k,j}(x) \end{cases}, \tag{8.12}$$

where $ag_i, ag_k \in Ag$.

The intensity of conflict between agents using a function of the distance between agents is also defined. The distance between agents ρ^x for the test object x is defined as: $\rho^x : Ag \times Ag \rightarrow [0, 1]$

$$\rho^x(ag_i, ag_k) = \frac{\sum_{v_j \in V^d} \phi^x_{v_j}(ag_i, ag_k)}{card\{V^d\}}, \tag{8.13}$$

where $ag_i, ag_k \in Ag$.

Agents $ag_i, ag_k \in Ag$ are in a friendship relation due to the object x, which is written $R^+(ag_i, ag_k)$, if and only if $\rho^x(ag_i, ag_k) < 0.5$. Agents $ag_i, ag_k \in Ag$ are in a conflict relation due to the object x, which is written $R^-(ag_i, ag_k)$, if and only if $\rho^x(ag_i, ag_k) \geq 0.5$.

Then disjoint clusters of the resource agents remaining in the friendship relations are created. The process of creating clusters in this approach is very similar to the hierarchical agglomerative clustering method and proceeds as follows. Initially, each resource agent is treated as a separate cluster. These two steps are performed until the stop condition, which is given in the first step, is met.

1. One pair of different clusters is selected (in the very first step a pair of different resource agents) for which the distance reaches a minimum value. If the selected value of the distance is less than 0.5, then agents from the selected pair of clusters are combined into one new cluster. Otherwise, the clustering process is terminated.
2. After defining a new cluster, the values of the distance between the clusters are recalculated. The following method for recalculating the values of the distance is used. Let $\rho^x : 2^{Ag} \times 2^{Ag} \rightarrow [0, 1]$, let D_i be a cluster formed from the merger of two clusters $D_i = D_{i,1} \cup D_{i,2}$ and let it be given a cluster D_j then

$$\rho^x(D_i, D_j) = \begin{cases} \frac{\rho^x(D_{i,1}, D_j) + \rho^x(D_{i,2}, D_j)}{2} & \text{if } \rho^x(D_{i,1}, D_j) < 0.5 \\ & \text{and } \rho^x(D_{i,2}, D_j) < 0.5 \\ \max\{\rho^x(D_{i,1}, D_j), \rho^x(D_{i,2}, D_j)\} & \text{if } \rho^x(D_{i,1}, D_j) \geq 0.5 \\ & \text{or } \rho^x(D_{i,2}, D_j) \geq 0.5 \end{cases} \tag{8.14}$$

The method that was presented above has a clearly defined stop condition. The stop condition is based on the assumption that one cluster should not contain two resource agents that are in a conflict relation due to the test object.

As before, a synthesis agent is defined for each cluster. However, the system's structure has changed from static to dynamic. This change was taken into account in the system's definition. A dispersed decision-making system with dynamically generated clusters is defined as:

$$WSD^{dyn}_{Ag} = \langle Ag, \{D_{ag} : ag \in Ag\}, \{As_x : x \text{ is a classified object}\}, \tag{8.15}$$

$$\{\delta_x : x \text{ is a classified object}\}\rangle,$$

where Ag is a finite set of resource agents; $\{D_{ag} : ag \in Ag\}$ is a set of decision tables of the resource agents; As_x is a finite set of synthesis agents defined for the clusters that are dynamically generated for test object x, $\delta_x : As_x \rightarrow 2^{Ag}$ is an injective function that each synthesis agent assigns to the cluster that is generated due to the classification of object x.

8.4.4   Dynamic Structure with Inseparable Clusters

In [17], the third approach was proposed. In this method, the definitions of the relations of friendship and conflict as well as a method for determining the intensity of the conflict between agents are the same as that described above, but the definition of a cluster is changed. This time no disjoint clusters are created. The cluster due to the classification of object x is the maximum, due to the inclusion relation, subset of resource agents $X \subseteq Ag$ such that

$$\forall_{ag_i, ag_k \in X} \ R^+(ag_i, ag_k). \tag{8.16}$$

Thus, the cluster is the maximum, due to inclusion relation, set of resource agents that remain in the friendship relation due to the object x.

As before, a synthesis agent is defined for each cluster and the definition of the dispersed decision-making system with dynamically generated clusters is used.

8.4.5   Dynamic Structure with Negotiations

In [17], the fourth approach was proposed. In this method, a dynamic structure is also used, but the process of creating clusters is more extensive and, as a consequence, the clusters are more complex and reconstruct and illustrate the views of the agents on the classification better.

The main differences between this approach and the previous method are as follows. Now, three types of relations between agents are defined: friendship, neutrality and conflict (previously, only two types were used). The clustering process consists of two stages (previously, only one stage process was used). In the first step, initial groups are created, which contain agents in a friendship relation. In the second stage, a negotiation stage, agents that are in neutrality relation are attached to the existing groups. In order to define the intensity of the conflict between agents, two functions are used: the distance function between the agents (was used in the previous method) and a generalized distance function.

The process of creating clusters is as follows. In the first step, the relations between the agents are defined.

Let p be a real number that belongs to the interval $[0, 0.5)$. Agents $ag_i, ag_k \in Ag$ are in a friendship relation due to the object x, which is written $R^+(ag_i, ag_k)$, if and only if $\rho^x(ag_i, ag_k) < 0.5 - p$. Agents $ag_i, ag_k \in Ag$ are in a conflict relation due to the object x, which is written $R^-(ag_i, ag_k)$, if and only if $\rho^x(ag_i, ag_k) > 0.5 + p$. Agents $ag_i, ag_k \in Ag$ are in a neutrality relation due to the object x, which is written $R^0(ag_i, ag_k)$, if and only if $0.5 - p \le \rho^x(ag_i, ag_k) \le 0.5 + p$.

Then, initial clusters are defined. The initial cluster due to the classification of object x is the maximum subset of resource agents $X \subseteq Ag$ such that

$$\forall_{ag_i, ag_k \in X} \ R^+(ag_i, ag_k). \tag{8.17}$$

After the first stage of clustering, a set of initial clusters and a set of agents that are not included in any cluster are obtained. In this second group of agents there are the agents that remained undecided. So those that are in a neutrality relation with the agents belonging to some initial clusters. In the second step, the negotiation stage, these agents play a key role.

As is well known, the goal of the negotiation process is to reach a compromise by accepting some concessions by the parties that are involved in a conflict situation. In the negotiation process, the intensity of the conflict is determined by using the generalized distance function. This definition assumes that during the negotiation, agents put the greatest emphasis on the compatibility of the ranks assigned to the decisions with the highest ranks. That is, the decisions that are most significant for the agent. The compatibility of ranks assigned to less meaningful decision is omitted. We define the function ϕ_G^x for test object x; $\phi_G^x : Ag \times Ag \to [0, \infty)$

$$\phi_G^x(ag_i, ag_j) = \frac{\sum_{v_l \in Sign_{i,j}} |r_{i,l}(x) - r_{j,l}(x)|}{card\{Sign_{i,j}\}}, \tag{8.18}$$

where $ag_i, ag_j \in Ag$ and $Sign_{i,j} \subseteq V^d$ is the set of significant decision values for the pair of agents ag_i, ag_j.

The generalized distance between agents ρ_G^x for the test object x is also defined; $\rho_G^x : 2^{Ag} \times 2^{Ag} \to [0, \infty)$

$$\rho_G^x(X, Y) = \begin{cases} 0 & \text{if } card\{X \cup Y\} \le 1, \\ \dfrac{\displaystyle\sum_{ag, ag' \in X \cup Y} \phi_G^x(ag, ag')}{card\{X \cup Y\} \cdot (card\{X \cup Y\} - 1)} & \text{else,} \end{cases} \tag{8.19}$$

where $X, Y \subseteq Ag$. As can be easily seen, the value of the generalized distance function for two sets of agents X and Y is equal to the average value of the function ϕ_G^x for each pair of agents ag, ag' that belong to the set $X \cup Y$. For each agent that is not attached to any cluster, the value of generalized distance function is calculated for this agent and every initial cluster. Then, the agent is included to all of the initial clusters for which the generalized distance does not exceed a certain threshold, which is set by the system's user. Moreover, agents without a coalition for which the value does not exceed the threshold are combined into a new cluster. Agents who are in a conflict relation are not connected in one cluster. After completion of the second stage of the process of clustering, the final form of the clusters is obtained.

As before, a synthesis agent is defined for each cluster and the definition of the dispersed decision-making system with dynamically generated clusters is used.

8.5   Description of the Experiments

In the experimental part, the method of attribute selection that is based on the rough set theory was used first for the local knowledge, and then the modified knowledge bases were used in the different approaches to define the dispersed system's structure. The results obtained in this way were compared with each other and with the results obtained for dispersed system and full sets of the conditional attributes.

8.5.1   Data Sets

We did not have access to dispersed data that are stored in the form of a set of local knowledge bases and therefore some benchmark data that were stored in a single decision table were used. The division into a set of decision tables was done for the data that was used.

The data from the UCI repository were used in the experiments – Soybean data set, Landsat Satellite data set, Vehicle Silhouettes data set, Dermatology data set and Audiology data set. The test sets for the Soybean, the Landsat Satellite and the Audiology data were obtained from the UCI repository. The Vehicle Silhouettes and the Dermatology data set were randomly divided into a training set consisting of 70% of the objects and the test set consisting of the remaining 30% of the objects. Discretization of continuous attributes was carried out for the Vehicle Silhouettes data set. Discretization was performed before the Vehicle Silhouettes data set was divided into a training set and a test set. All conditional attributes that were common to the local decision tables were discretized. This process was realized due to the approximated method of the aggregation of decision tables [30]. The frequency-based discretization method was used.

Table 8.1 presents a numerical summary of the data sets.

For all of the sets of data, the training set, which was originally written in the form of a single decision table, was dispersed, which means that it was divided into a set of decision tables. A dispersion in five different versions (with 3, 5, 7, 9 and 11 decision tables) was considered. The following designations were used:

- WSD_{Ag1}^{dyn} - 3 decision tables;
- WSD_{Ag2}^{dyn} - 5 decision tables;
- WSD_{Ag3}^{dyn} - 7 decision tables;

Table 8.1 Data set summary

Data set	# of objects in training set	# of objects in test set	# Conditional attributes	# Decision classes
Soybean	307	376	35	19
Landsat Satellite	4435	1000	36	6
Vehicle Silhouettes	592	254	18	4
Dermatology	256	110	34	6
Audiology	200	26	69	24

- WSD_{Ag4}^{dyn} - 9 decision tables;
- WSD_{Ag5}^{dyn} - 11 decision tables.

The numbers of the local decision tables were chosen arbitrarily. It seems obvious that the minimum number of tables that can be considered in the case of the creation of the coalitions is three. Then, the number of tables was increased by a certain constant in order to examine a certain group of divisions. The division into a set of decision tables was made in the following way. The author defined the number of conditional attributes in each of the local decision tables. Then, the attributes from the original table were randomly assigned to the local tables. As a result of this division, some local tables had common conditional attributes. The universes of the local tables were the same as the universe of the original table, but the identifiers of the objects were not stored in the local tables. A detailed description of the method for data dispersion can be found in [16].

Once again, it is worth emphasizing that a series of tests using the train-and-test method on dispersed data sets were conducted in this study. In the system, object identifiers are not stored in local knowledge bases. This approach is intended to reflect real situations in which local bases are collected by independent individuals. This means that it is not possible to check whether the same objects are stored in different local knowledge bases and it is not possible to identify which objects are the same. That, in turn, means that using the k-fold cross-validation method is impossible when preserving the generality. More specifically, when the cross-validation method is used, a test sample would have to be drawn from each local decision table. Under the adopted assumptions, it was not possible to identify whether the same objects were drawn from different local knowledge bases. Thus, it was not possible to generate one decision table (test set) from the selected set of test samples. That is why the train-and-test method was applied. This approach has the advantage that the experiments were performed on the same set of cases using four different methods.

8.5.2 Attribute Selection

In the first stage of the experiments for each local decision tables, the reducts of the set of conditional attributes were generated. As was mentioned before, for this purpose, the RSES program was used. The following settings of the RSES program were used: Discernibility matrix settings – Full discernibility, Modulo decision; Method – Exhaustive algorithm. For all of the analyzed data sets for each version of the dispersion (3, 5, 7, 9 and 11 decision tables) and for each local decision table, a set of reducts was generated separately. In order to generate reducts, an exhaustive algorithm with a full discernibility modulo decision was used.

Many reducts were generated for certain decision tables. For example, for the Landsat Satellite data set and a dispersed system with three local decision tables for one of the tables, 1,469 reducts were obtained and for another table 710 reducts were obtained. If more than one reduct was generated for a table, one reduct was randomly selected from the reducts that had the smallest number of attributes. Then, the local decision tables were modified in the following way. The set of attributes in the local decision table was restricted to the attributes that occurred in the selected reduct. The universe and the decision attribute in the table remained the same as before.

Table 8.2 shows the number of conditional attributes that were deleted from the local decision tables. In the table, the following designations were applied: $\#Ag$ – is the number of the local decision tables (the number of agents) and $\#A_{ag_i}$ – is the number of the deleted attributes from the set of the conditional attributes of the ith local table (of the ith agent).

As can be seen for the Landsat Satellite data set and the Vehicle Silhouettes data set and the dispersed system with 7, 9 and 11 local decision tables, the reduction of the set of conditional attributes did not cause any changes. Therefore, these systems will no longer be considered in the rest of the chapter.

8.5.3 Evaluation Measures and Parameters Optimization

The measures for determining the quality of classifications were:

- *estimator of classification error e* in which an object is considered to be properly classified if the decision class that is used for the object belonged to the set of global decisions that were generated by the system;

$$e = \frac{1}{card\{U_{test}\}} \sum_{x \in U_{test}} I(d(x) \notin \hat{d}_{WSD_{Ag}}(x)), \qquad (8.20)$$

where $I(d(x) \notin \hat{d}_{WSD_{Ag}}(x)) = 1$, when $d(x_i) \notin \hat{d}_{WSD_{Ag}}(x)$ and $I(d(x) \notin \hat{d}_{WSD_{Ag}}(x)) = 0$, when $d(x) \in \hat{d}_{WSD_{Ag}}(x)$; the test set is stored in a decision table $D_{test} = (U_{test}, \bigcup_{ag \in Ag} A_{ag}, d)$; $\hat{d}_{WSD_{Ag}}(x)$ is a set of global decisions generated by the dispersed decision-making system WSD_{Ag} for the test object x,

- *estimator of classification ambiguity error* e_{ONE} in which an object is considered to be properly classified if only one correct value of the decision was generated for this object;

Table 8.2 The number of conditional attributes removed as a result of the reduction of knowledge

Data set, #Ag	#A_{ag_1}	#A_{ag_2}	#A_{ag_3}	#A_{ag_4}	#A_{ag_5}	#A_{ag_6}	#A_{ag_7}	#A_{ag_8}	#A_{ag_9}	#$A_{ag_{10}}$	#$A_{ag_{11}}$
Soybean, 3	4	5	2	–	–	–	–	–	–	–	–
Soybean, 5	4	0	0	1	0	–	–	–	–	–	–
Soybean, 7	0	0	0	1	1	0	0	–	–	–	–
Soybean, 9	0	0	0	0	0	1	1	0	0	–	–
Soybean, 11	0	0	0	0	0	0	0	0	1	0	0
Landsat Satellite, 3	1	9	8	–	–	–	–	–	–	–	–
Landsat Satellite, 5	0	0	1	0	0	–	–	–	–	–	–
Landsat Satellite, 7	0	0	0	0	0	0	0	–	–	–	–
Landsat Satellite, 9	0	0	0	0	0	0	0	0	0	–	–
Landsat Satellite, 11	0	0	0	0	0	0	0	0	0	0	0
Vehicle Silhouettes, 3	2	4	2	–	–	–	–	–	–	–	–
Vehicle Silhouettes, 5	1	0	2	0	0	–	–	–	–	–	–
Vehicle Silhouettes, 7	0	0	0	0	0	0	0	–	–	–	–
Vehicle Silhouettes, 9	0	0	0	0	0	0	0	0	0	–	–
Vehicle Silhouettes, 11	0	0	0	0	0	0	0	0	0	0	0
Dermatology, 3	3	3	6	–	–	–	–	–	–	–	–
Dermatology, 5	0	0	2	0	3	–	–	–	–	–	–
Dermatology, 7	0	0	0	0	0	0	1	–	–	–	–
Dermatology, 9	0	0	0	0	0	0	0	1	0	–	–
Dermatology, 11	0	0	0	0	0	0	0	0	0	1	0
Audiology, 3	2	1	19	–	–	–	–	–	–	–	–
Audiology, 5	2	2	2	4	1	–	–	–	–	–	–
Audiology, 7	1	1	2	1	3	1	0	–	–	–	–
Audiology, 9	1	1	1	2	1	2	1	0	0	–	–
Audiology, 11	0	1	0	1	1	1	1	2	1	0	0

$$e_{ONE} = \frac{1}{card\{U_{test}\}} \sum_{x \in U_{test}} I(d(x) \neq \hat{d}_{WSD_{Ag}}(x)), \qquad (8.21)$$

where $I(d(x) \neq \hat{d}_{WSD_{Ag}}(x)) = 1$, when $\{d(x)\} \neq \hat{d}_{WSD_{Ag}}(x)$ and $I(d(x) \neq \hat{d}_{WSD_{Ag}}(x)) = 0$, when $\{d(x)\} = \hat{d}_{WSD_{Ag}}(x)$;

- *the average size of the global decisions sets $\overline{d}_{WSD_{Ag}}$ that was generated for a test set;*

$$\overline{d}_{WSD_{Ag}} = \frac{1}{card\{U_{test}\}} \sum_{x \in U_{test}} card\{\hat{d}_{WSD_{Ag}}(x)\}. \tag{8.22}$$

As was mentioned in Sect. 8.4, some parameters in the methods have been adopted. These parameters are as follows:

- m_1 – parameter that determines the number of relevant objects that are selected from each decision class of the decision table and are then used in the process of cluster generation. This parameter is present in all of the dynamic approaches;
- p – parameter that occurs in the definition of friendship, conflict and neutrality relations. This parameter is present in the last approach – dynamic structure with negotiations;
- m_2 – parameter that is used in the approximated method for the aggregation of decision tables. This parameter is present in all of the methods;
- ε – parameter of the DBSCAN algorithm. This parameter is present in all of the methods;

The values of these parameters were optimized according to the following schedule. At first, parameters m_1, p and m_2 were optimized. Of course, these parameters were optimized in a way that was adequate to the considered approach. For the static structure, only parameter m_2 was considered, for the dynamic structure with disjoint clusters and the dynamic structure with inseparable clusters parameters m_1 and m_2 were used and for the dynamic structure with negotiations, all of the above-mentioned parameters were optimized. For this purpose, parameter values $m_1, m_2 \in \{1, \ldots, 10\}$ and $p \in \{0.05, 0.1, 0.15, 0.2\}$ for the Soybean data sets, the Vehicle data set, the Dermatology data set and the Audiology data set and $m_1, m_2 \in \{1, \ldots, 5\}$ and $p \in \{0.05, 0.1, 0.15, 0.2\}$ for the Satellite data set were examined.

At this stage of experiments, in order to generate global decisions, the weighted voting method was used instead of the density-based algorithm. Then, the minimum value of the parameters m_1, p and m_2 were chosen, which resulted in the lowest value of the estimator of classification error on a test set.

In the second stage of the experiments, parameter ε was optimized. This was realized by using the optimal parameter values that were determined in the previous step. Parameter ε was optimized by performing a series of experiments with different values of this parameter that were increased from 0 by the value 0.0001. Then, a graph was created and the points that indicated the greatest improvement in the efficiency of inference were chosen. In the rest of the chapter, the results that were obtained for the optimal values of the parameter m_1, p, m_2 and ε are presented.

8.6 Experiments and Discussion

In this section, the results of the experiments for all of the considered methods are presented. First, a comparison for each of the approaches is made separately. The results obtained after the application of the attribute selection method are compared with the results obtained without the use of the attribute selection method. Then, a comparison of all of the approaches is made.

The results of the experiments are presented in Tables 8.3, 8.4, 8.5 and 8.6 – each table illustrates one approach. In each table, the results for both cases – with and without the use of the method of attribute selection – are given. The following information is given in the tables:

- the results in two groups — with attribute selection and without attribute selection,
- the name of the dispersed system (System),
- the selected, optimal parameter values (Parameters),
- the estimator of the classification error e,
- the estimator of the classification ambiguity error e_{ONE},
- the average size of the global decisions sets $\overline{d}_{WSD_{Ag}^{dyn}}$.

In each table the best results in terms of the measures e and $\overline{d}_{WSD_{Ag}^{dyn}}$ and for the given methods are bolded.

The discussion of the results was divided into separate subsections on the considered approaches.

8.6.1 Results for Static Structure

Table 8.3 shows the results for the approach with a static structure. As can be seen, for the Soybean data set, for some of the dispersed sets, the use of the method for attribute selection did not affect the quality of the classification. For the systems with three and eleven resource agents WSD_{Ag1}, WSD_{Ag5}, better results were obtained using the attribute selection. In the case of the Landsat Satellite data set and all of the considered dispersed sets, the replacement of the sets of attributes by reducts resulted in an improved quality of the classification.

It should also be noted that the biggest differences in the values of the measure e were observed for the Landsat Satellite data set and the dispersed system with three resource agents. In the case of the Vehicle Silhouettes data set, mostly poorer results were obtained after using attribute selection. In the case of the Dermatology data set, for some of the dispersed sets, the use of the method for attribute selection did not affect the quality of the classification. For the systems with five and nine resource agents WSD_{Ag2}, WSD_{Ag4}, better results were obtained using the attribute selection. For the Audiology data set, two times better results were obtained and two times worse results were obtained after applying attribute selection method.

Table 8.3 Results of experiments for static structure

System	Results with attribute selection					Results without attribute selection				
	Parameters		e	e_{ONE}	$\overline{d}_{WSD_{Ag}^{dyn}}$	Parameters		e	e_{ONE}	$\overline{d}_{WSD_{Ag}^{dyn}}$
	m_2	ε				m_2	ε			
Soybean data set										
WSD_{Ag1}	1	0.003	**0.024**	0.293	1.918	1	0.0025	0.027	0.295	2.005
WSD_{Ag2}	1	0.00625	0.037	0.332	2.104	1	0.00575	**0.035**	0.327	1.968
	1	0.00425	0.104	0.255	1.340	1	0.0035	**0.093**	0.242	1.33
WSD_{Ag3}	1	0.00575	0.008	0.306	1.928	1	0.00575	0.008	0.306	1.968
	1	0.00375	0.024	0.274	1.564	1	0.00375	0.024	0.274	1.559
WSD_{Ag4}	1	0.01075	0.058	0.321	1.739	1	0.0108	0.058	0.319	1.755
WSD_{Ag5}	1	0.0128	0.029	0.309	1.750	1	0.0128	0.029	0.309	1.755
	1	0.0084	**0.088**	0.253	1.287	1	0.0084	0.090	0.253	1.298
Landsat Satellite data set										
WSD_{Ag1}	5	0.0016	**0.021**	0.384	1.722	10	0.0014	0.042	0.401	1.757
	5	0.0008	**0.064**	0.262	1.282	10	0.0007	0.088	0.278	1.292
WSD_{Ag2}	6	0.0031	**0.008**	0.383	1.697	1	0.0031	0.012	0.396	1.744
	6	0.0012	**0.047**	0.211	1.222	1	0.0012	0.047	0.221	1.230
Vehicle Silhouettes data set										
WSD_{Ag1}	3	0.0023	0.165	0.508	1.516	20	0.00225	**0.106**	0.492	1.524
	3	0.0012	0.236	0.417	1.224	20	0.00125	**0.177**	0.378	1.236
WSD_{Ag2}	1	0.0036	0.173	0.524	1.543	17	0.0035	**0.154**	0.508	1.535
	1	0.0015	**0.240**	0.409	1.248	17	0.0015	0.244	0.406	1.224

(continued)

Table 8.3 (continued)

Dermatology data set	Results with attribute selection					Results without attribute selection				
WSD^{dyn}_{Ag1}	1	0.0001	0.018	0.018	1	1	0.0001	0.018	0.018	1
WSD^{dyn}_{Ag2}	1	0.0005	**0.064**	0.127	1.064	1	0.00075	0.073	0.136	1.064
WSD^{dyn}_{Ag1}	1	0.00097	0.154	0.462	1.846	1	0.00066	0.154	0.462	1.808
	1	0.00003	0.192	0.423	1.308	1	0.00003	0.192	0.423	1.308
WSD^{dyn}_{Ag2}	2	0.0041	0.308	0.500	1.846	2	0.0032	0.308	0.500	1.808
	2	0.0009	**0.385**	0.423	1.038	2	0.0001	0.423	0.423	1
WSD^{dyn}_{Ag3}	1	0.00066	0.231	0.654	1.462	1	0.00006	0.231	0.615	1.423
	1	0.00003	0.308	0.615	1.346	1	0.00003	**0.269**	0.615	1.385
WSD^{dyn}_{Ag4}	1	0.00211	0.115	0.462	1.769	1	0.0019	**0.077**	0.462	1.654
	1	0.0012	0.154	0.423	1.346	1	0.001	0.154	0.423	1.385
WSD^{dyn}_{Ag5}	1	0.0029	**0.269**	0.808	1.846	1	0.0024	0.308	0.808	1.769
	1	0.00003	0.385	0.654	1.308	1	0.0001	0.385	0.654	1.308

Table 8.4 Results of experiments for dynamic structure with disjoint clusters

System	Results with attribute selection						Results without attribute selection					
	Parameters		ε	e	e_{ONE}	$\bar{d}_{WSD_{Ag}^{dyn}}$	Parameters		ε	e	e_{ONE}	$\bar{d}_{WSD_{Ag}^{dyn}}$
	m_1	m_2					m_1	m_2				
Soybean data set												
WSD_{Ag1}^{dyn}	4	1	0.0126	**0.021**	0.295	2.394	1	1	0.0072	0.026	0.319	2.401
	4	1	0.0114	**0.035**	0.290	2.011	1	1	0.00645	0.064	0.287	2.082
WSD_{Ag2}^{dyn}	3	1	0.0145	**0.016**	0.311	1.721	4	1	0.01575	0.019	0.311	2.059
	3	1	0.0115	**0.035**	0.277	1.503	4	1	0.013	0.043	0.285	1.545
WSD_{Ag3}^{dyn}	6	1	0.0174	0.019	0.298	1.843	6	1	0.01875	**0.016**	0.306	2.008
	6	1	0.0135	0.021	0.277	1.588	6	1	0.0135	**0.019**	0.279	1.598
WSD_{Ag4}^{dyn}	7	1	0.0233	**0.011**	0.277	1.846						
	7	1	0.0167	**0.032**	0.218	1.535	1	1	0.006	0.043	0.261	1.529
WSD_{Ag5}^{dyn}	2	2	0.0201	**0.035**	0.324	1.761	1	4	0.005	0.037	0.327	1.838
	2	2	0.0024	0.069	0.258	1.279						
Landsat Satellite data set												
WSD_{Ag1}^{dyn}	1	6	0.0024	**0.027**	0.346	1.738	1	5	0.00225	0.033	0.330	1.736
WSD_{Ag2}^{dyn}	3	1	0.0056	0.012	0.421	1.797	1	3	0.0053	0.012	0.402	1.724
	3	1	0.0024	**0.041**	0.225	1.260	1	3	0.0022	0.047	0.215	1.235
Vehicle Silhouettes data set												
WSD_{Ag1}^{dyn}	1	1	0.0047	0.161	0.512	1.535	2	2	0.0051	**0.138**	0.476	1.528
	1	1	0.0015	0.244	0.413	1.252	2	2	0.0018	**0.209**	0.358	1.224
WSD_{Ag2}^{dyn}	1	3	0.0068	0.181	0.524	1.539	1	10	0.0069	0.181	0.516	1.528
	1	3	0.0034	0.248	0.421	1.248	1	10	0.003	**0.244**	0.421	1.236

(continued)

Table 8.4 (continued)

	Results with attribute selection						Results without attribute selection					
Dermatology data set												
WSD_{Ag1}^{dyn}	2	1	0.0001	0.018	0.018	1	1	1	0.0001	0.018	0.018	1
WSD_{Ag2}^{dyn}	2	1	0.0001	**0.009**	0.009	1	2	7	0.0001	0.018	0.018	1
WSD_{Ag3}^{dyn}	10	1	0.0015	0.027	0.045	1.018	10	1	0.0005	0.027	0.045	1.018
WSD_{Ag4}^{dyn}	5	2	0.0015	0.027	0.045	1.018	5	1	0.002	0.027	0.036	1.009
WSD_{Ag5}^{dyn}	5	1	0.0001	0.018	0.027	1.009	5	1	0.0001	0.018	0.027	1.009
Audiology data set												
WSD_{Ag1}^{dyn}	2	1	0.003	0.154	0.538	1.808	2	1	0.00179	0.154	0.538	1.808
	2	1	0.0015	0.192	0.462	1.308	2	1	0.0009	0.192	0.500	1.385
WSD_{Ag2}^{dyn}	1	1	0.0026	0.154	0.577	1.808	1	1	0.0001	0.154	0.577	1.808
	1	1	0.001	0.231	0.423	1.231	1	1	0.00111	0.231	0.462	1.308
WSD_{Ag3}^{dyn}							8	1	0.00288	**0.077**	0.346	1.423
	3	1	0.0026	0.115	0.346	1.308	8	1	0.00207	0.115	0.346	1.308
WSD_{Ag4}^{dyn}	2	1	0.004	**0.115**	0.423	1.500						
	2	1	0.0035	0.154	0.423	1.346	4	1	0.002775	0.154	0.385	1.346
WSD_{Ag5}^{dyn}	2	2	0.0046	**0.077**	0.500	1.654	3	3	0.00642	0.115	0.538	1.885
	2	2	0.002	**0.231**	0.423	1.231	3	3	0.0009	0.269	0.423	1.269

Table 8.5 Results of experiments for dynamic structure with inseparable clusters

System	Results with attribute selection						Results without attribute selection					
	Parameters		ε	e	e_{ONE}	$\bar{d}_{WSD_{Ag}^{dyn}}$	Parameters		ε	e	e_{ONE}	$\bar{d}_{WSD_{Ag}^{dyn}}$
	m_1	m_2					m_1	m_2				
Soybean data set												
WSD_{Ag1}^{dyn}	3	1	0.0087	0.069	0.245	1.617	2	1	0.00885	**0.019**	0.277	1.864
WSD_{Ag2}^{dyn}	3	1	0.015	0.016	0.314	1.731	3	1	0.01365	**0.008**	0.266	2
	3	1	0.012	0.032	0.285	1.545	3	1	0.01215	**0.016**	0.239	1.527
WSD_{Ag3}^{dyn}	6	1	0.0175	0.019	0.295	1.902	6	1	0.0188	**0.016**	0.298	2.016
	6	1	0.0135	0.021	0.266	1.590	6	1	0.0134	**0.019**	0.271	1.590
WSD_{Ag4}^{dyn}	7	1	0.0205	**0.019**	0.261	1.779	2	1	0.01775	0.027	0.287	1.761
	7	1	0.017	**0.032**	0.231	1.551	2	1	0.01625	0.035	0.242	1.519
WSD_{Ag5}^{dyn}	2	2	0.0225	0.045	0.309	1.614	2	2	0.01825	**0.035**	0.319	1.910
	2	2	0.0115	**0.080**	0.258	1.287	2	2	0.00825	0.082	0.255	1.274
Landsat Satellite data set												
WSD_{Ag1}^{dyn}	2	5	0.0034	**0.017**	0.413	1.757	1	4	0.0032	0.021	0.382	1.751
WSD_{Ag2}^{dyn}	2	2	0.00529	**0.014**	0.394	1.731	2	1	0.0052	0.019	0.402	1.739
	2	2	0.00235	**0.037**	0.237	1.268	2	1	0.0022	0.042	0.219	1.231
Vehicle Silhouettes data set												
WSD_{Ag1}^{dyn}	1	1	0.0049	0.154	0.524	1.520	2	2	0.0057	**0.126**	0.476	1.520
	1	1	0.0021	0.240	0.441	1.236	2	2	0.00285	**0.189**	0.382	1.232
WSD_{Ag2}^{dyn}	1	3	0.007	**0.177**	0.524	1.543	1	10	0.0069	0.181	0.516	1.528
	1	3	0.0034	0.248	0.421	1.248	1	10	0.003	0.248	0.421	1.232

(continued)

Table 8.5 (continued)

	Results with attribute selection						Results without attribute selection					
Dermatology data set												
WSD^{dyn}_{Ag1}	1	1	0.0001	0.018	0.018	1	1	1	0.0001	0.018	0.018	1
WSD^{dyn}_{Ag2}	2	1	0.0001	**0.009**	0.009	1	2	7	0.0001	0.018	0.018	1
WSD^{dyn}_{Ag3}	2	1	0.002	0.036	0.055	1.018	5	1	0.002	**0.027**	0.045	1.018
WSD^{dyn}_{Ag4}	5	2	0.002	0.027	0.045	1.018	5	1	0.002	0.027	0.036	1.009
WSD^{dyn}_{Ag5}	5	1	0.001	0.018	0.027	1.009	4	2	0.001	0.018	0.036	1.018
Audiology data set												
WSD^{dyn}_{Ag1}	2	1	0.003	0.154	0.538	1.808	2	1	0.00179	0.154	0.538	1.808
	2	1	0.0015	0.192	0.462	1.308	2	1	0.0009	0.192	0.500	1.385
WSD^{dyn}_{Ag2}	3	1	0.0039	0.115	0.462	1.692	1	1	0.00219	**0.077**	0.577	1.923
	3	1	0.0014	0.231	0.423	1.231	1	1	0.00108	**0.192**	0.462	1.269
WSD^{dyn}_{Ag3}	2	1	0.0045	0.115	0.462	1.731	8	1	0.00288	**0.077**	0.346	1.423
	2	1	0.0023	0.154	0.346	1.231	8	1	0.00208	**0.115**	0.346	1.308
WSD^{dyn}_{Ag4}	4	1	0.0044	0.154	0.423	1.769	1	2	0.00117	**0.038**	0.346	1.423
	4	1	0.0001	0.231	0.346	1.154	1	2	0.00109	**0.077**	0.346	1.346
WSD^{dyn}_{Ag5}	4	2	0.004	0.077	0.500	1.731	2	2	0.0036	0.077	0.500	1.885
	4	2	0.0012	0.231	0.462	1.385	2	2	0.0022	0.231	0.462	1.385

Table 8.6 Results of experiments for dynamic structure with negotiations

System	Results with attribute selection							Results without attribute selection						
	Parameters				e	e_{ONE}	$\overline{d}_{WSD_{Ag}^{dyn}}$	Parameters				e	e_{ONE}	$\overline{d}_{WSD_{Ag}^{dyn}}$
	m_1	p	m_2	ε				m_1	p	m_2	ε			
Soybean data set														
WSD_{Ag1}^{dyn}	4	0.05	1	0.0033	0.059	0.274	1.811	5	0.1	1	0.0088	**0.019**	0.266	1.697
WSD_{Ag2}^{dyn}	5	0.1	2	0.0037	0.037	0.378	2.082	4	0.05	1	0.0144	**0.008**	0.290	2.082
	5	0.1	2	0.0024	0.053	0.309	1.566	4	0.05	1	0.0122	**0.021**	0.258	1.529
WSD_{Ag3}^{dyn}	6	0.05	1	0.00222	0.013	0.295	1.989	2	0.05	4	0.0156	0.013	0.301	1.899
	6	0.05	1	0.00147	**0.024**	0.255	1.566	2	0.05	4	0.0117	0.032	0.277	1.572
WSD_{Ag4}^{dyn}								4	0.1	2	0.0174	0.024	0.293	1.822
	8	0.2	1	0.00204	**0.008**	0.253	1.561	1	0.05	3	0.0103	0.043	0.242	1.521
WSD_{Ag5}^{dyn}	10	0.15	1	0.00252	0.035	0.340	1.904	2	0.05	2	0.0225	0.035	0.322	1.875
	10	0.15	1	0.00171	**0.045**	0.274	1.388	2	0.05	2	0.0123	0.080	0.263	1.303
Landsat Satellite data set														
WSD_{Ag1}^{dyn}	1	0.05	1	0.00201	**0.009**	0.445	1.788	4	0.05	4	0.0029	0.022	0.390	1.786
WSD_{Ag2}^{dyn}	2	0.05	2	0.00174	**0.008**	0.390	1.702	1	0.05	3	0.0046	0.011	0.367	1.618
	2	0.05	2	0.00084	**0.034**	0.230	1.260	1	0.05	3	0.0024	0.040	0.220	1.237
Vehicle Silhouettes data set														
WSD_{Ag1}^{dyn}	2	0.05	3	0.0042	0.157	0.504	1.508	9	0.05	2	0.0038	**0.122**	0.390	1.516
	2	0.05	3	0.0023	0.228	0.437	1.272	9	0.05	2	0.002	**0.185**	0.409	1.272
WSD_{Ag2}^{dyn}	3	0.05	3	0.0044	0.193	0.524	1.531	1	0.05	10	0.0043	**0.177**	0.559	1.531
	3	0.05	3	0.0024	0.260	0.441	1.240	1	0.05	10	0.0022	0.260	0.453	1.228
Dermatology data set														
WSD_{Ag1}^{dyn}	10	0.2	1	0.0001	0.018	0.018	1	1	0.05	1	0.0001	0.018	0.018	1

(continued)

Table 8.6 (continued)

	Results with attribute selection							Results without attribute selection						
WSD^{dyn}_{Ag2}	1	0.05	1	0.0001	0.018	0.018	1	1	0.05	7	0.0001	0.018	0.018	1
WSD^{dyn}_{Ag3}	2	0.05	1	0.004	0.036	0.109	1.082	1	0.2	9	0.001	0.018	**0.018**	1.027
WSD^{dyn}_{Ag4}	5	0.05	1	0.002	**0.027**	0.036	1.009	5	0.05	1	0.002	0.036	0.045	1.009
WSD^{dyn}_{Ag5}	4	0.2	1	0.002	**0.018**	0.036	1.018	5	0.05	1	0.004	0.027	0.055	1.027

Audiology data set

	Results with attribute selection							Results without attribute selection						
WSD^{dyn}_{Ag1}	2	0.05	1	0.0015	0.192	0.462	1.308	1	0.3	1	0.00069	**0.154**	0.462	1.615
								1	0.3	1	0.000015	0.192	0.423	1.308
WSD^{dyn}_{Ag2}	3	0.15	2	0.0036	0.115	0.500	1.692	1	0.2	2	0.00099	**0.077**	0.462	1.808
	3	0.15	2	0.0025	0.154	0.423	1.385	1	0.2	2	0.00078	0.154	0.385	1.308
WSD^{dyn}_{Ag3}	3	0.2	1	0.002	0.115	0.385	1.423	5	0.1	1	0.002844	**0.077**	0.346	1.462
	3	0.2	1	0.0001	0.154	0.231	1.077	1	0.05	1	0.00195	0.154	0.385	1.308
WSD^{dyn}_{Ag4}	6	0.15	1	0.0001	0.192	0.308	1.154	1	0.1	2	0.001248	**0.115**	0.385	1.462
								1	0.05	1	0.000725	**0.154**	0.346	1.231
WSD^{dyn}_{Ag5}	4	0.1	1	0.0044	0.115	0.500	1.577	2	0.05	2	0.0036	0.115	0.462	1.577
	4	0.1	1	0.0035	0.192	0.423	1.385	5	0.2	1	0.00249	0.192	0.423	1.308

8.6.2 Results for Dynamic Structure with Disjoint Clusters

Table 8.4 shows the results for the approach with a dynamic structure and disjoint clusters. This time, for the Soybean data set, in most of the dispersed sets, the use of the method for attribute selection improved the quality of the classification (the only exception was the system with seven resource agents WSD_{Ag3}^{dyn}). In the case of the Landsat Satellite data set and all of the considered dispersed sets, the replacement of the sets of attributes by reducts resulted in an improved the quality of the classi-fication. However, this time differences in the values of the measure e were not as noticeable as before. In the case of the Vehicle Silhouettes data set, mostly poorer results were obtained after using attribute selection. In the case of the Dermatology data set, the use of the method for attribute selection did not affect the quality of the classification. For the Audiology data set for two of the dispersed sets (with nine and eleven resource agents WSD_{Ag4}, WSD_{Ag5}), better results were obtained.

8.6.3 Results for Dynamic Structure with Inseparable Clusters

Table 8.5 shows the results for the approach with a dynamic structure and inseparable clusters. As can be seen, for the Soybean data set and most of the dispersed sets, poorer results were obtained using the method for attribute selection (the only exceptions are the systems with nine and eleven resource agents WSD_{Ag4}^{dyn}, WSD_{Ag5}^{dyn}). In the case of the Landsat Satellite data set and all of the considered dispersed sets, the replacement of the sets of attributes by reducts resulted in an improved quality of the classification. In the case of the Vehicle Silhouettes data set and the system with three resource agents WSD_{Ag1}^{dyn}, poorer results were obtained after using attribute selection. In the case of the Dermatology data set, better results were obtained once (for the system with five WSD_{Ag2}^{dyn}) and worse results were obtained once (for the system with seven WSD_{Ag3}^{dyn}). For the Audiology data set and most of the dispersed sets, poorer results were obtained.

8.6.4 Results for Dynamic Structure with Negotiations

Table 8.6 shows the results for the approach with a dynamic structure and negotia-tions. In this case, for the Soybean data set, the Landsat Satellite data set and the Dermatology data set, for most of the considered dispersed sets, better results were obtained when applying the method of attribute selection. The only exceptions were the systems for the Soybean data set with three and five resource agents WSD_{Ag1}^{dyn}, WSD_{Ag2}^{dyn} and the system with seven resource agents WSD_{Ag3}^{dyn} for the Dermatology data set. The biggest differences in the values of the measure e were observed for the Landsat Satellite data set and the dispersed system with three resource agents and for the Soybean data set and the dispersed system with seven resource agents. In the case of the Vehicle Silhouettes data set and the Audiology data set, poorer results were obtained after using attribute selection.

8.6.5 Comparison of All Methods

A graph was created in order to give a detailed comparison of the results with and without attribute selection. Figure 8.1 presents the values of the estimator of classification error for the results given in Tables 8.3, 8.4, 8.5 and 8.6. The results for four approaches to creating a system's structure are marked on the graph. Only the results obtained for larger average size of the global decisions sets are presented on the graph. For each of the dispersed sets, the results with using the method of attribute selection are given on the left side and the results without the use of attribute selection are presented on the right side of the graph.

The following conclusions and observations can be made based on Fig. 8.1. Considering all of the presented results, it can be observed that:

- the method with a dynamic structure and negotiations obtained the best results for eleven dispersed sets (for the Soybean data set and systems WSD_{Ag1}^{dyn}, WSD_{Ag2}^{dyn} and WSD_{Ag4}^{dyn}, for the Landsat Satellite data set and both systems that were considered, for the Dermatology data set and systems WSD_{Ag1}^{dyn}, WSD_{Ag3}^{dyn}, WSD_{Ag4}^{dyn} and WSD_{Ag5}^{dyn}, for the Audiology data set and systems WSD_{Ag1}^{dyn} and WSD_{Ag2}^{dyn});
- the method with a dynamic structure and inseparable clusters obtained the best results for eleven dispersed sets (for the Soybean data set and systems WSD_{Ag1}^{dyn} and WSD_{Ag2}^{dyn}, for the Dermatology data set and systems WSD_{Ag1}^{dyn}, WSD_{Ag2}^{dyn}, WSD_{Ag4}^{dyn} and WSD_{Ag5}^{dyn}, for the Audiology data set and all systems);
- the method with a static structure received the best results for seven dispersed sets (for the Soybean data set and systems WSD_{Ag3}, WSD_{Ag5}, for the Landsat Satellite data set and system WSD_{Ag2}^{dyn}, for the Vehicle Silhouettes data set and both systems that were considered, for the Dermatology data set and the Audiology data set and system WSD_{Ag1}^{dyn});
- the method with a dynamic structure with disjoint clusters obtained the best results for seven dispersed sets (for the Dermatology data set and systems WSD_{Ag1}^{dyn}, WSD_{Ag2}^{dyn}, WSD_{Ag4}^{dyn} and WSD_{Ag5}^{dyn}, for the Audiology data set and systems WSD_{Ag1}^{dyn}, WSD_{Ag3}^{dyn} and WSD_{Ag5}^{dyn}).

Fig. 8.1 Comparison of the
results with and without
attribute selection

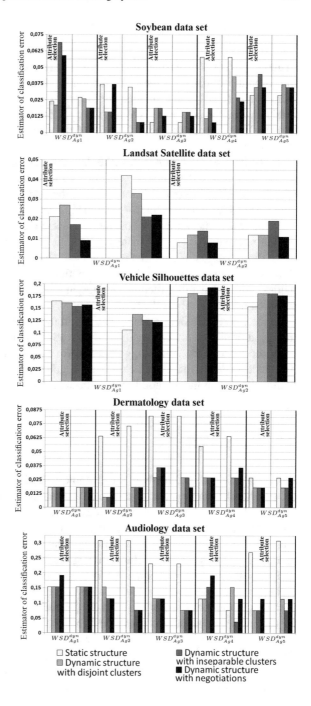

8.6.6 Comparison for Data Sets

In the case of the Soybean data set and dynamic approaches, the selection of attributes improved the efficiency of the classification for dispersed sets with a greater number of resource agents; for a smaller number of resource agents, poorer results were received. For the Soybean data set, the application of the selection of attributes had the least impact on the efficiency of the classification in the method with a static structure.

In the case of the Landsat Satellite data set, the use of the method for attribute selection improved the efficiency of the classification for all of the considered methods. The smallest differences in the values of the measure e were noted for the method with a dynamic structure and inseparable clusters.

In the case of the Vehicle Silhouettes data set, better results were obtained without the use of the method for attribute selection. It is suspected that the reason for this is the discretization of continuous attributes, which had to be made for the Vehicle Silhouettes data set. Only for the system with five resource agents WSD_{Ag2}^{dyn} and the method with a dynamic structure and inseparable clusters did the use of the attribute selection method improve the result.

In the case of the Dermatology data set, in most cases, the selection of attributes improved or did not affect the efficiency of the classification.

In the case of the Audiology data set, for a dynamic structure with inseparable clusters and a dynamic structure with negotiations, better results were obtained without the use of the method for attribute selection.

8.7 Conclusion

Research in which the method for attribute selection that is based on the rough set theory was employed in a dispersed system is presented here. Four different approaches to the creation of a dispersed system's structure were considered: the approach with a static structure, the approach with a dynamic structure and disjoint clusters, the approach with a dynamic structure and inseparable clusters and the approach with a dynamic structure and negotiations.

In the experimental part of the chapter, a presentation of the executed experiments and their results for five data sets from the UCI Repository were given. The results with and without the use of the method for attribute selection were compared for each approach separately. It was noted that the method for attribute selection did not improve the quality of the classification for all approaches and data sets (e.g. the method with a dynamic structure and inseparable clusters and the Soybean data set; the Vehicle Silhouettes data set). However, in most cases, better results were obtained after the application of this method.

References

1. Baron, G.: Analysis of multiple classifiers performance for discretized data in authorship attribution. In: Czarnowski, I., Howlett, R.J., Jain, L.C. (eds.) Intelligent Decision Technologies 2017: Proceedings of the 9th KES International Conference on Intelligent Decision Technologies (KES-IDT 2017) – Part II, pp. 33–42. Springer International Publishing, Cham (2018). https://doi.org/10.1007/978-3-319-59424-8_4
2. Cano, A., Ventura, S., Cios, K.J.: Multi-objective genetic programming for feature extraction and data visualization. Soft Computing **21**(8), 2069–2089 (2017). https://doi.org/10.1007/s00500-015-1907-y
3. Cichocki, A., Mandic, D.P., Phan, A.H., Caiafa, C.F., Zhou, G., Zhao, Q., Lathauwer, L.D.: Tensor decompositions for signal processing applications from two-way to multiway component analysis. CoRR (2014). arXiv:1403.4462
4. Gatnar, E.: Multiple-Model Approach to Classification and Regression. PWN, Warsaw (2008)
5. Jackowski, K., Krawczyk, B., Woźniak, M.: Improved adaptive splitting and selection: the hybrid training method of a classifier based on a feature space partitioning. Int. J. Neural Syst. **24**(03), 1430,007 (2014). https://doi.org/10.1142/S0129065714300071
6. Krawczyk, B., Woźniak, M.: Dynamic classifier selection for one-class classification. Knowl. Based Syst. **107**, 43–53 (2016). https://doi.org/10.1016/j.knosys.2016.05.054
7. Kuncheva, L.I.: Combining Pattern Classifiers Methods and Algorithms. Wiley, New York (2004)
8. Kuncheva, L.I.: A bound on kappa-error diagrams for analysis of classifier ensembles. IEEE Trans. Knowl. Data Eng. **25**(3), 494–501 (2013). https://doi.org/10.1109/TKDE.2011.234
9. Kuncheva, L.I., Bezdek, J.C., Duin, R.P.: Decision templates for multiple classifier fusion: an experimental comparison. Pattern Recognit. **34**(2), 299–314 (2001). https://doi.org/10.1016/S0031-3203(99)00223-X
10. Müller, J.P., Fischer, K.: Application impact of multi-agent systems and technologies: a survey. In: Shehory, O., Sturm, A. (eds.) Agent-Oriented Software Engineering: Reflections on Architectures, Methodologies, Languages, and Frameworks, pp. 27–53. Springer, Berlin (2014). https://doi.org/10.1007/978-3-642-54432-3_3
11. Ng, K.C., Abramson, B.: Probabilistic multi-knowledge-base systems. Appl. Intell. **4**(2), 219–236 (1994). https://doi.org/10.1007/BF00872110
12. Nguyen, S.H., Bazan, J., Skowron, A., Nguyen, H.S.: Layered learning for concept synthesis. In: Peters, J.F., Skowron, A., Grzymała-Busse, J.W., Kostek, B., Świniarski, R.W., Szczuka, M.S. (eds.) Transactions on Rough Sets I, pp. 187–208. Springer, Berlin (2004). https://doi.org/10.1007/978-3-540-27794-1_9
13. Oliveira, L.S., Morita, M., Sabourin, R.: Feature selection for ensembles using the multi-objective optimization approach. In: Jin, Y. (ed.) Multi-Objective Machine Learning, pp. 49–74. Springer, Berlin (2006). https://doi.org/10.1007/3-540-33019-4_3
14. Pawlak, Z.: Rough Sets. Int. J. Comput. Inf. Sci. **11**, 341–356 (1982)
15. Polikar, R.: Ensemble based systems in decision making. IEEE Circuits Syst. Mag. **6**(3), 21–45 (2006)
16. Przybyła-Kasperek, M., Wakulicz-Deja, A.: Application of reduction of the set of conditional attributes in the process of global decision-making. Fundam. Inf. **122**(4), 327–355 (2013). https://doi.org/10.3233/FI-2013-793
17. Przybyła-Kasperek, M., Wakulicz-Deja, A.: A dispersed decision-making system - the use of negotiations during the dynamic generation of a system's structure. Inf. Sci. 288 (C), 194–219 (2014). https://doi.org/10.1016/j.ins.2014.07.032
18. Przybyła-Kasperek, M., Wakulicz-Deja, A.: Global decision-making system with dynamically generated clusters. Inf. Sci. **270**, 172–191 (2014). https://doi.org/10.1016/j.ins.2014.02.076
19. Przybyła-Kasperek, M., Wakulicz-Deja, A.: Global decision-making in multi-agent decision-making system with dynamically generated disjoint clusters. Appl. Soft Comput. **40**, 603–615 (2016). https://doi.org/10.1016/j.asoc.2015.12.016

20. Rogova, G.: Combining the results of several neural network classifiers. In: Yager, R.R., Liu, L. (eds.) Classic Works of the Dempster–Shafer Theory of Belief Functions, pp. 683–692. Springer, Berlin (2008). https://doi.org/10.1007/978-3-540-44792-4_27
21. Schneeweiss, C.: Distributed Decision Making. Springer, Berlin (2003)
22. Schneeweiss, C.: Distributed decision making-a unified approach. Eur. J. Oper. Res. **150**(2), 237–252 (2003)
23. Shoemaker, L., Banfield, R.E., Hall, L.O., Bowyer, K.W., Kegelmeyer, W.P.: Using classifier ensembles to label spatially disjoint data. Inf. Fusion **9**(1), 120–133 (2008). https://doi.org/10.1016/j.inffus.2007.08.00 (Special issue on Applications of Ensemble Methods)
24. Skowron, A.: Rough Set Exploration System. http://logic.mimuw.edu.pl/rses/. Accessed 01 March 2017
25. Skowron, A., Jankowski, A., Świniarski, R.W.: Foundations of rough sets. In: Kacprzyk, J., Pedrycz, W. (eds.) Springer Handbook of Computational Intelligence, pp. 331–348. Springer, Berlin (2015). https://doi.org/10.1007/978-3-662-43505-2_21
26. Ślęzak, D., Janusz, A.: Ensembles of bireducts: towards robust classification and simple representation. In:. Kim, T.H, Adeli, H., Ślęzak, D., Sandnes, F.E., Song, X., Chung, K.I., Arnett, K.P. (eds.) Future Generation Information Technology: Third International Conference, FGIT 2011 in Conjunction with GDC 2011, Jeju Island, Korea, December 8–10, 2011. Proceedings, pp. 64–77. Springer, Berlin, Heidelberg (2011). https://doi.org/10.1007/978-3-642-27142-7_9
27. Ślęzak, D., Widz, S.: Is it important which rough-set-based classifier extraction and voting criteria are applied together? In: Szczuka, M., Kryszkiewicz, M., Ramanna, S., Jensen, R., Hu Q. (eds.) Rough Sets and Current Trends in Computing: 7th International Conference, RSCTC 2010, Warsaw, Poland, June 28-30,2010. Proceedings, pp. 187–196. Springer, Berlin, Heidelberg (2010). https://doi.org/10.1007/978-3-642-13529-3_21
28. Słowiński, R., Greco, S., Matarazzo, B.: Rough-set-based decision support. In: Burke, E.K., Kendall, G. (eds.) Search Methodologies: Introductory Tutorials in Optimization and Decision Support Techniques, pp. 557–609. Springer, Boston (2014). https://doi.org/10.1007/978-1-4614-6940-7_19
29. Stasiak, B., Mońko, J., Niewiadomski, A.: Note onset detection in musical signals via neural-network-based multi-odf fusion. Int. J. Appl. Math. Comput. Sci. **26**(1), 203–213 (2016)
30. Wakulicz-Deja, A., Przybyła-Kasperek, M.: Hierarchical multi-agent system. In: Recent Advances in Intelligent Information Systems, pp. 615–628. Academic Publishing House EXIT (2009)
31. Wang, S., Pedrycz, W., Zhu, Q., Zhu, W.: Subspace learning for unsupervised feature selection via matrix factorization. Pattern Recognit. **48**(1), 10–19 (2015). https://doi.org/10.1016/j.patcog.2014.08.004
32. Wróblewski, J.: Ensembles of classifiers based on approximate reducts. Fundam. Inf. **47**(3–4), 351–360 (2001)
33. Wu, Y., Zhang, A.: Feature selection for classifying high-dimensional numerical data. In: Proceedings of the 2004 IEEE Computer Society Conference on Computer Vision and Pattern Recognition, 2004. CVPR 2004, vol. 2, p. II. IEEE (2004)

Chapter 9
Feature Selection Approach for Rule-Based Knowledge Bases

Agnieszka Nowak-Brzezińska

Abstract The subject-matter of this study is knowledge representation in rule-based knowledge bases. The two following issues will be discussed herein: *feature selection* as a part of mining knowledge bases from a knowledge engineer's perspective (it is usually aimed at completeness analysis, consistency of the knowledge base and detection of redundancy and unusual rules) as well as from a domain expert's point of view (domain expert intends to explore the rules with regard to their optimization, improved interpretation and a view to improve the quality of knowledge recorded in the rules). In this sense, exploration of rules, in order to select the most important knowledge, is based, in a great extent, on the analysis of similarities across the rules and their clusters. Building the representatives for created clusters of rules bases on the analysis of the left-hand sides of this rules and then selection of the best descriptive once. Thus we may treat this approach as a *feature selection* process.

Keywords Knowledge base · Inference algorithms · Cluster analysis · Outlier detection · Rules representatives · Rules visualization

9.1 Introduction

Feature selection have become the focus of much research in areas of application for which datasets with tens or hundreds of thousands of variables are available. These areas include text processing of Internet documents or gene expression array analysis, and many others. The objective of variable selection is three-fold: improving the prediction performance of the predictors, providing faster and more cost-effective predictors, and providing a better understanding of the underlying process that generated the data. There are many potential benefits of variable and feature selection: facilitating data visualization and data understanding, reducing the measurement and

A. Nowak-Brzezińska (✉)
Institute of Computer Science, University of Silesia in Katowice, Będzińska 39,
41-200 Sosnowiec, Poland
e-mail: agnieszka.nowak@us.edu.pl

© Springer International Publishing AG 2018
U. Stańczyk et al. (eds.), *Advances in Feature Selection for Data and Pattern Recognition*, Intelligent Systems Reference Library 138,
https://doi.org/10.1007/978-3-319-67588-6_9

storage requirements, reducing training and utilization times, defying the curse of dimensionality to improve prediction performance [6, 8]. Clustering has long been used for feature construction. The idea is to replace a group of "similar" variables by a cluster centroid, which becomes a feature [18]. However, when we take the results of the clustering process (clusters), and then, using different techniques we select descriptors which represent the clusters in the best possible way, we may call it *feature selection* approach. The idea proposed in this paper corresponds to such an approach. The main author's motivation for interesting in *feature selection* methods was the necessity of finding efficient techniques for managing large set of rules in the rule-based knowledge bases (*KBs*). Last thirty years has brought an enormous interest in integrating database and knowledge-based system (*KBS*) technologies in order to create an infrastructure for modern advanced applications, as *KBs* which consist of database systems extended with some kind of knowledge, usually expressed in the form of *rules*. The size of the *KBs* is constantly increasing in volume and type of the data saved in rules is more complex, thus this kind of knowledge requires methods for its better organization. If there is no structure in which rules would be organized in, the system is unfortunately inefficient. Managing large *KBs* is important from both knowledge engineer and domain expert point of view. A knowledge engineer can, as a result, take control of the verification of the *KBs* correctness in order to maximize the efficiency of the decision support system to be developed, while a domain expert can consequently take care of the completeness of the gathered information. Therefore the methods of selecting the most crucial information from rules or their clusters help to manage the *KBs* and ensure high efficiency of decision support systems based on the knowledge represented in the form of rules.

The rest of the chapter is organized as follows: in Sect. 9.2 the general information about the motivation of the author's scientific goals is presented, including the related works and proposed approach. It contains the description of the similarity analysis as well as the pseudocode of the agglomerative hierarchical clustering algorithm used for creating groups of rules. It also includes the introduction of the methods for creating the representatives. Section 9.3 describes the proposed methods of selection the features for best representation of created clusters of rules, based on the Rough Set Theory. It contains definition of the *lower* and *upper approximation*-based approaches. The results of the experiments are included in Sect. 9.4. The summarization of the research goals and results of the experiments fills the Sect. 9.5.

9.2 Feature Selection Methods for Rule Mining Processes

This chapter focuses mainly on selecting subsets of features that are useful to build a good predictor or to be a good descriptor of the domain knowledge. Having a set of n examples $\{x_k, y_k\}$ $(k = 1, \ldots, n)$ consisting of m input variables x_k, i $(i = 1, \ldots, m)$ and usually one output variable y_k, feature selection methods use variable ranking techniques with a defined scoring function computed from the values x_k, i and y_k, $k = 1, \ldots, n$. Feature selection methods divide into wrappers, filters, and embedded methods. Wrappers utilize the learning machine of interest as a black box to score subsets of variable according to their predictive power. Filters select subsets of variables as a preprocessing step, independently of the chosen predictor. Embedded methods perform variable selection in the process of training and are usually specific to given learning machines [6, 8].The subject of the analysis are rules - a specific type of data. *Rules* are logical statements (implications) of the „*if* condition$_1$ & ... & condition$_n$ *then* conclusion" type. It seems that there is no simpler way to express the knowledge of experts and to make the rules easily understood by people not involved in the expert system building. They may be given apriori by the domain expert or generated automatically by using one of many dedicated algorithms for rule induction. An example of rule generated automatically from the dataset `contact lenses` [11] is as follows:

```
(tear-prod-rate=reduced)=>(contact-lenses=none[12])  12
```

and it should be read as: *if (tear-prod-rate=reduced)* **then** *(contact-lenses=none)* which is covered by 12 of instances in the original dataset (50% of instances cover this rule because the number of all instances in this dataset equals 24).

Rules have been extensively used in knowledge representation and reasoning as they are very space efficient representation [10]. However, there appears a problem when the number of rules is too high. Whenever we have large number of data to analyze it is necessary to use an efficient technique for managing it. The example is dividing those data into smaller number of groups. If some data are more similar than other elements in the set, it is possible to build groups from them. In result, in a given group there are only elements for which within similarity is greater than some threshold. In the literature, it is possible to find numerous results of research interest in methods for managing large sets of data (also the rules) based on the clustering approach as well as joining and reducing rules [10]. Unfortunately, the related works, possible to find in the literature, are usually based on the specific type of rules, called *association rules*, and have many limitations in relation to the rules that are the subject of the author's research interest (an exemplar limitation is that fact that association rules are generated automatically from data while the author analyze *KBs* with rules that are possible to be given by domain experts).The author presents an idea, in which, based on the similarity analysis, rules with similar premises are merged in order to form a group. In this approach only the left hand side of the rules is analyzed, but it is possible to take into account also the right side of the rules. The results of the author's previous research in this subject are included in [13].

In this study the author wants to show the way of the exploration of complex *KBs* with groups of similar rules manipulating the forms of their representatives. The role of the representatives of the rules' clusters is very important. Knowing the representatives of the clusters of rules the user of the *KB* has got the general knowledge about every group (representative is a kind of the description for a given group).Process of using the feature selection approach in the context of the rules and their clusters is as follows. At the beginning *KB* contains N rules: $KB = \{r_1, r_2, \ldots, r_i, \ldots, r_N\}$. After clustering process, *KB* becomes a set of g groups of rules $R = \{R_1, R_2, \ldots, R_l, \ldots, R_g\}$. For such groups we need to build representatives. This is a main step of the feature selection process. By using a defined criteria we try to select, from every group of rules, premises (pairs *(attribute, value)*) which describe the groups in the best possible way. It is not a trivial task as there are different number of premises in the left hand side of the rules, rules may create a queue, rules my not have at least one common premise etc. Thus the process of selection the features that describe the rules in a given group in an optimal way is so important and worth to analyze.

The author spent a lot of time analyzing and proposing methods for rules' and their clusters' representation. It is very important to choose the form of the representative properly as this influences the results of further analysis. *Feature selection* process is based here on the *Rough Set Theory* (*RST*) (described in Sect. 9.3). *RST* allows to make a choice between *lower* and *upper approximation* approach for a given set, which is in this way a set of rules' clusters representatives. *Lower approximation* approach provides a general description while the *upper approximation* approach results in creating more detailed description for created groups of rules. To create such representatives we make selection of the descriptors (pairs of the attributes and their values) from the left hand side of the rules to represent the whole group of rules. It is very important to find the most efficient descriptions as they influence on the overall efficiency of the knowledge extraction process, realized by every *decision support system* (*DSS*).

9.2.1 Significance of Rules Mining Process

Exploration of *KBs* can run in different ways, depending on who is to be the recipient of the results. If it is carried out by the knowledge engineer it usually results in knowing the rules which are unusual, redundant or inconsistent. If the number of rules is too high, the exploration of *KB* may result in dividing the rules in the smaller pieces (groups). Thanks to this, the knowledge engineer can manage the set of the rules in an efficient way. If the knowledge exploration is carried out from the user point of view, it is usually focused on the time and the form of discovering new knowledge from a knowledge hidden in rules. If the knowledge mining process concerns the domain expert it aims in finding a knowledge useful for improving the quality of the knowledge already being saved in the set of rules. Thanks to mining of *KBs*, domain experts can get information about frequent and rare (unusual) patterns in rules, what in turn can speed up the process of deeper analysis of the field.

This study presents the aspects of the *KB*'s exploration especially from the knowledge engineer and domain expert point of view. Rules in a given *KB* can be given apriori by a domain expert or generated automatically from the data. In both cases the domain expert may wish to analyze the knowledge hidden in rules and if the number of rules is too big, this process can be inefficient. Exploration of *KB* in the context of clusters of rules' representation allows the domain expert to know in what way (if only) the knowledge is divided, whether it contains many groups of different rules, or a few groups of many rules. The first case corresponds to a situation in which the domain of the expert knowledge is dispersed into many aspects (i.e. many varied medical types of diseases), while the second one to the situation when the knowledge is focused on one (or a few) direction, and probably there is a need to explore the knowledge of the expert in order to induce new rules, which would describe the new cases. Both of the cases allows the domain expert to improve the knowledge by its deeper or wider exploration. The knowledge engineer may wish to know if there are some unusual rules, redundant rules or if there are any inconsistency hidden in rules. It is possible to get such an information by using clustering techniques. By analyzing the created structure of the groups of rules, the knowledge engineer can see much more information than it is possible to see when the rules are not clustered. Achieving small number of large clusters means that the knowledge in a given domain is focused on one or few pieces of it, and maybe it is necessary to analyze it more carefully in order to discover any new important features which could make the rules more varied. If there are many small clusters of rules it probably means that there are a small number of rules describing different types of the cases (saved in each cluster of rules). Each of these cases should be verified because it affects the quality of the knowledge in a given *DSS*. All the issue related to the process of clustering the rules and verifying the created structure are described in detail in Sect. 9.2.5.

From the domain expert point of view, knowledge exploration process is realized inter alia by the clusters of rules' representatives, which allow the domain expert to gain the full description of the knowledge hidden in rules. By creating the representative for every cluster of rules it is possible know the character of the domain knowledge very fast. An approach proposed in this research, based on two types of the representation: more general and more detailed, allows the domain expert, for example, to find the attributes (with their values) which appear in every rule (or the majority of the rules) and describe every cluster of rules. Such an attribute is not valuable to keep in the cluster's description. All the details of the methods used to create the rules' representatives are included in Sect. 9.3.

9.2.2 Feature Selection in the Rules and Their Clusters

In machine learning and statistics, *feature selection* is the process of selecting a subset of relevant features (variables or predictors) in order to simplify the representation of knowledge in a given system. It also assumes that the data contains many features that are either redundant or irrelevant, and can thus be removed without incurring much loss of information. The presented approach contains selection for the representatives of the rules clusters in *KB* the premises which match the defined requirements. Neither of the lengths of the representatives: too big nor too small are not welcome. Representatives should not contain premises which are common for majority of created clusters because they will not allow for distinction between groups of rules. Therefore it is so important to build representatives properly.

9.2.3 Related Works

An interesting and effective approaches to the problem with managing large sets of rules and necessity of their partitioning can be found in [9, 12, 15, 16, 18], where known systems like CHIRA, XTT2 or D3RJ are described. In CHIRA, rules are aggregated if their premises are built from the same conditional attributes or if the premises of one rule are a subset of the premises of the other [16]. In [15] rules that lie close to each other are joined if their joint causes no deterioration in accuracy of the obtained rule. XTT2 provides a modularized rule base, where rules working together are placed in one context corresponding to a single decision table [12] while D3RJ is a method which produces more general rules, where each descriptor can endorse the subset of values [9]. In most of these tools, a global set of rules is partitioned by the system designer into several parts in an arbitrary way. Unfortunately, none of these tools can be simply used to mine *KBs* described at the beginning of this chapter. They use different approaches than clustering based on similarity analysis, therefore it is not possible to compare the results of using author's implementation and these tools, as they require different input data format.

The representation of clusters in literature focuses on so called *centroid* or *medoid* point. But this representation does not match for the representation of groups of rules with labels containing different number of pairs (*attribute*, *value*) in its conditional part. It is impossible to count the center point of a cluster of rules. We rather need to construct a symbolic description for each cluster, something to represent the information that makes rules inside a cluster similar to each other and that would convey this information to the user. Cluster labeling problems are often present in modern text and Web mining applications with document browsing interfaces [17]. The authors present the idea of so called Description Comes First (*DCF*) approach which changes the troublesome conventional order of a typical clustering algorithm. It separates the processes of candidate cluster label discovery (responsible for collecting all phrases potentially useful for creating good cluster label) and data clustering. Out of all candidate labels only these which are „supported" (most similar) to cluster centroids discovered in parallel are selected. In this study, instead of finding a centroid point of a given cluster we try to find a set of features that match a given condition (appropriate frequency) for being a good representative.

9.2.4 Clustering the Rules Based on the Similarity Approach

An approach presented in this research is based on both clustering the rules and creating the representatives for them. Subject of analysis are rules. They may be formed using both type of the data representation: quantitative and qualitative. The majority of clustering approaches are based on the assumption of measuring the distance between analyzed objects and merging those with the smallest distance. Many of the implementations are based on the calculating the distances between the data, what means that, if the dataset contains the data in the other type of representation (qualitative), it is impossible to measure the distance for such data and usually they are overlooked in the analysis. In case of mixed type of the analyzed data, it is necessary to use measures which enable to compare the data without calculating the distances between them—and the measures which allow for this are usually similarity-based measures.

A large number of similarity coefficients have been proposed in the literature [2, 4], from which three were implemented and further analyzed:

- simple matching coefficient (SMC),
- Jaccard measure,
- Gower measure.

In all similarity measures, described in this work, similarity S between two rules r_i and r_j can be denoted as weighted sum of similarities s_k considering k-th common attribute of these rules. This can be written as Eq. (9.1):

$$S(r_i, r_j) = \sum_{k:a_k \in (A(r_i) \cap A(r_j))} w_k s_k(r_{ik}, r_{jk}), \tag{9.1}$$

where $A(r)$ is set of attributes of rule r, w_k is the weight assigned to k-th attribute and r_{ik} and r_{jk} are values of k-th attributes of i-th and j-th rule respectively.

Simple matching coefficient (SMC) [13] is the simplest measure of similarity considered in this work. It handles continuous attributes the same way it does with categorical attributes, namely (Eq. (9.2)):

$$s(r_{ik}, r_{jk}) = \begin{cases} 1 & \text{if } r_{ik} = r_{jk}, \\ 0 & \text{otherwise.} \end{cases} \tag{9.2}$$

In this case, however, overall similarity of rules S is simple sum, as weight w_k of each attribute is equal to 1. Due to that fact this similarity measure tends to favor rules with more attributes. More advanced form of this measure is *Jaccard index* [3]. It eliminates aforementioned drawback of *SMC* by setting weight $w_k = \frac{1}{Card(A(r_i) \cup A(r_j))}$, where $Card : V \to \mathbb{N}$ is the cardinality of a set.

Last measure described in this work is *Gower index* [5]—the most complicated one as it handles categorical data differently from numerical data. Similarity considering categorical data is count the same way as in case of the *Jaccard* or the *SMC* measures. Similarity of continuous attributes can be denoted as following (Eq. (9.3)):

$$s(r_{ik}, r_{jk}) = \begin{cases} 1 - \frac{|r_{ik} - r_{jk}|}{range(a_k)} & \text{if } range(a_k) \neq 0, \\ 1 & \text{otherwise,} \end{cases} \tag{9.3}$$

where $range(a_k) = max(a_k) - min(a_k)$ is range of k-th common attribute.

In the experiments, all measures described in this subsection were used as a parameter of the clustering algorithm. The author wanted to check their influence on resultant structure of clustering. It was shown that e.g., some of them tends to generate structures with larger number of ungrouped rules and that different similarity measures influences average length of cluster representative, thus allowing one to consider *KB* partitioning basing on cluster representatives length.

9.2.5 Clustering the Rules

The proposed concept assumes division of a given *KB* into coherent subgroups of rules. This section presents the basic concepts and notations of clustering the rules from a given *KB*. Let us assume that *KB* is a single set $KB = \{r_1, \ldots, r_i, \ldots, r_N\}$ without any order of N rules. Each rule $r_i \in KB$ is stored as a Horn's clause defined as:

$$r_i : p_1 \wedge p_2 \wedge \ldots \wedge p_m \rightarrow c,$$

where p_f is f-th literal (a pair of an attribute and its value) (a, v_i^a) $(f = 1, 2, \ldots, m)$. Attribute $a \in A$ may be a conclusion of rule r_i as well as a part of the premises.

In mathematical meaning, clustering the rules is a collection of subsets $\{R_d\}_{d \in D}$ of R such that: $R = \bigcup_{d \in D} R_d$ and if $d, l \in D$ and $d \neq l$ then $R_d \cap R_l = \emptyset$. The subsets of rules are non-empty and every rule is included in one and only one of the subsets. The partition of rules $PR \subseteq 2^R : PR = \{R_1, R_2, \ldots, R_g\}$, where: g is the number of groups of rules creating the partition PR and R_g is g-th group of rules generated by the partitioning strategy (PS). It is a general concept, however from the author's research point of view the most interesting is the *similarity based partition*, which finds pairs of the most similar rules iteratively. The process terminates if the similarity is no longer at least T or if the number of created clusters match a given condition. In result the g groups of rules $R_1, R_2, \ldots, R_l, \ldots, R_g$ is achieved. At the beginning of the clustering process there are only single rules to combine in groups, but when two rules r_i, r_j create a group R_l we need to know how to calculate the similarity between two groups R_{l_1}, R_{l_2} or between a group R_l and a single rule r_i. The similarity measures presented above work properly for single rules as well as for their groups. The similarity of groups R_{l_1} and R_{l_2} is determined as the similarity between their representatives $Representative(R_{l_1})$ and $Representative(R_{l_2})$, as follows:

$$sim(R_{l_1}, R_{l_2}) = \frac{|Representative(R_{l_1}) \cap Representative(R_{l_2})|}{|Representative(R_{l_1}) \cup Representative(R_{l_2})|}.$$

The efficiency of the created structure of objects, when using clustering approach, depends strongly on the type of the clustering. There are two main types: hierarchical and partitional-based techniques. The author wanted the results of the clustering to be resistant to the outliers in rules, and to make the results independent on the user requirements, thus the hierarchical techniques (*Agglomerative Hierarchical Clustering* algorithm *AHC*) was chosen to being implemented [7, 19]. The author needed to made small modifications in the classical version of this algorithm. Clustering algorithm proposed by the author looks for two the most similar rules (related to their premises) and combine them iteratively. Very often there are *KBs* in which many rules are totally dissimilar (do not have at least one common premise). That is why it was natural to form a termination condition somehow. Usually, the resultant structure contains both some number of groups of rules and a set of small groups of rules with only one (singular) rule inside (they will be called *ungrouped rules*). At the beginning each object is considered a cluster itself (or one may say that each object is placed within a cluster that consists only of that object). The pseudocode of the *AHC* algorithm [7] is presented as follows.

Algorithm 9.1: Pseudocode of the *AHC* algorithm

1. **Require** stop condition *sc*, ungrouped set of rules $KB = \{r_1, r_2, \ldots, r_i, \ldots, r_N\}$;
2. **Ensure** $R = \{R_1, R_2, \ldots, R_l, \ldots, R_g\}$ grouped tree-like structure of objects;
3. **procedure** *Classic AHC Algorithm*(*sc,KB*)
4. **begin**
5. **var** M - similarity matrix that consists of every clusters pair similarity value;
6. Place each rule r_i from *KB* into a separate cluster.;
7. Build similarity matrix *M*;
8. Using *M* find the most similar pair of clusters and merge them into one;
9. Update *M*;
10. **if** *sc* was met **then**
11.   end the procedure;
12. **else**
13.   REPEAT from step 3.
14. **endif**
15.   **RETURN** the resultant structure.
16. **end procedure**

The most important step is the second one, in which the similarity matrix *M* is created based on the selected similarity measure and a pair of the two most similar rules (or groups of rules) are merged. In this step two parameters are given by a user: the similarity measure and the clustering method. The author decides to compare in this research results of the clustering rules based on changing the clustering parameters. In this work clustering is stopped when given number of clusters is generated.

When the clustering process is finished, the process of forming the representatives—a kind of feature selection approach—begins. Every created clusters of rules is labeled with the two type of the representatives (described in details in Sect. 9.3). Thanks to such a solution, it is possible to mine the knowledge hidden in clusters of rules by analyzing their representatives.

It is worth to mention at this moment, that too many clustering parameters used in clustering process (like the number of desired clusters, the similarity measure, the clustering method) finally result in varied structures of the created hierarchy of clusters of rules. The resultant structure of clusters of rules can be evaluated by the analysis of different parameters, from which the following are the most informative: the number of clusters, the size of the biggest cluster, the number of ungrouped objects (dissimilar to the other). The knowledge that is explored from the created clusters of rules depends on the method of creating their representatives. If too restrictive techniques are used then it is possible to achieve an empty representatives, while too small requirements can results in achieving too long clusters' representatives, which are difficult to analyze by the user. All these parameters are the subject of the experiments taken in this study. The results of the experiments are presented in Sect. 9.4.

Apart from the usual choice of similarity functions, linkage criterion must be selected (since a cluster consists of multiple objects, there are multiple candidates to compute the distance to). The most popular methods (known also as inter-cluster similarity measure) are *Single Link* (*SL*), *Complete Link* (*CoL*), *Average Link* (*AL*) and *Centroid Link* (*CL*). They were described by the author in [14].

9.2.6   Cluster's Representative

It is very important for data to be presented in the most friendly way. Clustering a set of objects is not only about the dividing the objects into some number of groups, taking into account their similarity. It is also about giving the informative descriptions to the created clusters.

In the literature, the most often used form of the representative for a created cluster of objects, is the method based on the idea of so called *centroid* point. In mathematics and physics it is the point at the center of any shape, sometimes called center of area or center of volume. The co-ordinates of the centroid are the average (arithmetic mean) of the co-ordinates of all the points of the shape. For a shape of uniform density, the centroid coincides with the center of mass which is also the center of gravity in a uniform gravitational field.

Sometimes the centroid is replaced by the so called *medoid*, which is nothing more than the representative of a cluster with a data set whose average dissimilarity to all the objects in the cluster is minimal. Medoids are similar in concept to means or centroids, but medoids are always members of the data set. Medoids are most commonly used on data when a mean or centroid cannot be defined.

When there are a multi-dimensional space of the knowledge representation, it is difficult to built the representative which would be suitable for all the data from a given cluster. When the data, for which the representative is being created, are complex and there are many of varied data in a given cluster, a good technique to built the representative is taking into account only those information which is met in the majority of data from this particular group.

There are numerous ways for achieving different type of the representatives.

In author's previous research the representative was created in the following way. Having as input data: cluster C, and a threshold value t [%], find in the cluster's attributes set A only these attributes that can be found in t rules and put them in set A'. Then for each attribute $a \in A'$ check if a is symbolic or numeric. If it is symbolic then count modal value, if numeric - average value and return the attribute-value pairs from A' as cluster's representative. The representative created this way contains attributes that are the most important for the whole cluster (they are common enough). This way the examined group is well described by the minimal number of variables. However the author saw the necessity to analyze more methods for creating clusters' representatives and their influence on the resultant structure efficiency. Thus, in this study other approaches, called by the author as the *lower* and the *upper approximation* based representatives, are presented.

9.3   Rough Set Approach in Creating Rules' Clusters Representatives

The similarity between rules in *KB* is measured using the approach based on the notions of *indiscernibility relation*, which is a central concept in Rough Set Theory (*RST*). The *indiscernibility relation* allows to examine the similarity of objects [1]. Objects are considered indiscernible if they share the same values for all the attributes. In the proposed approach, this strict requirement of indiscernibility defined in canonical *RST* is relaxed. Because the clustering is done based on the assumption that objects are similar rather than identical (when we use indiscernibility relation for similarity measure) we may still deem two objects indiscernible, even if attributes that discern them exist, but their number is low. The larger the value of a similarity coefficient, the more similar the objects are.

This section presents the basic notion of the indiscernibility relation and both *lower* and *upper approximation* which inspired the author in the research on rules clustering.

Arbitrary rules $r_i, r_j \in KB$ are *indiscernible* by *PR* ($r_i \; \widetilde{PR} \; r_j$) if and only if r_i and r_j have the same value on all elements in *PR*:

$$IND(PR) = \{(r_i, r_j) \in KB \times KB : \forall_{a \in PR} \, a(r_i) = a(r_j)\}.$$

Obviously, the *indiscernibility relation*, associated with *PR*, is an *equivalence* relation on *PR*. As such, it induces rules partition, denoted *PR**, which is a set of all equivalence classes of the *indiscernibility relation*. It does not necessarily contain a subset of attributes. It may also include the criterion of creating groups of rules, i.e. groups of rules with at least *m* number of premises or groups of rules with premises containing a specific attribute or particular pair (attribute, value).

Indiscernibility relation *IND(C)* related to the premises of rules (*C*) (here denoted as $cond(r_i)$ and $cond(r_j)$) is as follows:

$$IND(C) = \{(r_i, r_j) \in R \times R : cond(r_i) = cond(r_j)\},$$

and it means that rules r_i and r_j are *indiscernible* by *PR* (identical).

Extending *indiscernibility relation IND(C)* to the similarity context, we may say that indiscernibility relation *IND(C)* related to the premises of rules can be defined in the following way:

$$IND(C) = \{(r_i, r_j) \in R \times R : sim(cond(r_i), cond(r_j)) \geq T\}$$

where *T* is some given threshold for minimal similarity.

9.3.1 Lower and Upper Approximation Approach

The essence of *RST* is that it is a very general technique for approximating a given set and can be very useful while forming the groups' representatives. Clusters' representatives influence the quality of rules partition (cohesion and separation of the created groups) and the efficiency of the inference process. Many different approaches can be found to determine representatives for groups of rules: (i) the set of all premises of rules that form group R_l, (ii) the conjunction of the selected premises of all rules included in a given group R_l as well as (iii) the conjunction of the premises of all rules included in a given group R_l. An approach inspired by *RST* with *lower* and *upper approximations* (details are included in the next subsection) enables managing the process of finding unusual rules and exacting relevant rules or the most relevant rules (when it is impossible to find exact rules).

9.3.1.1 The Definition of the Lower and Upper Approximation

If X denotes a subset of universe element U ($X \subset U$) then the *lower approximation* of X in B ($B \subseteq A$) denoted as $\underline{BX}$, is defined as the union of all these elementary sets which are contained in X: $\underline{BX} = \{x_i \in U \,|\, [x_i]_{IND(B)} \subset X\}$.

Upper approximation $\overline{BX}$ is the union of these elementary sets, which have a non-empty intersection with X: $\overline{BX} = \{x_i \in U \,|\, [x_i]_{IND(B)} \cap X \neq \emptyset\}$.

According to the definition of the *lower* and *upper approximation* of a given set we may say that the *lower approximation* forms representatives (for groups of rules) which contain only premises of rules which are present in every rule in a given group, while the *upper approximation* contains all premises which at least once were present in any of the rules in this group. It allows for defining for every group R_l the following two representatives:

- $\underline{Representative(R_l)}$ - for *lower approximation* set of the cluster's of rules R_j representatives,
- $\overline{Representative(R_l)}$ - for *upper approximation*.

The *lower approximation* set of a representative of group R_l is the union of all these literals from premises of rules, which certainly appear in R_l, while the *upper approximation* is the union of the literals that have a non-empty intersection with the subset of interest. The definition of the *lower approximation* of $Representative(R_l)$ is as follows:

$$\underline{Representative(R_l)} = \bigcup \{p_f : \forall_{r_i \in R_l} \quad p_f \in cond(r_i)\},$$

and it contains all premises which definitely form all rules in this group, while the *upper approximation* contains premises of rules which possibly describe this group, and it is as follows:

$$\overline{Representative(R_l)} = \bigcup \{p_f : \exists_{r_i \in R_l} \quad p_f \in cond(r_i)\}.$$

There are rules, which have some premises common with some (not all) of the rules in group R_l and that is why these premises do not form the *lower approximation*.

9.3.1.2 The Example of Feature Selection Process for the Representatives of the Clusters of Rules

Let us assume that the *KB* contains the following rules:

$r_1 : (a, 1) \wedge (b, 1) \wedge (c, 1) \rightarrow (dec, A)$
$r_2 : (a, 1) \rightarrow (dec, B)$
$r_3 : (d, 1) \rightarrow (dec, A)$
$r_4 : (d, 1) \wedge (e, 1) \rightarrow (dec, C)$
$r_5 : (a, 1) \wedge (d, 1) \rightarrow (dec, B)$

After using the *AHC* algorithm for clustering the rules two groups R_1 containing rules r_3, r_4 and R_1 containing rules r_1, r_2, r_5 are created. The groups' representatives for clusters R_1 and R_2 are as follows:

- for a cluster $R_1 = \{r_3, r_4\}$ the feature selection process results in creating the following representatives:

  – $Representative(R_1) = \{(d, 1)\}$,
  – $\overline{Representative(R_1)} = \{(d, 1), (e, 1)\}$,

- for a cluster $R_2 = \{r_1, r_2, r_5\}$ he following representatives are selected:

  – $\underline{Representative(R_2)} = \{(a, 1)\}$,
  – $\overline{Representative(R_2)} = \{(a, 1), (b, 1), (c, 1), (d, 1)\}$.

It means that literal $(d, 1)$ appears in every rule in this group while $(e, 1)$ makes rule r_4 distinct from r_3. In other words, $(d, 1)$ certainly describes every rule in group R_1 while $(e, 1)$ possibly describes this group (there is at least one rule which contains this literal in the conditional part).

9.3.2 KbExplorer - A Tool for Knowledge Base Exploration

It is worth to mention that all the author's research in the area of *DSS* was focused on the implementation of inference algorithms for rules and clusters of rules. In result, the `kbExplorer` system (http://kbexplorer.ii.us.edu.pl/) was created. It is an expert system shell that allows for flexible switching between different inference methods based on knowledge engineer preferences.The user has the possibility of creating *KB*s using a special creator or by importing a *KB* from a given data source. The format of the *KB*s enables working with a rule set generated automatically based on the rough set theory as well as with rules given apriori by the domain expert. The *KB* can have one of the following file formats: XML, RSES, TXT. It is possible to define attributes of any type: nominal, discrete or continuous. There are no limits for the number of rules, attributes, facts or the length of the rule.

9.4 Experiments

The goal of the experiments was to analyze varied *KBs* in order to explore the knowledge hidden in them. The aim of the author was to check whether the process of clustering and creating the representatives of created clusters of rules enables to mine useful knowledge. Having a hierarchy of the clusters of rules, with their representatives, it should be possible to extract a *meta*-knowledge about the knowledge stored in a given *KB*. Such a *meta*-knowledge may be treated as a kind of information (data) selected from a given *KB*. The process is not trivial, therefore a short explanation is given in next subsection additionally.

9.4.1 From Original Data to the Rules

The experiments were carried out on four different *KBs* from the UCI Machine Learning Repository [11]. Original datasets were used in order to generate the decision rules using *RSES* software and *LEM*2 algorithm [1]. The group of algorithms noticed as *LEM* (Learning from Examples Module) contains two versions. Both are components of the data mining system *LERS* (Learning from Examples using Rough Sets) based on some rough set definitions. For this particular research rules were generated using *LEM*2 version, based on the idea of blocks of attribute-value pairs, but there is no restriction for the method of rules induction. It is most frequently used since in most cases it gives better results than other algorithms. In general, *LEM*2 computes a local covering and then converts it into a rule set. Then the *kbExplorer* system is used to analyze the similarities of rules and to create the structure of clusters of rules. The system allows to use different clustering parameters like similarity measures, clustering methods, the number of created clusters as well as the method responsible for creating the representatives for clusters. The analysis of the influence of all this parameters on the efficiency of knowledge mining process was the main goal of the experiments. The next subsection presents the results of the experiments. The analysis of the results as well as the summarization of the presented research are included in the end of this section.

9.4.2 The Results of the Experiments

The analyzed datasets have got different characteristics: taking into account the number of attributes, as well as the type of the attributes, and the number of rules.

Table 9.1 The general characteristic of the analyzed *KBs*

$p < 0,05$	*AttrN*	*RulesN*	*NodesN*	*ClusterN*
Diabetes.rul	9	490	957 ± 12.94	23.0 ± 12.94
Heart disease.rul	14	99	188 ± 5.48	$10,0 \pm 5.48$
Nursery.rul	9	867	1704 ± 18.05	30.0 ± 18.05
Weater symbolic.rul	5	5	7 ± 1.42	3.0 ± 1.42

Table 9.2 The number of the experiments

	Similarity measure			
Representative	Gower	SMC	Jaccard	Total
lowerApprox	160	160	160	480
upperApprox	160	160	160	480
Total	160	320	320	960

The smallest number of attributes was 5, the greatest 14. The smallest number of rules was 5, the greatest 490. All the details of analyzed datasets are included in Table 9.1. The meaning of the columns in this table is as follows:

- *AttrN* - number of different attributes occurring in premises or conclusions of rules in given knowledge base,
- *RulesN* - number of rules in examined knowledge base,
- *ClustersN* - number of nodes in dendrogram representing resultant structure.

Unless it was not specified otherwise, all the values in all Tables contains the *mean* value $\pm$ the *standard deviation*. High *standard deviation* values result from different clustering parameters, and therefore, big differences in values of parameters that had been analyzed.

The number of starts of the clustering algorithm that had been made equal 960 (see Table 9.2).

Taking into account two methods of creating the representatives (*lower* and *upper approximation* based approaches) for three different similarity measures (*Gower*, *Jaccard* and the *SMC* measure) there were 160 executions of the clusterings for a given similarity measure and the method for creating the representatives. Therefore all of the combinations results in 960 in total. Taking into account the two methods of creating representatives and four clustering methods (*SL, CL, CoL* and *AL*) there were 120 runs of the clustering processes for each option. 160 or 120 here are the number of cases for all the examined datasets.The subject of the research is not related to the classification problem. There is no need to provide a validation process. The goal of the research was to show the method of selection information from the knowledge hidden in rule-based *KBs*. The method is based on analyzing the similarity between

Table 9.3 The values of the clustering parameters for analyzed *KBs*

$p < 0,05$	BCS	BRL	U	emptyR
Diabetes.rul	279.41 ± 162.84	85.50 ± 97.58	4.21 ± 7.73	6.67 ± 2.36
Heart disease.rul	45.80 ± 30.84	22.29 ± 23.30	2.19 ± 3.57	31.34 ± 27.98
Nursery.rul	378.90 ± 277.24	4.5 ± 3.51	3.79 ± 7.95	85.13 ± 87.13
Weater symbolic.rul	3.0 ± 1.42	2.15 ± 1.24	2.20 ± 1.73	0.07 ± 1.91
Mean±SD	151.31 ± 201.48	30.88 ± 57.68	2.93 ± 5.63	31.29 ± 49.82

the rules and clustering similar rules. Labeling the created groups of rules by the representatives is a kind of *feature selection* process.

The goal of the first experiment was to analyze the characteristics of different *KBs* in context of the most interesting parameters for clustering approach as the number of rules in the biggest cluster, the length of the representative of the biggest cluster etc. The results of this experiment are included in Table 9.3. There are following parameters described in that table: *U* - number of singular clusters in resultant structure of grouping, *BRL* - biggest representative length (number of descriptors in biggest cluster's representative) and *BCS* - biggest cluster size (number of rules in the cluster that contains the most of them).

The performance of different similarity measures was evaluated in the context of knowledge mining using information like: the number of rules clusters (*ClustersN*), the number of ungrouped rules (*U*), the sizes of the biggest cluster (*BCS*) and the length of the most specific representative (*BRL*). More specific means more detailed, containing a higher number of descriptors. The optimal structure of KBs with rules clusters should contain the well separated groups of rules, and the number of such groups should not be too high. Moreover, the number of ungrouped rules should be minimal. Creating an optimal description of each cluster (representative) is very important because they are used further to select a proper group (and reject all the others) in the inference process, in order to mine knowledge hidden in rules (by accepting the conclusion of the given rule as a true fact). The most important experiment was focused on analyzing the influence of the chosen representative's type on the examined clustering parameters (*U*, *BCS*, *BRL*, etc.) as well as on the frequency of an appearance of empty representative, which is the unexpected situation and it is obvious that clustering with empty representatives is not a good clustering, cause we loose an information about a group when we do not know its representative.

Table 9.4 The frequency of the empty representatives vs. Representative's method

$p < 0.05$	siE	anE	Total
lApprox	449(93.54%)	31(6.46%)	480(50%)
uApprox	336(70%)	144(30%)	480(50%)
Total	785(81.77%)	175(18.23%)	960(100%)

Table 9.4 shows that in case of using *lower approximation-based* representative (*lApprox*) in 93.54% of cases the representative sometimes was empty, and only in 6.46% always was nonempty. In case of using *upper approximation-based* representative (*uApprox*) in 30% of all clusterings examined during the experiments, the representative was always nonempty. Symbols *siE* and *anE* have the following meaning: *siE* - "sometimes the representative is empty", while *anE* - "the representative is always nonempty". This shows that the method that is used in the process of creating the representatives influence, for example, on the frequency of the cases in which the representatives are empty and thus uninformative.

Table 9.5 presents the values of the examined clustering parameters (*U*, *BCS*, *BRL*, etc.) when using *lower* or *upper approximation*-based representatives.

In case of the *lower approximation*-based representative (which is a method of creating the short descriptions of clusters of rules) the length of the biggest representative as well as the size of the biggest representative are tens times smaller than in case of using the *upper approximation*-based representative. The number of ungrouped rules is few times smaller when using the *lower approximation*-based representative instead of the *upper approximation*-based representative. Therefore, if we do not want to achieve too many outliers in *KB*, we should use the *lower approximation*-based method.

The last, but not least, result of the experiments is presented in Table 9.6, where it is possible to compare the values of clustering parameters (*U*, *BCS*, *BRL*, etc.) for both cases: clusterings with possible empty representative and clusterings for which the representative of each cluster is always nonempty.

It is clearly seen that in case of clusterings without empty representatives the size of the biggest cluster is greater than in case of clusterings with possible empty representatives. What can be interesting is the fact that the number of ungrouped rules (outliers) is twice greater when the representative is always nonempty. The reason is the following: greater number of ungrouped rules is achieved when the

Table 9.5 The values of the clustering parameters for *lower* and *upper approximation*-based method

$p < 0.05$	BCS	BRL	U	emptyR
lApprox	144.2 ± 194.8	1.04 ± 0.3	1.8 ± 2.7	57.2 ± 60.7
uApprox	158.4 ± 207.9	60.7 ± 69.8	4.01 ± 7.3	6.0 ± 4.6
Total	151.3 ± 201.5	30.9 ± 57.7	2.9 ± 5.6	31.29 ± 49.8

Table 9.6 The values of the clustering parameters for cases of empty and nonempty representative

$p < 0.05$	BCS	BRL	U	emptyR
siE	143.1 ± 179.7	36.6 ± 62.4	2.6 ± 5.1	38.6 ± 52.5
anE	188.1 ± 276.9	5.2 ± 2.9	4.2 ± 7.6	-1.0 ± 0.0
Total	151.3 ± 201.5	30.9 ± 57.7	2.9 ± 5.6	31.3 ± 49.8

condition for merging rules or clusters of rules is no longer fulfilled. Such a situation takes place when the rules or their clusters have no common part, and in consequence the representatives of the clusters which would be created for such elements are being empty.

9.4.3   The Summary of the Experiments

The results of the experiments allow to see that clustering parameters influence significantly on the efficiency of the process of mining knowledge in a given *KB*. They decide about the form and the size of the cluster and therefore they provide different forms of cluster's representation. Experiments have confirmed that the representatives created by using *lower approximation-based representatives* result in achieving much smaller length of the biggest representative as well as the size of the biggest representative than in case of using the *upper approximation-based representative*. The method of creating representatives also affects the ability to detect outliers in rules or rules clusters (the number of ungrouped rules is few times smaller when using the *lower approximation*-based representative instead of the *upper approximation*-based representative).

9.5   Conclusion

The author introduces a method for *KB*'s exploration, based on the rules clustering idea inspired by the *rough set theory* (*RST*). The proposed approach contains smart representation of knowledge, based on general or detailed form of clusters' descriptions. This research presents the results of the experiments which was based on the comparison of using the two proposed approaches for different *KBs* in the context of varied parameters connected with clustering of rules and knowledge mining approaches. An exploration of rules, in order to select the most important knowledge, is based, in a great extent, on the analysis of similarities across the rules and clusters and making use of the opportunities arising from cluster analysis algorithms. An exploration of clusters of rules based on the *Rough Set Theory* approach can be treated as a way of feature selection from the rules that form a given cluster, and thus the method of creating the representatives of rules clusters is so important. The proper selection of all available clustering parameters (in this study three different similarity measures, four clustering methods as well as a given number of clusters to create are available) and methods of creating groups' representatives (two defined methods for creating the representatives: *lower* and *upper approximation*-based approach were considered in the experiments) enables efficient selection of the most important information from the knowledge hidden in *KBs* that contain even very large number of rules.

The results of the experiments confirmed that the *lower approximation* based representative's method (the methods based on the rough set theory are examined in this research) is relevant to the more general descriptions of the clusters of rules, the highest frequency of the empty representatives, and the smallest number of ungrouped rules, while the *upper approximation*-based representative's method more often causes greater length of the biggest representative and the higher number of ungrouped rules.

In future the author plans to extend the analysis by using larger *KB* with higher number of rules and/or attributes. It is also planed to check whether using other clustering algorithms, than the analyzed for this research, leads to gain the expected efficiency.

References

1. Bazan, J.G., Szczuka, M.S., Wróblewski, J.: A new version of rough set exploration system. In: Alpigini, J.J., Peters, J.F., Skowron, A., Zhong, N. (eds.) Rough Sets and Current Trends in Computing. Lecture Notes in Computer Science 2475, Springer, Berlin, Heidelberg, pp. 397–404 (2002)
2. Boriah, S., Chandola, V., Kumar, V.: Similarity measures for categorical data: A comparative evaluation. Proceedings of the 8th SIAM International Conference on Data Mining. pp. 243–254, (2008)
3. Finch, H.: Comparison of distance measures in cluster. analysis with dichotomous data. Journal of Data Science 3. Ball State University. pp. 88–100 (2005)
4. Goodall, D.: A new similarity index based on probability. Biometrics. **22**, 882–907 (1966)
5. Gower, J.: A general coefficient of similarity and some of its properties. Biometrics. **27**, 857–871 (1971)
6. Guyon, I.: An introduction to variable and feature selection. J. Mach. Learn. Res. **3**, 1157–1182 (2003)
7. Jain, A.K., Dubes, R.C.: Algorithms for clustering data. Prentice Hall, New Jersey (1988)
8. Kumar, V., Minz, S.: Feature selection: A literature review. Smart Comput. Rev. **4**(3), 211–229 (2014)
9. Latkowski, R., Mikołajczyk, M.: Data decomposition and decision rule joining for classification of data with missing values. Lecture Notes in Artificial Intelligence 3066, Springer Verlag, pp. 254–263 (2004)
10. Lee, O., Gray, P.: Knowledge base clustering for kbs maintenance. Lecture Notes in Artificial Intelligence 3066, Springer Verlag. Journal of Software Maintenance and Evolution **10**(6), 395–414 (1998)
11. Lichman, M.: Uci machine learning repository (2013). http://archive.ics.uci.edu/ml
12. Nalepa, G., Ligeza, A., Kaczor, K.: Overview of knowledge formalization with XTT2 rules. Rule-Based Reasoning, Programming, and Applications, LNCS 6826, Springer Verlag. pp. 329–336 (2011)
13. Nowak-Brzezińska, A.: Mining rule-based knowledge bases inspired by rough set theory. Fundam. Inform. **148**, 35–50 (2016)
14. Nowak-Brzezińska, A., Rybotycki, T.: Visualization of medical rule-based knowledge bases. J. Med. Inform. Technol. **24**, 91–98 (2015)
15. Pindur, R., Susmaga, R., Stefanowski, J.: Hyperplane aggregation of dominance decision rules. Fundam. Inform. **61**, 117–137 (2004)
16. Sikora, M., Gudyś, A.: Chira-convex hull based iterative algorithm of rules aggregation. Fundam. Inform. **123**, 143–170 (2013)

17. Stefanowski, J., Weiss, D.: Comprehensible and accurate cluster labels in text clustering. In: Large Scale Semantic Access to Content (Text, Image, Video, and Sound), RIAO '07, pp. 198–209 (2007)
18. Toivonen, H., Klemettinen, M., Ronkainen, P., Hatonen, K., Mannila, H.: Pruning and grouping discovered association rules. Workshop Notes of ECML Workshop on Statistics, Machine Learning, and Knowledge Discovery in Databases. pp. 47–52 (1995)
19. Wierzchoń, S., Kłopotek, M.: Algorithms of Cluster Analysis. IPI PAN, Warsaw, Poland (2015)

Part III
Image, Shape, Motion, and Audio Detection and Recognition

Chapter 10
Feature Selection with a Genetic Algorithm for Classification of Brain Imaging Data

Annamária Szenkovits, Regina Meszlényi, Krisztian Buza, Noémi Gaskó, Rodica Ioana Lung and Mihai Suciu

Abstract Recent advances in brain imaging technology, coupled with large-scale brain research projects, such as the BRAIN initiative in the U.S. and the European Human Brain Project, allow us to capture brain activity in unprecedented details. In principle, the observed data is expected to substantially shape our knowledge about brain activity, which includes the development of new biomarkers of brain disorders. However, due to the high dimensionality, the analysis of the data is challenging, and selection of relevant features is one of the most important analytic tasks. In many cases, due to the complexity of search space, evolutionary algorithms are appropriate to solve the aforementioned task. In this chapter, we consider the feature selection task from the point of view of classification tasks related to functional magnetic resonance imaging (fMRI) data. Furthermore, we present an empirical comparison of

A. Szenkovits (✉) · N. Gaskó · R.I. Lung · M. Suciu
Centre for the Study of Complexity, Babeş-Bolyai University, str. Kogalniceanu, Nr.1, Cluj-Napoca, Romania
e-mail: szenkovitsa@cs.ubbcluj.ro

N. Gaskó
e-mail: gaskonomi@cs.ubbcluj.ro

R.I. Lung
e-mail: rodica.lung@econ.ubbcluj.ro

M. Suciu
e-mail: mihai-suciu@cs.ubbcluj.ro

R. Meszlényi
Department of Cognitive Science, Budapest University of Technology and Economics, Egry József utca 1, Budapest 1111, Hungary
e-mail: meszlenyi.regina@ttk.mta.hu

R. Meszlényi
Brain Imaging Centre, Research Centre for Natural Sciences, Hungarian Academy of Sciences, Magyar tudósok krt. 2, Budapest 1117, Hungary

K. Buza
Knowledge Discovery and Machine Learning, Institute für Informatik III, Rheinische Friedrich-Wilhelms-Universität Bonn, Römerstr. 164, 53117 Bonn, Germany
e-mail: buza@biointelligence.hu

© Springer International Publishing AG 2018
U. Stańczyk et al. (eds.), *Advances in Feature Selection for Data and Pattern Recognition*, Intelligent Systems Reference Library 138, https://doi.org/10.1007/978-3-319-67588-6_10

conventional LASSO-based feature selection and a novel feature selection approach designed for fMRI data based on a simple genetic algorithm.

Keywords Functional magnetic resonance imaging (fMRI) · Functional connectivity · Classification · Feature selection · Mild cognitive impairment · Biomarker

10.1 Introduction

Advanced brain imaging technology allows us to capture brain activity in unprecedented details. The observed data is expected to substantially shape our knowledge about the brain, its disorders, and to contribute to the development of new biomarkers of its diseases, including *mild cognitive impairment* (MCI). MCI represents a transitional state between the cognitive changes of normal aging and very early dementia and is becoming increasingly recognized as a risk factor for Alzheimer disease (AD) [14].

The brain may be studied at various levels. At the neural level, the anatomy of neurons, their connections and spikes may be studied. For example, neurons responding to various visual stimulii (such as edges) as well as neurons recognizing the direction of audio signals have been identified [49, 50]. Despite these spectacular results, we note that *C. Elegans*, having only 302 neurons in total, is the only species for which neural level connections are fully described [43]. Furthermore, in this respect, *C. Elegans* is extremely simple compared with many other species. For example, the number of neurons in the human brain is approximately 100 billion and each of them has up to 10,000 synapses [16]. Imaging neural circuits of that size is difficult due to various reasons, such as diversity of cells and limitations of traditional light microscopy [27].

While neurons may be considered as the elementary components of the brain, at the other end, psychological studies focus on the behavior of the entire organism. However, due to the large number of neurons and their complex interactions, it is extremely difficult to establish the connection between low-level phenomena (such as the activity of neurons) and high-level observations referring to the behavior of the organism. In fact, such connections are only known in exceptional cases: for example, deprivation to visual stimuli causes functional blindness due to modified brain function, in other words: the brain does not "learn" how to see, if no visual input is provided [20, 46].

In order to understand how the brain activity is related to phenotypic conditions, many recent studies follow an "explicitly integrative perspective" resulting in the emergence of the new domain of "network neuroscience" [2]. While there are brain networks of various spatial and temporal scale, in this study we focus on networks describing *functional connectivity* between brain regions. Two regions are said to be functionally connected if their activations are synchronized. When measuring the activity of a region, usually, the aggregated activity of millions of neurons is captured as function of time. Spatial and temporal resolution (number of regions and frequency of measurements) depend on the experimental technology. In case of functional magnetic resonance imaging (fMRI), the spatial resolution is currently around fifty-thousand voxels (i.e., the brain activity is measured in approximately fifty-thousand cubic areas of the brain), while the temporal resolution is between 0.5 and 2 s. Roughly speaking, the raw data consists of approximately fifty-thousand time-series, each one corresponding to one of the voxels in which the brain activity is measured.

When analyzing fMRI data, after the necessary preprocessing steps, voxels are organized into regions of interest (ROIs). The time series of the voxels belonging to a ROI are aggregated into a single time series representing the activation of the ROI as function of time. Next, functional connectivity strength between ROIs can be calculated. This process is illustrated in Fig. 10.1. Traditionally, functional connectivity strength is described with linear correlation coefficients between the time series associated with the ROIs. However, according to recent results, more complex relationships can be represented with other time series distance measures such as dynamic time warping (DTW) [30].

The functional brain network can be represented with the connectivity matrix of the aforementioned ROIs, i.e., by the functional connectivity strength between each pair of ROIs. Functional connectivity matrices show remarkable similarity between subjects, however some connectivity patterns may be characteristic to various conditions and disorders such as gender, age, IQ, and schizophrenia, addiction or cognitive impairment, see e.g. [28, 29] and the references therein.

While the connectivity features can be seen as a compact representation of brain activity compared with the raw data, the dimensionality of the resulting data is still very high, making feature selection essential for subsequent analysis. Therefore, in this study we consider the task of feature selection, which can be described as follows: given a set of features, the goal is to select a subset of features that is appropriate for the subsequent analysis tasks, such as classification of instances. In many cases, due to the complexity of search space, evolutionary algorithms are appropriate to solve this task.

In the last decade, extensive research has been performed on evolutionary algorithms for feature selection. In this chapter, we will consider the feature selection task with special focus on its applications to functional magnetic resonance imaging (fMRI) data. Furthermore, we will present an empirical comparison of conventional feature selection based on the "Least Absolute Shrinkage and Selection Operator" (LASSO) and a novel feature selection approach designed for fMRI data based on a minimalistic genetic algorithm (mGA). Finally, we point out that feature selection is essential for the successful classification of fMRI data which is a key task in developing new biomarkers of brain disorders.

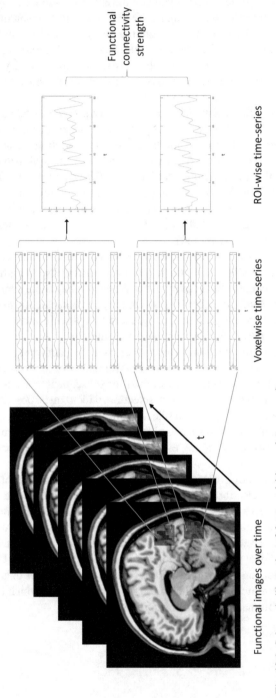

Fig. 10.1 Schematic illustration of data acquisition and preprocessing

10.2 Materials and Methods

In this section we provide details of the dataset and pre-processing (Sect. 10.2.1), the feature selection methods we consider: LASSO (Sect. 10.2.2) and mGA (Sect. 10.2.3). Subsequently, we describe our experimental protocol in Sect. 10.2.4.

10.2.1 Data and Preprocessing

We used publicly available data from the Consortium for Reliability and Reproducibility (CoRR) [52], in particular, the LMU 2 and LMU 3 datasets [3, 4]. The datasets contain fourty-nine subjects (22 males, age (mean $\pm$ SD): 68.6 $\pm$ 7.3 years, 25 diagnosed with mild cognitive impairment (MCI)), each subject participated at several resting-state fMRI measurement sessions, thus the total number of measurement sessions is 146. In each measurement session, besides high resolution anatomical images, 120 functional images were collected over 366 sec, resulting in time-series of length 120 for every brain voxel. The dataset was collected at the Institute of Clinical Radiology, Ludwig-Maximilians-University, Munich, Germany. For further details on how the data was obtained we refer to the web page of the data: http://fcon_1000.projects.nitrc.org/indi/CoRR/html/lmu_2.html.

Preprocessing of the raw data includes motion-correction, identification of gray matter (GM) voxels, and the application of low-pass and high-pass filters. For a detailed description of the preprocessing pipeline, see [30].

We used the Willard functional atlas of FIND Lab, consisting of 499 functional regions of interest (ROIs) [34] to obtain 499 functionally meaningful averaged blood-oxygen-level dependent (BOLD) signals in each measurement. This allows us to calculate ROI-based functional connectivity as follows: we calculate the pairwise dynamic time warping (DTW) distances [36] between the aforementioned 499 BOLD signals, resulting in $499 \times 498/2 = 124251$ connectivity features. We obtained these connectivity features for each of the 146 measurement sessions, leading to a dataset of 146 instances and 124251 features. From the publicly available phenotypic data, mild cognitive impairment (MCI) was selected as classification target.

Given the relatively low amount of instances, selection of relevant features (i.e., the ones that are characteristic for the presence or absence of mild cognitive impairment) is a highly challenging task to any of the feature selection techniques.

10.2.2 Least Absolute Shrinkage and Selection Operator

The Least Absolute Shrinkage and Selection Operator is widely used for the analysis of high-dimensional data, including brain imaging data. Therefore, we review this technique next.

We are given a dataset $\mathbf{X} \in \mathbb{R}^{N \times d}$ containing N instances with d features. Each instance x_i (the i-th row of $\mathbf{X}$) is associated with a label y_i, the vector y contains all the labels: $y = (y_1, \ldots, y_N)$. We aim at finding a function $f(x)$ in the form $f(x) = \sum_{j=1}^{d} \theta^{(j)} x^{(j)}$, where $x^{(j)}$ is the j-th component of vector x, and $\forall j : \theta^{(j)} \in \mathbb{R}$, so that the function fits the data, i.e., $f(x_i) \approx y_i$ for all (x_i, y_i) pairs. For simplicity, we use $\theta = (\theta^{(1)}, \ldots, \theta^{(d)})$ to denote the vector of all the parameters.

The above task can be considered as an ordinary least squares (OLS) regression problem, where the objective is to find the parameter vector θ^* that minimizes the sum of squared errors:

$$\theta^* = \arg \min_{\theta} \frac{1}{N} ||y - \mathbf{X}\theta||_2^2. \tag{10.1}$$

In the case when $N \geq d$ (and matrix $\mathbf{X}$ has a full column rank), i.e., when we have more training instances than features, the optimal θ^* vector exists and is unique. However, if the number of features exceeds the number of available training instances, matrix $\mathbf{X}$ loses its full column rank, therefore the solution is not unique anymore. In this case, the model tends to overfit the dataset $\mathbf{X}$, in the sense that it fits not only to the "regularities" of the data, but also to measurement noise while it is unlikely to fit to unseen instances of a new dataset $\mathbf{X}'$.

To be able to choose the "correct" parameter vector from the numerous possibilities, one must assume some knowledge about the parameters, and use that in the optimization. The most common solution of this problem is to add a regularization term to the function we try to minimize, called *objective function*. In case of the well-known ridge-regression method [17] we assume that the Euclidean-norm (L_2-norm) of the θ vector is small, and the objective is to find the parameter vector θ^* that minimizes the sum of squared errors and the regularization term:

$$\theta^* = \arg \min_{\theta} \left(\frac{1}{N} ||y - \mathbf{X}\theta||_2^2 + \lambda ||\theta||_2^2 \right), \tag{10.2}$$

where $\lambda \in \mathbb{R}$ is a hyperparameter controlling the regularization, i.e., in case of $\lambda = 0$ the method is equivalent to the OLS, but as we increase the λ value, the L_2-norm of the θ vector has to decrease to minimize the objective function.

However, in cases when we can hypothesize that only *some* of all the available features have influence on the labels, the above regularization based on the L_2-norm of θ may not lead to an appropriate solution θ^*. In particular, ridge-regression method results in low absolute value weights for the features, while it tends to "distribute" the weights between features, i.e., in most cases almost all of the features will receive nonzero weights. In contrast, regularization with L_1-norm usually results in zero weights for many features, therefore this method can be seen as a feature selection technique: it selects those features for which the associated weights are not zero, while the others are not selected. This method is called LASSO [42].

Formally, LASSO's objective is to find the parameter vector θ^* that minimizes the sum of squared errors and the L_1 regularization term:

$$\theta^* = \arg\min_{\theta} \left(\frac{1}{N} ||y - \mathbf{X}\theta||_2^2 + \lambda ||\theta||_1 \right), \tag{10.3}$$

where $\lambda \in \mathbb{R}$ is a hyperparameter controlling the sparsity of the resulting model, i.e. the number of weights that are set to zero.

As mentioned above, in brain imaging studies the number of instances is usually much lower than the number of features, therefore LASSO may be used. Indeed, it has been shown to be successful for classification tasks related to brain networks [28, 29, 35] in cases where the number of features is 10–50 times larger than the number of instances. In [28, 29] the features selected by LASSO, i.e., functional connectivity strength values, did not only lead to good classification performance, but the selected connections showed remarkable stability through the rounds of cross-validation and resulted in well-interpretable networks that contain connections that differ the most between groups of subjects. The task we consider in this study is even more challenging compared with the tasks considered in [28, 29], because the number of features is more than 800 times higher than the number of available instances.

10.2.3 Minimalist Genetic Algorithm for Feature Selection

Evolutionary Algorithms (EAs) represent a simple and efficient class of nature inspired optimization techniques [10, 18]. In essence, EAs are population based stochastic techniques that mimic natural evolution processes to find the solution for an optimization problem. The main advantages of these approaches are their low complexity in implementation, their adaptability to a variety of problems without having specific domain knowledge, and effectiveness [13, 31, 37].

There are many variants of EAs with different operators (e.g. selection, recombination, mutation, etc.) and control parameters such as population size, crossover and mutation probabilities [12]. A set of random solutions is evolved with the help of some variation operators and only good solutions will be kept in the population with the help of selection for survivor operators. Solutions are evaluated by using a fitness function constructed depending on the problem. In most cases it represents the objective of the optimization problem, and guides the search to optimal solutions by preserving during the selection process solutions with high (for maximization problems) fitness values. Based on the encoding and operators used, the main classes of evolutionary algorithms are considered to be: Genetic algorithms, Genetic programming (tree structure encoding), Evolution strategies (real-valued encoding), and Evolutionary programming (typically uses only mutation) [12].

Genetic algorithms (GAs) use binary encoding and specific mutation and crossover operators [12]. A potential solution evolved within a GA is referred to as an individual, encoded as a chromosome which is composed of genes - and represented in its simplest form as a string of 0 and 1. Encoding/decoding of an individual is problem dependent.

As the feature selection problem aims at selecting the relevant features from a large number of features [45], evolutionary algorithms seem to be an appropriate choice for tackling it because they can deal with the search space generated by the high number of features [21, 45]. Evaluation is performed by using either the classification accuracy or some convex combination between various indicators, such as classification accuracy, number of features, overall classification performance, class specific accuracy, and Pearson correlation [6, 39, 44, 47]. Multi-objective optimization algorithms have also been used to find trade-offs between the number of features and classification accuracy as two conflicting objectives [19, 25].

Genetic Algorithms are a natural choice for approaching the feature selection problem due to the encoding used: in the binary string each value shows if the corresponding feature is selected or not. Consequently, there are many studies that employ various variants of genetic algorithms for solving feature selection problems in fields such as computer science, engineering, medicine, biochemistry, genetics and molecular biology, chemistry, decision sciences, physics and astronomy, pharmacology, toxicology and pharmaceutics, chemical engineering, business, management and accounting, agricultural and biological sciences, materials science, earth and planetary sciences, social sciences and humanities. For example, in [24] a genetic algorithm is used to select the best feature subset from 200 time series features and use them to detect premature ventricular contraction - a form of cardiac arrhythmia. In [7] a genetic algorithm was used to reduce the number of features in a complex speech recognition task and to create new features on machine vision task. In [33], a genetic algorithm optimizes a weight vector used to scale the features.

Recently, genetic algorithms have been used for feature selection and classification of brain related data. Problems approached include brain tumor classification [15, 26], EEG analysis [22, 23], and seizure prediction [9]. Genetic algorithms are designed as stand-alone feature selection methods [32] or as part of a more complex analysis combined with simulated annealing [23, 26], neural networks [9, 15, 41] or support vector machines [22, 26, 41]. To the best of our knowledge, genetic algorithms have not yet been used for feature selection on fMRI data.

In the following we propose a minimalist version of a genetic algorithm for mining features in the brain data. The *Minimalist Genetic Algorithm* (mGA) evolves one binary string encoded individual by using only uniform mutation for a given number of generations in order to improve the classification accuracy while restricting the number of selected features. In what follows, the mGA is described in detail.

Encoding

mGA uses bit string representation of length $L = 124251$ – equal to the total number of features – where 1 means that a feature is selected, and 0 means that the certain feature is not selected:

$$\underbrace{010\ldots100}_{\text{length: }L=124251}.\qquad(10.4)$$

Initialization

In the first generation a random initial solution is generated by assigning each gene a value of 1 with probability p_{in}. The probability p_{in} controls number of selected features in the initial individual, it is problem dependent, and thus can be set according to domain knowledge.

Evaluation of the fitness function

The aim of the search is to maximize the classification accuracy while avoiding over-fitting. Thus we construct a fitness function based on accuracy, considering also as a penalization the proportion of the selected features to the length of the chromosome (individual). By combining both accuracy and number of selected features, we hope to achieve a trade-off between them and to avoid overfitting. In particular, the fitness f of individual I is computed as:

$$f(I) = A(I) - \frac{n_I}{L}, \tag{10.5}$$

where $A(I)$ is the classification accuracy and n_I is the number of features selected in I (the number of 1 s in the binary representation). We divide the number of selected features n_I by the total length of the chromosome to keep the two terms in Eq. (10.5) in [0, 1] and maintain a balance between them.

Mutation

mGA uses the uniform random mutation for variation by which each gene is modified with a probability p_m. In detail, for each gene a random number between 0 and 1 is generated following a uniform distribution; if this value is less than the given mutation probability p_m, the value of the gene is modified (from 1 to 0 or conversely).

In the following example, the second gene of individual I in the left is modified from 0 to 1 as the second random number generated is 0.001:

$$I = 0\mathbf{1} \ldots 10 \xrightarrow[rand():0.236, \mathbf{0.001}, \ldots, 0.385, 0.798]{} I' = 0\mathbf{0} \ldots 10. \tag{10.6}$$

Outline of mGA

The main steps of mGA are outlined in Algorithm 1. First, a random initial solution I is generated. Then, the following steps are repeated until the stopping criterion is met: (i) creation of an offspring I' by the application of mutation to the individual I representing the current solution, (ii) if the offspring has a higher fitness value, we keep it as the new current solution, otherwise we do not change the current solution. The process stops after a predefined number of iterations, denoted by $MaxGen$.

Parameters

mGA uses the following parameters: p_{in}: the probability of each gene being set to 1 when generating the random individual in the first generation, p_m: the probability used for uniform mutation and the number of generations ($MaxGen$).

Algorithm 10.1: Outline of mGA

Initialize random individual (I);
Evaluate I;
$nrGen = 0$;
while $nrgen < MaxGen$ **do**
    Apply mutation to $I \rightarrow I'$;
    Evaluate I';
    if $f(I') > f(I)$ (I' better than I) **then**
        $I = I'$;
    end if
    $nrgen + +$;
end while

Table 10.1 mGA Parameter settings

Parameter	p_m	p_{in}	$MaxGen$
Value	0.004	0.004	3000

10.2.4 Experimental Set-Up

The goal of our experiments was to compare the mGA with LASSO in terms of their ability to select relevant features. In order to objectively compare the selected feature sets, and to quantitatively assess their quality, we aimed to classify subjects according to the presence or absence of mild cognitive impairment (MCI) using the selected features. In other words: we evaluated both algorithms indirectly in context of a classification task. Next, we describe the details of our experimental protocol.

As measurements from the same subjects are not independent, we performed our experiments according to the leave-one-subject-out cross-validation schema. As our dataset contains measurements from 49 subjects, we had 49 rounds of cross-validation. In each round, the instances belonging to one of the subjects were used as *test data*, while the instances of the remaining 48 subjects were used as *training data*. We performed feature selection with both methods (i.e., the mGA and LASSO) using the training data.[1] Subsequently, both sets of selected features were evaluated.

For LASSO, we set the parameter λ to 0.005 in order to restrict the number of selected features to be around the number of instances (154 ± 5.1).[2] The parameter values used for mGA are presented in Table 10.1. The values of p_m and p_{in} were set empirically to limit the number of selected features (starting with approx. 500 and increasing only if better sets are generated) in order to avoid overfitting caused by selecting unnecessarily large feature sets.

In order to assess the quality of a selected feature set, we classified test instances with a simple 1-nearest neighbor [1] classifier with Euclidean distance based on the

---

[1] The accuracy for the fitness function of mGA was calculated solely on the training data. In particular we measured the accuracy of a nearest neighbor classifier in an *internal* 5-fold cross-validation on the training data.

[2] We note that $\lambda = 0.001$ and $\lambda = 0.0001$ led to very similar classification accuracy. For simplicity, we only show the results in case of $\lambda = 0.005$ in Sect. 10.3.

selected features only. That is: we restrict test and training instance vectors to the selected feature set, and for every test instance x we search for its *nearest neighbor* in the training data. With nearest neighbor of the test instance x we mean the training instance that is the closest to x according to the Euclidean distance. The predicted label of the test instance x is the label of its nearest neighbor.

For the task-based evaluation of feature sets in context of the classification according to MCI described above, in principle, one could use various classifiers such as recent variants of neural networks [40] or decision rules [51], however, we decided to use the nearest neighbor classifier, because our primary goal is to compare the set of selected features and nearest neighbor relies highly on the features. Furthermore, nearest neighbor is well-understood from the theoretical point of view and works surprisingly well in many cases [5, 8, 11].

To demonstrate how our classifiers perform compared to the chance level, we generated 100 000 random labeling with "coin-flipping" (50–50% chance of generating the label corresponding to the presence or absence of MCI), and calculated the accuracy values of these random classifications. The 95th percentile of this random classifier's accuracy-distribution is 56.8%. Therefore, we treat the classification as significantly better than random "coin-flipping" if its accuracy exceeds the threshold of 56.8%.

10.3   Results

Classification accuracy

The classification accuracy of 1-nearest neighbour classifier based on LASSO-selected and mGA-selected feature sets are presented in Fig. 10.2.

For both LASSO-selected and mGA-selected feature sets, the classification accuracy is significantly better than the accuracy of "coin-flipping". Furthermore, in this task, where the number of features is extremely high compared with the number of instances, the mGA-based feature selection clearly outperforms LASSO-based feature selection in terms of classification accuracy.

Fig. 10.2 Classification accuracy using features selected by LASSO and mGA

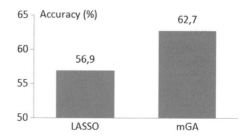

Table 10.2 Mean ± standard deviation of selected feature set sizes through the 49 rounds of cross-validation with the two methods

LASSO	mGA
154.3 ± 5.1	775.4 ± 249.4

Stability of the selected features

A good feature selection algorithm should result not only in good classification performance, but an interpretable feature set as well. The two approaches we consider in this study, LASSO and mGA, differ greatly in the number of selected features (see Table 10.2). The LASSO algorithm selects about 154 features with a low standard deviation, while the mGA algorithm chooses about 5 times more features with large standard deviation.

The selected feature sets can be examined from the point of view of stability, i.e., we can calculate how many times each feature was selected during the 49 rounds of cross-validation (Fig. 10.3a). As features describe connections between brain ROIs, we can also calculate how many times each ROI was selected through the cross-validation (Fig. 10.3b).

In terms of the stability of features, the difference between the two algorithms is undeniable (Fig. 10.3a). Clearly, the feature set selected by LASSO is very stable compared with the mGA-selected features, as there are 47 connections that were selected in at least 40 rounds (about 80%) out of all the 49 rounds of cross-validation, while in case of the mGA algorithm, there is no feature that was selected more than 6 times. Interestingly, the distinction between the two algorithms almost disappears if we consider the stability of selected ROIs. In Fig. 10.3b one can see that both algorithms identify a limited number (less than five) hub ROIs, that have considerably more selected connections, than the rest of the brain.

Runtime

In our experiments, the runtime of a single evaluation in mGA was ≈3.5 s on an Intel® Core™ i7-5820K CPU @ 3.30GHz × 12, 32 GB RAM, with the settings from Table 10.1. The total runtime is influenced by the parameter settings ($MaxGen$, p_m and p_{in}) as they control the number of evaluations and the number of selected features involved in the evaluation of the classification accuracy in (10.5). Using the MATLAB-implementation of LASSO, on average, ≈4 s were needed for the selection of features in each round of the cross-validation. As mGA requires multiple evaluations, the total runtime of LASSO is much lower than that of mGA.

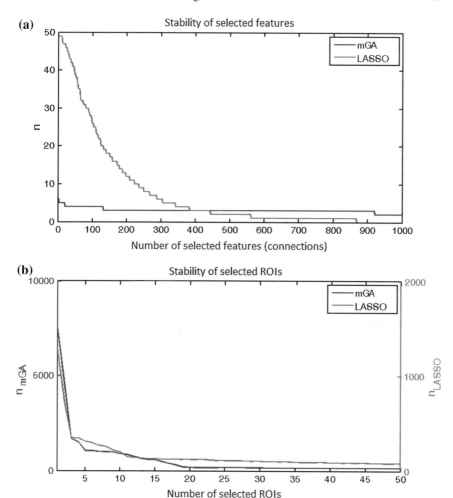

Fig. 10.3 a, The stability of selected features for LASSO and mGA. Considering the 49 rounds of leave-one-subject-out cross-validation, for each $n = 1, \ldots, 49$, we count how many features appear at least n-times among the selected features. The vertical axis shows n, while the horizontal axis displays the number of features that appear at least n-times among the selected features. **b**, The stability of selected ROIs for LASSO and mGA. Considering all 49 rounds of cross-validation, we count how many times each ROIs is selected in total (the selection of a feature corresponds to the selection of two ROIs; as several features are associated with the same ROI, a ROI may be selected multiple times: e.g., if both features $f_{1,2} = (r_1, r_2)$ and $f_{1,3} = (r_1, r_3)$ were selected in all the 49 rounds of cross-validation, and no other features were selected, ROI r_1 would appear $49 \times 2 = 98$ times in total, while r_2 and r_3 would appear 49 times in total). The left and right vertical axes show n_{mGA} and n_{LASSO}, while the horizontal axis shows how many ROIs were selected at least n_{mGA}-times and n_{LASSO}-times, respectively. (As mGA selects about 5 times more features in each round of the cross-validation, there is a scaling factor of 5 between the two vertical axes.)

10.4 Discussion

The results show that even in the case of extremely high number of features (number of features is more than 800 times higher than the number of instances), both LASSO and mGA algorithms are able to classify subjects according to presence or absence of MCI significantly better than "coin flipping". In terms of accuracy, mGA clearly outperformed LASSO. In contrast, the set of features selected by LASSO is substantially more stable. With respect to the stability of selected ROIs, in our experiments, the two algorithms resulted in very similar characteristics.

The differences regarding stability may be attributed to inherent properties of the algorithms: if two features are similarly useful for fitting the model to the data, but one of them is slightly better than the other, due to its regularization term, LASSO tends to select the better feature, while mGA may select any of them with similar probabilities.

The most frequently selected ROIs can be important from a clinical point of view as well. The top five ROIs of the two algorithms show a significant overlap (see Table 10.3).

The top 20% of ROIs are visualized in Fig. 10.4a, b. One can note that while the most important ROIs i.e. the hubs of the two methods are the same, the LASSO based map is more symmetric, i.e. it respects strong homotopic (inter-hemispheric) brain connections, while the mGA based map shows clear left hemisphere dominance. Most importantly, several out of the top 20% ROIs selected by both the LASSO and the mGA, have been reported in meta-studies examining Alzheimer's disease and MCI [38, 48].

Table 10.3 The top five selected ROIs in case of LASSO and mGA

	LASSO			mGA	
	ROI ID	Region		ROI ID	Region
1	143	Cuneal cortex	1	143	Cuneal cortex
2	144	Occipital pole, Lingual gyrus	2	144	Occipital pole, Lingual gyrus
3	147	Left lateral occipital cortex, superior division	3	145	Right precentral gyrus
4	149	Cerebellum	4	24	Left frontal pole
5	145	Right precentral gyrus	5	146	Frontal medial cortex, subcallossal cortex

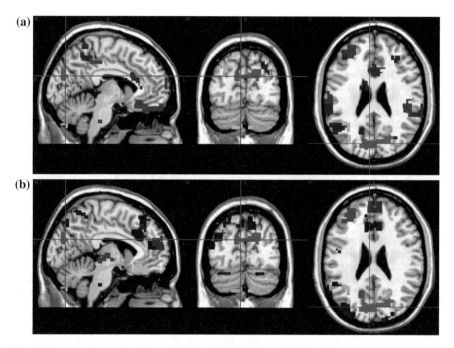

Fig. 10.4 The top 20% of selected ROIs using LASSO (**a**), and mGA (**b**). Warm colors represent more frequently chosen ROIs (hubs)

10.5   Conclusions

The analysis of brain imaging data is a challenging task, because of the high dimensionality. Selecting the relevant features is a key task of the analytic pipeline. We considered two algorithms, the classic LASSO-based feature selection, and a novel genetic algorithm (mGA) designed for the analysis of functional magnetic resonance imaging (fMRI) data. We compared them in context of the recognition of mild cognitive impairment (MCI) based on fMRI data. In terms of classification accuracy, the features selected by mGA outperformed the features selected by LASSO. According to our observations, the set of features selected by LASSO is more stable over multiple runs. Nevertheless, both methods provide meaningful information about the data, confirming the search potential of genetic algorithms and providing a starting point to further and deeper analyses of brain imaging data by heuristic methods.

Acknowledgements This work partially was supported by a grant of the Romanian National Authority for Scientific Research and Innovation, CNCS-UEFISCDI, project number PN-II-RU-TE-2014-4-2332 and the National Research, Development and Innovation Office (Hungary), project number: NKFIH 108947 K.

References

1. Altman, N.S.: An introduction to kernel and nearest-neighbor nonparametric regression. Am. Stat. **46**(3), 175–185 (1992)
2. Bassett, D.S., Sporns, O.: Network neuroscience. Nat. Neurosci. **20**(3), 353–364 (2017)
3. Blautzik, J., Keeser, D., Berman, A., Paolini, M., Kirsch, V., Mueller, S., Coates, U., Reiser, M., Teipel, S.J., Meindl, T.: Long-term test-retest reliability of resting-state networks in healthy elderly subjects and patients with amnestic mild cognitive impairment. J. Alzheimer's Dis. **34**(3), 741–754 (2013)
4. Blautzik, J., Vetter, C., Peres, I., Gutyrchik, E., Keeser, D., Berman, A., Kirsch, V., Mueller, S., Pöppel, E., Reiser, M., et al.: Classifying fmri-derived resting-state connectivity patterns according to their daily rhythmicity. NeuroImage **71**, 298–306 (2013)
5. Buza, K., Nanopoulos, A., Schmidt-Thieme, L.: Time-series classification based on individualised error prediction. In: 13th International Conference on Computational Science and Engineering, pp. 48–54. IEEE (2010)
6. Canuto, A.M.P., Nascimento, D.S.C.: A genetic-based approach to features selection for ensembles using a hybrid and adaptive fitness function. In: The 2012 International Joint Conference on Neural Networks (IJCNN), pp. 1–8 (2012). https://doi.org/10.1109/IJCNN.2012.6252740
7. Chang, E.I., Lippmann, R.P.: Using genetic algorithms to improve pattern classification performance. In: Proceedings of the 1990 Conference on Advances in Neural Information Processing Systems 3. NIPS-3, pp. 797–803. Morgan Kaufmann Publishers Inc., San Francisco, CA, USA (1990)
8. Chen, G.H., Nikolov, S., Shah, D.: A latent source model for nonparametric time series classification. In: Advances in Neural Information Processing Systems, pp. 1088–1096 (2013)
9. D'Alessandre, M., Vachtseyanos, G., Esteller, R., Echauz, J., Sewell, D., Litt, B.: A systematic approach to seizure prediction using genetic and classifier based feature selection. In: International Conference on Digital Signal Processing, DSP, vol. 2 (2002). https://doi.org/10.1109/ICDSP.2002.1028162
10. De Jong, K.: Evolutionary Computation: A Unified Approach. MIT Press, Bradford Book (2006)
11. Devroye, L., Gyorfi, L., Krzyzak, A., Lugosi, G.: On the strong universal consistency of nearest neighbor regression function estimates. Ann. Stat. 1371–1385 (1994)
12. Eiben, A.E., Smith, J.E.: Introduction to Evolutionary Computing, 2nd edn. Springer Publishing Company, Incorporated (2015). https://doi.org/10.1007/978-3-662-44874-8
13. de la Fraga, L.G., Coello Coello, C.A.: A review of applications of evolutionary algorithms in pattern recognition. In: Wang, P.S.P. (ed.) Pattern Recognition, Machine Intelligence and Biometrics, pp. 3–28. Springer Berlin, Heidelberg (2011). https://doi.org/10.1007/978-3-642-22407-2_1
14. Grundman, M., Petersen, R.C., Ferris, S.H., Thomas, R.G., Aisen, P.S., Bennett, D.A., Foster, N.L., Jack Jr., C.R., Galasko, D.R., Doody, R., et al.: Mild cognitive impairment can be distinguished from Alzheimer disease and normal aging for clinical trials. Arch. Neurol. **61**(1), 59–66 (2004)
15. Gwalani, H., Mittal, N., Vidyarthi, A.: Classification of brain tumours using genetic algorithms as a feature selection method (GAFS). In: ACM International Conference Proceeding Series, vol. 25–26, August (2016). https://doi.org/10.1145/2980258.2980318
16. Herculano-Houzel, S.: The human brain in numbers: a linearly scaled-up primate brain. Front. Hum. Neurosci. **3**, 31 (2009)
17. Hoerl, A.E., Kennard, R.W.: Ridge regression: biased estimation for nonorthogonal problems. Technometrics **12**(1), 55–67 (1970)
18. Holland, J.H.: Adaptation in Natural and Artificial Systems. MIT Press, Cambridge, MA, USA (1992)
19. de la Hoz, E., de la Hoz, E., Ortiz, A., Ortega, J., Martínez-Álvarez, A.: Feature selection by multi-objective optimisation: application to network anomaly detection by hierarchical self-organising maps. Knowl. Based Syst. **71**, 322–338 (2014). https://doi.org/10.1016/j.knosys.2014.08.013

20. Hyvärinen, J., Carlson, S., Hyvärinen, L.: Early visual deprivation alters modality of neuronal responses in area 19 of monkey cortex. Neurosci. Lett. **26**(3), 239–243 (1981)
21. de la Iglesia, B.: Evolutionary computation for feature selection in classification problems. Wiley Interdiscip. Rev. Data Min. Knowl. Discov. **3**(6), 381–407 (2013). https://doi.org/10. 1002/widm.1106
22. Jalili, M.: Graph theoretical analysis of Alzheimer's disease: discrimination of AD patients from healthy subjects. Inf. Sci. **384** (2017). https://doi.org/10.1016/j.ins.2016.08.047
23. Ji, Y., Bu, X., Sun, J., Liu, Z.: An improved simulated annealing genetic algorithm of EEG feature selection in sleep stage. In: 2016, Asia-Pacific Signal and Information Processing Association Annual Summit and Conference. APSIPA 2016 (2017). https://doi.org/10.1109/ APSIPA.2016.7820683
24. Kaya, Y., Pehlivan, H.: Feature selection using genetic algorithms for premature ventricular contraction classification. In: 2015 9th International Conference on Electrical and Electronics Engineering (ELECO), pp. 1229–1232 (2015). https://doi.org/10.1109/ELECO.2015.7394628
25. Khan, A., Baig, A.: Multi-objective feature subset selection using non-dominated sorting genetic algorithm. J. Appl. Res. Technol. **13**(1), 145–159 (2015). https://doi.org/10.1016/ S1665-6423(15)30013-4
26. Kharrat, A., Halima, M., Ben Ayed, M.: MRI brain tumor classification using Support Vector Machines and meta-heuristic method. In: International Conference on Intelligent Systems Design and Applications, ISDA, vol. 2016, June (2016). https://doi.org/10.1109/ISDA.2015. 7489271
27. Lichtman, J.W., Denk, W.: The big and the small: challenges of imaging the brain's circuits. Science **334**(6056), 618–623 (2011)
28. Meszlényi, R., Peska, L., Gál, V., Vidnyánszky, Z., Buza, K.: Classification of fmri data using dynamic time warping based functional connectivity analysis. In: Signal Processing Conference (EUSIPCO), 2016 24th European, pp. 245–249. IEEE (2016)
29. Meszlényi, R., Peska, L., Gál, V., Vidnyánszky, Z., Buza, K.A.: A model for classification based on the functional connectivity pattern dynamics of the brain. In: Third European Network Intelligence Conference, pp. 203–208 (2016)
30. Meszlényi, R.J., Hermann, P., Buza, K., Gál, V., Vidnyánszky, Z.: Resting state fmri functional connectivity analysis using dynamic time warping. Front. Neurosci. **11**, 75 (2017)
31. Michalewicz, Z., Dasgupta, D. (eds.): Evolutionary Algorithms in Engineering Applications, 1st edn. Springer-Verlag New York Inc, Secaucus, NJ, USA (1997)
32. Noori, F., Qureshi, N., Khan, R., Naseer, N.: Feature selection based on modified genetic algorithm for optimization of functional near-infrared spectroscopy (fNIRS) signals for BCI. In: 2016 2nd International Conference on Robotics and Artificial Intelligence, ICRAI 2016 (2016). https://doi.org/10.1109/ICRAI.2016.7791227
33. Raymer, M.L., Punch, W.F., Goodman, E.D., Kuhn, L.A., Jain, A.K.: Dimensionality reduction using genetic algorithms. IEEE Trans. Evol. Comput. **4**(2), 164–171 (2000). https://doi.org/ 10.1109/4235.850656
34. Richiardi, J., Altmann, A., Milazzo, A.C., Chang, C., Chakravarty, M.M., Banaschewski, T., Barker, G.J., Bokde, A.L., Bromberg, U., Büchel, C., et al.: Correlated gene expression supports synchronous activity in brain networks. Science **348**(6240), 1241–1244 (2015)
35. Rosa, M.J., Portugal, L., Hahn, T., Fallgatter, A.J., Garrido, M.I., Shawe-Taylor, J., Mourao-Miranda, J.: Sparse network-based models for patient classification using fmri. Neuroimage **105**, 493–506 (2015)
36. Sakoe, H., Chiba, S.: Dynamic programming algorithm optimization for spoken word recognition. IEEE Trans. Acoust. Speech Signal Process. **26**(1), 43–49 (1978)
37. Sanchez, E., Squillero, G., Tonda, A.: Industrial Applications of Evolutionary Algorithms. Springer-Verlag Berlin Heidelberg (2012). https://doi.org/10.1007/978-3-642-27467-1
38. Schroeter, M.L., Stein, T., Maslowski, N., Neumann, J.: Neural correlates of Alzheimer's disease and mild cognitive impairment: a systematic and quantitative meta-analysis involving 1351 patients. Neuroimage **47**(4), 1196–1206 (2009)

39. da Silva, S.F., Ribeiro, M.X., João do E.S. Batista Neto, J., Traina-Jr., C., Traina, A.J.: Improving the ranking quality of medical image retrieval using a genetic feature selection method. Decis. Support Syst. **51**(4), 810 – 820 (2011). https://doi.org/10.1016/j.dss.2011.01.015. (Recent Advances in Data, Text, and Media Mining & Information Issues in Supply Chain and in Service System Design)
40. Stańczyk, U.: On performance of DRSA-ANN classifier. In: International Conference on Hybrid Artificial Intelligence Systems, pp. 172–179. Springer (2011)
41. Tajik, M., Rehman, A., Khan, W., Khan, B.: Texture feature selection using GA for classification of human brain MRI scans. Lecture Notes in Computer Science, vol. 9713. Springer International Publishing, Switzerland (2016)
42. Tibshirani, R.: Regression shrinkage and selection via the lasso. J. R. Stat. Soc. Ser. B (Methodol.)267–288 (1996)
43. White, J.G., Southgate, E., Thomson, J.N., Brenner, S.: The structure of the nervous system of the nematode caenorhabditis elegans. Philos. Trans. R. Soc. Lond. B Biol. Sci. **314**(1165), 1–340 (1986)
44. Winkler, S.M., Affenzeller, M., Jacak, W., Stekel, H.: Identification of cancer diagnosis estimation models using evolutionary algorithms: a case study for breast cancer, melanoma, and cancer in the respiratory system. In: Proceedings of the 13th Annual Conference Companion on Genetic and Evolutionary Computation, GECCO'11, pp. 503–510. ACM, New York, NY, USA (2011). https://doi.org/10.1145/2001858.2002040
45. Xue, B., Zhang, M., Browne, W.N., Yao, X.: A survey on evolutionary computation approaches to feature selection. IEEE Trans. Evol. Comput. **20**(4), 606–626 (2016). https://doi.org/10.1109/TEVC.2015.2504420
46. Yaka, R., Yinon, U., Rosner, M., Wollberg, Z.: Pathological and experimentally induced blindness induces auditory activity in the cat primary visual cortex. Exp. Brain Res. **131**(1), 144–148 (2000)
47. Yang, J., Honavar, V.G.: Feature subset selection using a genetic algorithm. IEEE Intell. Syst. **13**(2), 44–49 (1998). https://doi.org/10.1109/5254.671091
48. Yang, J., Pan, P., Song, W., Huang, R., Li, J., Chen, K., Gong, Q., Zhong, J., Shi, H., Shang, H.: Voxelwise meta-analysis of gray matter anomalies in Alzheimer's disease and mild cognitive impairment using anatomic likelihood estimation. J. Neurol. Sci. **316**(1), 21–29 (2012)
49. Ye, C.Q., Poo, M.M., Dan, Y., Zhang, X.H.: Synaptic mechanisms of direction selectivity in primary auditory cortex. J. Neurosci. **30**(5), 1861–1868 (2010)
50. Yoshor, D., Bosking, W.H., Ghose, G.M., Maunsell, J.H.: Receptive fields in human visual cortex mapped with surface electrodes. Cereb. Cortex **17**(10), 2293–2302 (2007)
51. Zielosko, B., Chikalov, I., Moshkov, M., Amin, T.: Optimization of decision rules based on dynamic programming approach. In: Faucher, C., Jain, L.C. (eds.) Innovations in Intelligent Machines-4: Recent Advances in Knowledge Engineering, pp. 369–392. Springer (2014)
52. Zuo, X.N., Anderson, J.S., Bellec, P., Birn, R.M., Biswal, B.B., Blautzik, J., Breitner, J.C., Buckner, R.L., Calhoun, V.D., Castellanos, F.X., et al.: An open science resource for establishing reliability and reproducibility in functional connectomics. Sci. Data **1** (2014)

Chapter 11
Shape Descriptions and Classes of Shapes. A Proximal Physical Geometry Approach

James Francis Peters and Sheela Ramanna

Abstract This chapter introduces the notion of classes of shapes that have descriptive proximity to each other in planar digital 2D image object shape detection. A finite planar shape is planar region with a boundary (shape contour) and a non-empty interior (shape surface). The focus here is on the triangulation of image object shapes, resulting in maximal nerve complexes (MNCs) from which shape contours and shape interiors can be detected and described. An MNC is collection of filled triangles (called 2-simplexes) that have a vertex in common. Interesting MNCs are those collections of 2-simplexes that have a shape vertex in common. The basic approach is to decompose an planar region containing an image object shape into 2-simplexes in such a way that the filled triangles cover either part or all of a shape. After that, an unknown shape can be compared with a known shape by comparing the measurable areas of a collection of 2-simplexes covering both known and unknown shapes. Each known shape with a known triangulation belongs to a class of shapes that is used to classify unknown triangulated shapes. Unlike the conventional Delaunay triangulation of spatial regions, the proposed triangulation results in simplexes that are filled triangles, derived by the intersection of half spaces, where the edge of each half space contains a line segment connected between vertices called

This research has been supported by the Natural Sciences and Engineering Research Council of Canada (NSERC) discovery grants 194376, 185986 and Instituto Nazionale di Alta Matematica (INdAM) Francesco Severi, Gruppo Nazionale per le Strutture Algebriche, Geometriche e Loro Applicazioni grant 9 920160 000362, n.prot U 2016/000036.

J.F. Peters (✉) · S. Ramanna
Computational Intelligence Laboratory, ECE Department, University of Manitoba, Winnipeg, MB R3T 5V6, Canada
e-mail: James.Peters3@umanitoba.ca

J.F. Peters
Faculty of Arts and Sciences, Department of Mathematics, Adiyaman University, 02040 Adiyaman, Turkey

S. Ramanna
Department of Applied Computer Science, University of Winnipeg, Winnipeg, MB R3B 2E9, Canada
e-mail: s.ramanna@uwinnipeg.ca

sites (generating points). A straightforward result of this approach to image geometry is a rich source of simple descriptions of plane shapes of image objects based on the detection of nerve complexes that are maximal nerve complexes or MNCs. The end result of this work is a proximal physical geometric approach to detecting and classifying image object shapes.

Keywords Class of shapes · Descriptive proximity · Physical geometry · Shape detection · 2-simplex · Triangulation

11.1 Introduction

This chapter introduces a proximal physical geometric approach to detecting and classifying image object shapes. K. Borsuk was one of the first to suggest studying sequences of plane shapes in his theory of shapes [6]. Both spatial and descriptive forms of representation will be considered [30]. *Spatial representation* can be considered in two ways: (i) one where a space is given and its characteristics are studied via geometry and topology (ii) a space is approximated using some form of tool [17]. *Descriptive representation* starts with probe functions that map features of objects to numbers in $\mathbb{R}^n$ [30]. In a descriptive representation, the simplicial complexes are a result of nerve constructions of observations (objects) in the feature space. A nerve $N(C)$ in a finite collection of sets C is a simplicial complex with vertices of sets in C and with simplices corresponding to all non-empty intersections among these sets. Various forms of geometric nerves are usually collections known as simplicial complexes in a normed linear space (for details see, e.g., [13], [30, Sect. 1.13]).

In geometry and topology, a **simplex** is the convex hull (smallest closed convex set) which contains a nonempty set of vertices [4, Sect. I.5, p. 6]. A *convex hull* of a set of vertices A (denoted by conv A) in n dimensions is the intersection of all convex sets containing conv A [43]. A *convex set* A contains all points on each line segment drawn between any pair of points contained in A. For example, a 0-simplex is a vertex, a 1-simplex is a straight line segment and a 2-simplex is a triangle that includes the plane region which it bounds (see Fig. 11.1). In general, a k-simplex is a polytope with $k + 1$ vertices. A *polytope* is an n-dimensional point set P that is an intersection of finitely many closed half spaces in the Euclidean space $\mathbb{R}^n$. So, for example, a 2-simplex is a 2-dimensional polytope (i.e., a filled triangle) in the Euclidean plane represented by $\mathbb{R}^2$. A collection of simplexes is called a **simplicial complex** (briefly, *complex*).

Delaunay triangulations, introduced by B.N Delone (Delaunay) [12], represent discrete triangular-shaped approximation of continuous space. This representation partitions the space into regions. This partitioning (or decomposition) is made possible by a popular method called the Voronoï diagram [40]. A variation of Edelsbrunner–Harer nerves which are collections of Voronoï regions and maximal nerve complexes also called maximal nucleus clusters and its application in visual arts can be found in [36]. A proximity space [10] setting for MNCs makes it possible to investigate the strong closeness of subsets in MNCs as well as the spatial and descriptive closeness of MNCs themselves.

Unlike the conventional Delaunay triangulation of spatial regions, the proposed triangulation introduced in this chapter results in collections of 2-simplexes that are filled triangles, derived by the intersection of half spaces, where the edge of each half space contains a line segment connected between vertices called sites (generating points). Each MNC consists of a central simplex called the **nucleus**, which is surrounded by adjacent simplexes. The description of MNCs starts with easily determined features of MNC filled triangles, namely, centroids, areas, lengths of sides. Also of interest are the number of filled triangles that have the MNC nucleus as a common vertex. Notice that a raster colour image MNC nucleus is a pixel with measurable features such as diameter and colour channel intensities. Basically, a *plane shape* such as a rabbit shadow shape is a hole with varying contour length and varying numbers of points in its interior. Another interesting feature is Rényi entropy [36–38] which can be used to measure the information level of these nerve complexes. The setting for image object shapes is a descriptive proximity space that facilitates comparison and classification of shapes that are descriptively close to each other. A practical application of the proposed approach is the study of the characteristics of surface shapes. This approach leads to other applications such as the detection of object shapes in sequences of image frames in videos. This chapter highlights the importance of strong descriptive proximities between triangulated known and unknown shapes in classifying shapes in digital images. For an advanced study of triangulated image object shapes, see, e.g., M.Z. Ahmad and J.F. Peters [2] and J.F. Peters [34]. For an advanced study of descriptive proximities that are relevant in image object shape detection, see A. Di Concilio, C. Guadagni, J.F. Peters and S. Ramanna [11].

11.2   Introduction to Simplicial Complexes

Shapes can be quite complex when they are found in digital images. By covering a part or all of a digital image with simplexes, we simplify the problem of describing object shapes, thanks to the known geometric features of either individual simplices or simplicial complexes. An important form of simplicial complex is a collection

Fig. 11.1 k-simplices

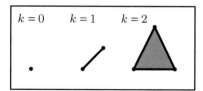

of simplexes called a **nerve**. The study of nerves was introduced by P. Alexandroff [3], elaborated by K. Borsuk [5], J. Leray [20], and a number of others such as M. Adamaszek et al. [1], E.C. de Verdière et al. [39], H. Edelsbrunner and J.L. Harer [14]. The problem of describing complex shapes is further simplified by extracting feature values from nerves that are embedded in simplicial complexes covering spatial regions. This is essentially a point-free geometry approach introduced by [32].

K. Borsuk also observed that every polytope can be decomposed into a finite sum of elementary simplexes, which he called brics. A **polytope** is the intersection of finitely many closed half spaces [42]. This leads to a simplicial complex K covered by simplexes $\Delta_1, \ldots, \Delta_n$ such that the nerve of the decomposition is the same as K [5]. Briefly, a **simplex** $\Delta(S)$ is the convex hull of a set of points S, i.e., the smallest convex set containing S. Simplexes in this chapter are restricted to vertices, line segments and filled triangles in the Euclidean plane, since our main interest is in the extraction of features of simplexes superimposed on planar digital images.

A planar simplicial complex K is a **nerve**, provided the simplexes in K have a common intersection (called the nucleus of the nerve). A nerve of a simplicial complex K (denoted by $\mathrm{Nrv}\,K$) in the triangulation of a plane region is defined by

$$\mathrm{Nrv}\,K = \left\{ \Delta \subseteq K : \bigcap \Delta \neq \varnothing \right\} \ \text{(Nerve)}$$

In other words, the simplexes in a nerve have proximity to each other, since they share the nucleus. The **nucleus** of a nerve is the set of points common to the simplexes in a nerve. Nerves can be constructed either spatially in terms of contiguous simplexes with a common nucleus or descriptively in terms of simplexes with parts that have descriptions that match the description of the nucleus. In the descriptive case, the nucleus of a nerve $\mathrm{Nrv}\,K$ equals the descriptive intersection of the simplexes in the nerve, i.e., the set of all points common in nerve simplexes and that have matching descriptions. In either case, the nucleus is a source of information common to the simplexes of a nerve. Examples of a pair of adjacent nerves in the triangulation of a digital image are shown in Fig. 11.2.

11.2.1 Examples: Detecting Shapes from Simplicial Complexes

A **nerve** is a simplicial complex in which the simplexes have have a common vertex, i.e., the simplexes of a nonempty intersection. If the simplexes in a nerve are convex sets, then the nerve is homotopy equivalent to the union of the simplexes. For an expository introduction to Borsuk's notion of shapes, see K. Borsuk and J. Dydak [7]. Borsuk's initial study of shapes has led to a variety of applications in science (see, e.g., the shape of capillarity droplets in a container by F. Maggi and C. Mihaila [22] and shapes of 2D water waves and hydraulic jumps by M.A. Fontelos, R. Lecaros,

Fig. 11.2 Overlapping
nerves

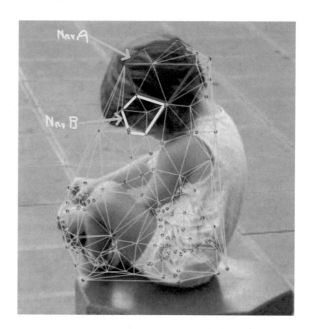

J.C. López-Rios and J.H. Ortega [16]). Shapes from a physical geometry perspective
with direct application in detecting shapes of image objects are introduced in [32].

Image object shape detection and object class recognition are of great interest in
Computer Vision. For example, basic shape features can be represented by boundary
fragments and shape appearance represented by patches of an auto shadow shape
in a traffic video frame. This is the basic approach to image object shape detection
by A. Opelt, A. Pinz and A. Zisserman in [24]. Yet another recent Computer Vision
approach to image object shape detection reduces to the problem of finding the
contours of an image object, which correspond to object boundaries and symmetry
axes. This is the approach suggested by I. Kokkinos and A. Yuille in [19]. We now
give a simple example of shape detection by composition from simplicial complexes.

Example 11.1 Let *S* be a set of vertices as shown in Fig. 11.3a. A shape contour
is constructed by drawing straight line segments between neighbouring vertices as
shown in Fig. 11.3b. In this case, the contour resembles a rabbit, a simplicial complex
constructed from a sequence of connected line segments. Then selected vertices are

Fig. 11.3 Simplex vertices (**a**)
and simplicial complex (**b**)

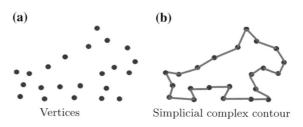

Vertices   Simplicial complex contour

Fig. 11.4 Augmented
simplicial complex (**a**) and
one of its nerves (**b**)

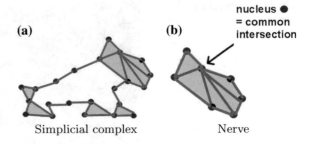

Simplicial complex   Nerve

triangulated to obtain three sets of filled triangles, each set with a common vertex (tail, feet and head in Fig. 11.4a). The triangulated head of the rabbit is an example of a nerve containing simplexes that have a common • as shown in Fig. 11.4b. ∎

Algorithm 11.1: Basic 2-simplex-based shape detection method

Input   : Read digital image Img containing shape sh A.
Output: MNCs on triangulated shapes sh A.
1 *Identify vertices V in sh Img that includes bdy(sh$A$) vertices*;
2 /* **N.B.** Vertices in bdy(sh$A$) and int(sh$A$) lead to shape complex shcx A*/ ;
3 *Triangulate V on Img (constructs cx Img)*;
4 *Identify nerve complexes on cx Img*;
5 *Look for maximal nerve complexes NrvK(p) (**MNCs**) on cx Img*;
6 /* **N.B.** Focus on NrvK(p) with vertex $p \in$ shcx A */ ;
7 *Compare each NrvK(p) on sh A with known triangulated shapes*;
8 /* *i.e.,* Compare shape contours and interiors. */ ;
9 *Check if NrvK(p) is descriptively close to nerves on shapes*;
10 /* *i.e.,* Classify sample shapes based on detected descriptively near nerves. */ ;

The basic steps in a 2-simplex-based shape detection method are represented visually in Figs. 11.3 and 11.4. Let cx$K$, sh$A$ denote **complex** K and **shape** A, respectively. Further, let bdy(sh$A$), int(sh$A$) denote the set of points in **boundary** of sh$A$ and the set of points in the **interior** of sh$A$, respectively. A triangulated shape constructs a shape complex A (denoted by shcx A). That is, a **shape complex** is a collection of 2-simplexes covering a shape. A shape complex shcx A is a cover of a shape sh$A$, provided

sh$A \subseteq$ shcx A, (shape complex shcx A covers (contains) the shape A).

Let p be a vertex in complex sh$A$ and let Nrv$A(p)$ be a nerve complex in which p is the vertex common to the filled triangles (2-simplexes) in the nerve. The basic shape detection steps are summarized in Algorithm 11.1.

In the study of object shapes in a particular image, a good practice is to consider each object shape as a member of a class of shapes. A *class of shapes* is a set of shapes with matching features. The perimeters of clusters of overlapping nerves in triangulated images provide good hunting grounds for shapes (see, e.g., the emergence of a head shape in the clusters of overlapping nerves in Fig. 11.2). Then image object shape detection reduces to checking whether the features of a particular image object shape match the features of a representative in a known class of shapes. In other words, a particular shape A is a member of a known class $\mathscr{C}$, provided the feature values of shape A match up with feature values of a representative shape in class $\mathscr{C}$. For example, an obvious place to look for a comparable shape feature is shape perimeter length. Similar shapes have similar perimeter length. If we add shape edge colour to the mix of comparable shape features, then colour image object shapes start dropping into different shape classes, depending on the shape perimeter length and perimeter edge colour of each shape.

11.3 Preliminaries of Proximal Physical Geometry

This section briefly introduces an approach to the detection of image object shape geometry, descriptive proximity spaces useful in reasoning about descriptions of image object shapes, and Edelsbrunner–Harer nerve complexes.

11.3.1 Image Object Shape Geometry

In the context of digital images, computational geometry provides a basis for the construction and analysis of various types of mesh overlays on images. In this chapter, the focus is on a variant of **Delaunay triangulation**, which is a covering of a digital image with filled triangles with non-intersecting interiors. A filled triangle is defined in terms of the boundary and the interior of set.

Let $A \, \delta \, B$ indicate that the nonempty sets A and B are near (close to) each other in a space X. The **boundary** of A (denoted bdy$A$) is the set of all points that are near A and near A^c [23, Sect. 2.7, p. 62]. The **closure** of A (denoted by cl$A$) is defined by

$$\text{cl}A = \{x \in X : x \, \delta \, A\} \text{ (Closure of } A \text{)}.$$

An important structure is the **interior** of A (denoted int$A$), defined by int$A$ = cl$A$ − bdy$A$. Let p, q, r be points in the space X. A **filled triangle** (denoted by fil$\Delta(pqr)$) is defined by

$$\text{fil}\Delta(pqr) = \text{int}\Delta(pqr) \cup \text{bdy } \Delta(pqr) \text{(filled triangle)}.$$

When it is clear from the context that simplex triangles are referenced, we write $\Delta(pqr)$ or ΔA or simply Δ, instead of fil$\Delta(pqr)$. Since image object shapes tend to irregular, the shapes of known geometric shapes in a filled triangle covering of an image give a more precise view of the shapes of image objects. Thanks to the known properties of triangles (e.g., uniform shape, sum of the interior angles, perimeter, area, lengths of sides), object shapes can be described in a very accurate fashion. A sample digital image geometry algorithm useful in triangulating a digital image is given in Algorithm 11.2.

Algorithm 11.2: Digital Image Geometry via Mesh Covering Image

Input   : Read digital image img.
Output: Mesh $\mathcal{M}$ covering an image.
1 $MeshSite \leftarrow MeshGeneratingPointType$;
2 $img \longmapsto MeshSitePointCoordinates$;
3 $S \leftarrow MeshSitePointCoordinates$;
4 /* S contains MeshSitePointType coordinates used as mesh generating points (seeds or sites). */ ;
5 $Vertices \leftarrow \{p, q, r\} \subset S$;
6 /* $Vertices$ contains sets of three neighbouring points in S. */ ;
7 $\mathcal{M} \leftarrow \bigcup fil\Delta(pqr)$ for all neighbouring $p, q, r \in Vertices$;
8 $\mathcal{M} \longmapsto img$;
9 /* Use $\mathcal{M}$ to gain information about image geometry. */ ;

11.3.2 Descriptions and Proximities

This section briefly introduces two basic types of proximities, namely, traditional *spatial proximity* and the more recent *descriptive proximity* in the study of computational proximity [30]. Nonempty sets that have **spatial proximity** are close to each other, either asymptotically or with common points. Sets with points in common are strongly proximal. Nonempty sets that have **descriptive proximity** are close, provided the sets contain one or more elements that have matching descriptions. A commonplace example of descriptive proximity is a pair of paintings that have matching parts such as matching facial characteristics, matching eye, hair, skin colour, or matching nose, mouth, ear shape. Each of these proximities has a strong form. A **strong proximity** embodies a special form of tightly twisted nextness of nonempty sets. In simple terms, this means sets that share elements, have strong prox-

imity. For example, the shaded parts **◖**, **◣**, **◗** in Fig. 11.5 contain points in the Euclidean plane that are shared 5by the tightly twisted (overlapping) shapes A and B.

The following example is another illustration of strongly near shapes.

Example 11.2 The pair of triangles A and B in Fig. 11.6 are strongly near, since these triangles have a common edge. The assumption is that interior of each triangle is filled with points from the underlying image.   ■

Proximities are nearness relations. In other words, a *proximity* between nonempty sets is a mathematical expression that specifies the closeness of the sets. A **proximity space** results from endowing a nonempty set with one or more proximities. Typically, a proximity space is endowed with a common proximity such as the proximities from Čech [9], Efremovič [15], Lodato [21], and Wallman [41], or the more recent descriptive proximity [25–27].

A pair of nonempty sets in a proximity space are *near (close to each other)*, provided the sets have one or more points in common or each set contains one or more points that are sufficiently close to each other. Let X be a nonempty set, $A, B, C \subset X$. Čech [9] introduced axioms for the simplest form of proximity δ_C, which satisfies

Čech Proximity Axioms [9, Sect. 2.5, p. 439]

(P1) $\varnothing \not\delta A, \forall A \subset X.$
(P2) $A \, \delta \, B \Leftrightarrow B \delta A.$
(P3) $A \cap B \neq \varnothing \Rightarrow A \delta B.$
(P4) $A \, \delta \, (B \cup C) \Leftrightarrow A \, \delta \, B$ or $A \, \delta \, C.$ ■

The proximity δ_L satisfies the Čech proximity axioms and

Lodato Proximity Axiom [21]

Fig. 11.5 Tightly twisted, overlapping shapes

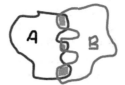

Fig. 11.6 Strongly near filled triangles

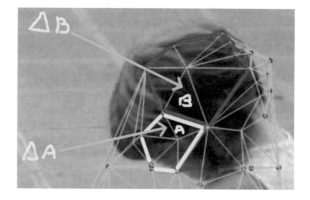

(a)   **(b)**

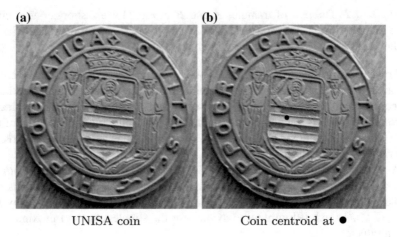

UNISA coin   Coin centroid at ●

Fig. 11.7 UNISA coin image (**a**) and UNISA coin image centroid ● (**b**)

(**P5**$_L$) $A \ \delta_L \ B$ and $\{b\} \ \delta_L \ C$ for each $b \in B \ \Rightarrow \ A \ \delta_L \ C$.   ∎

11.3.3 Descriptive Proximities

In the run-up to a close look at extracting features of triangulated image objects, we
first consider descriptive proximities. There are two basic types of *object features*,
namely, *object characteristic* and *object location*. For example, an object characteristic feature of a picture point is colour. And object location of a region is the geometric
centroid, which is the center of mass of the region. Let X be a set of points in a $n \times m$
rectangular 2D region containing points with coordinates (x_i, y_i), $i = 1, \ldots, n$ in
the Euclidean plane. Then, for example, the coordinates x_c, y_c of the centroid of a
2D region in the Euclidean space $\mathbb{R}^2$ are

$$x_c = \frac{1}{n} \sum_{i=1}^{n} x_i, \ y_c = \frac{1}{m} \sum_{i=1}^{m} y_i.$$

Example 11.3 (*Digital image centroid*). In Fig. 11.7, the black dot ● indicates the
location of a digital image centroid. In Fig. 11.7a, the digital image shown a UNISA
coin from a 1982 football tournament in Salerno, Italy. In Fig. 11.7b, the location of
the centroid of the UNISA coin is identified with ●. ∎

Each object characteristic feature of a concrete point or region has a real value
that is extracted by a probe ϕ which is a mapping $\phi : X \longrightarrow \mathbb{R}$. Let 2^X be the family
of subsets of X in the Euclidean plane $\mathbb{R}^2$, A a plane region in 2^X and let (x, y) be

the coordinates of the center of mass (centroid) of A. Also, let p be a point with coordinates (x_1, y_1). Each **object location** extracted by a location probe ϕ_L is a mapping

$$\phi_L : 2^X \longrightarrow \mathbb{R} \times \mathbb{R} \text{ (Location of a region centroid)},$$

$$e.g., \; phi_L(A) = (x, y) \text{ (Region } A \text{ centroid Coordinates)}.$$

$$\phi_L : X \longrightarrow \mathbb{R} \times \mathbb{R} \text{ (Location of a point)},$$

$$e.g., \; phi_L(p) = (x_1, y_1) \text{ (Coordinates of point } p).$$

This means that each planar point or singleton set region with n-features has a description defined by a feature vector in an $n + 2$-dimensional feature space. Let $\Phi(A), \Phi(p)$ denote feature vectors for a singleton set region A and point p, respectively, in a space X. Then

$$\Phi(A) = (\phi_L(A), \phi_1(A), \ldots, \phi_n(A)) \text{ (Region feature vector with location)},$$

$$e.g., \; \Phi(A) = ((x_M, y_M), \phi_1(A), \ldots, \phi_n(A)) \text{ (Region } A \text{ feature vector)},$$

$$\text{with the centroid of } A \text{ at location } ((x_M, y_M)).$$

$$\Phi(p) = (\phi_L(p), \phi_1(p), \ldots, \phi_n(p)) \text{ (Region feature vector that includes location)},$$

$$e.g., \; \Phi(p) = ((x, y), \phi_1(p), \ldots, \phi_n(p)) \text{ (Point , } p \text{ feature vector)},$$

$$\text{with the centroid of } p \text{ at location } ((x, y)).$$

Descriptive proximities resulted from the introduction of the descriptive intersection pairs of nonempty sets.

Descriptive Intersection [27] **and** [23, Sect. 4.3, p. 84].

(Φ) $\Phi(A) = \{\Phi(x) \in \mathbb{R}^n : x \in A\}$, set of feature vectors.

($\underset{\Phi}{\cap}$) $A \underset{\Phi}{\cap} B = \{x \in A \cup B : \Phi(x) \in \Phi(A) \& \in \Phi(x) \in \Phi(B)\}$. ∎

The descriptive proximity δ_Φ was introduced in [25–27]. Let $\Phi(x)$ be a feature vector for $x \in X$, a nonempty set of non-abstract points such as picture points. $A \, \delta_\Phi \, B$ reads A is descriptively near B, provided $\Phi(x) = \Phi(y)$ for at least one pair of points, $x \in A, y \in B$. The proximity δ in the Čech, Efremovič, and Wallman proximities is replaced by δ_Φ. Then swapping out δ with δ_Φ in each of the Lodato axioms defines a descriptive Lodato proximity that satisfies the following axioms.

Descriptive Lodato Axioms [28, Sect. 4.15.2]

(dP0) $\varnothing \, \not{\delta_\Phi} \, A, \forall A \subset X.$

(dP1) $A \, \delta_\Phi \, B \Leftrightarrow B \, \delta_\Phi \, A.$

(dP2) $A \underset{\Phi}{\cap} B \neq \varnothing \Rightarrow A \, \delta_\Phi \, B.$

(dP3) $A \, \delta_\Phi \, (B \cup C) \Leftrightarrow A \, \delta_\Phi \, B \text{ or } A \, \delta_\Phi \, C.$

(dP4) $A \, \delta_\Phi \, B \text{ and } \{b\} \, \delta_\Phi \, C \text{ for each } b \in B \Rightarrow A \, \delta_\Phi \, C.$ ∎

Nonempty sets A, B in a proximity space X are *strongly near* (denoted $A \overset{\wedge}{\delta} B$), provided the sets share points. Strong proximity $\overset{\wedge}{\delta}$ was introduced in [29, Sect. 2] and completely axiomatized in [35] (see, also, [18, Sect. 6 Appendix]). The descriptive strong proximity $\overset{\wedge}{\delta}_\varphi$ is the descriptive counterpart of $\overset{\wedge}{\delta}$.

Definition 11.1 Let X be a topological space, $A, B, C \subset X$ and $x \in X$. The relation $\overset{\wedge}{\delta}_\varphi$ on the family of subsets 2^X is a *descriptive strong Lodato proximity*, provided it satisfies the following axioms.

Descriptive Strong Lodato proximity [28, Sect. 4.15.2]

(dsnN0) $\varnothing \ \overset{\wedge}{\not\delta}_\varphi \ A, \forall A \subset X$, and $X \ \overset{\wedge}{\delta}_\varphi \ A, \forall A \subset X$

(dsnN1) $A \ \overset{\wedge}{\delta}_\varphi \ B \Leftrightarrow B \ \overset{\wedge}{\delta}_\varphi \ A$

(dsnN2) $A \ \overset{\wedge}{\delta}_\varphi \ B \Rightarrow A \underset{\varphi}{\cap} B \neq \varnothing$

(dsnN3)   If $\{B_i\}_{i \in I}$ is an arbitrary family of subsets of X and $A \ \overset{\wedge}{\delta}_\varphi \ B_{i*}$ for some $i^* \in I$ such that $\mathrm{int}(B_{i*}) \neq \varnothing$, then $A \ \overset{\wedge}{\delta}_\varphi \ (\bigcup_{i \in I} B_i)$

(dsnN4) $\mathrm{int}A \underset{\varphi}{\cap} \mathrm{int}B \neq \varnothing \Rightarrow A \ \overset{\wedge}{\delta}_\varphi \ B$ ∎

When we write $A \ \overset{\wedge}{\delta}_\varphi \ B$, we read A is *descriptively strongly near* B. The notation $A \ \overset{\wedge}{\not\delta}_\varphi \ B$ reads A is not descriptively strongly near B. For each *descriptive strong proximity*, we assume the following relations:

(dsnN5) $\Phi(x) \in \Phi(\mathrm{int}(A)) \Rightarrow x \ \overset{\wedge}{\delta}_\varphi \ A$

(dsnN6) $\{x\} \ \overset{\wedge}{\delta}_\varphi \ \{y\} \Leftrightarrow \Phi(x) = \Phi(y)$ ∎

So, for example, if we take the strong proximity related to non-empty intersection of interiors, we have that $A \ \overset{\wedge}{\delta}_\varphi \ B \Leftrightarrow \mathrm{int}A \underset{\varphi}{\cap} \mathrm{int}B \neq \varnothing$ or either A or B is equal to X, provided A and B are not singletons; if $A = \{x\}$, then $\Phi(x) \in \Phi(\mathrm{int}(B))$, and if B is also a singleton, then $\Phi(x) = \Phi(y)$.

Example 11.4 (*Descriptive strong proximity*) Let X be a space of picture points represented in Fig. 11.8 with red, green or blue colors and let $\Phi : X \to \mathbb{R}$ be a description of X representing the color of a picture point, where 0 stands for red (r), 1 for green (g) and 2 for blue (b). Suppose the range is endowed with the topology given by $\tau = \{\varnothing, \{r, g\}, \{r, g, b\}\}$. Then $A \ \overset{\wedge}{\delta}_\varphi \ E$, since $\mathrm{int}A \cap \mathrm{int}E \neq \varnothing$. Similarly, $B \ \overset{\wedge}{\not\delta}_\varphi \ C$, since $\mathrm{int}B \underset{\varphi}{\cap} \mathrm{int}C \neq \varnothing$.   ∎

Remark 11.1 (*Importance of strong descriptive proximities*) Strong proximities are important in the detection of similar shapes, since the focus here is on the features of those shape regions that are covered either partly or completely by maximal nerve complexes. In effect, the focus here is on the interiors of shapes that have matching descripionts. This is the basic requirement for strong descriptive proximity (see Axiom (**dsnN5**)). In support of our interest in shape interiors, Algorithm 11.1 provides the foundation for the detection of strong descriptive proximities between known and unknown shapes.   ∎

11.3.4   Edelsbrunner–Harer Nerve Simplexes

Let $\mathscr{F}$ denote a collection of nonempty sets. An Edelsbrunner–Harer *nerve* of $\mathscr{F}$ [14, Sect. III.2, p. 59] (denoted by Nrv$\mathscr{F}$) consists of all nonempty sub-collections whose sets have a common intersection and is defined by

$$\mathrm{Nrv}.\mathscr{F} = \left\{ X \subseteq \mathscr{F} : \bigcap X \neq \varnothing \right\}.$$

A natural extension of the basic notion of a nerve arises when we consider adjacent polygons and the closure of a set. Let A, B be nonempty subset in a topological space X. The expression $A \, \delta \, B$ (A near B) holds true for a particular proximity that we choose, provided A and B have nonempty intersection, i.e., $A \cap B \neq \varnothing$. Every nonempty set has a set of points in its interior (denoted int$A$) and a set of boundary points (denoted bdy$A$). A nonempty set is *open*, provided its boundary set is empty, i.e., bdy$A \neq \varnothing$. Put another way, a set A is open, provided all points $y \in X$ sufficiently close to $x \in A$ belong to A [8, Sect. 1.2]. A nonempty set is *closed*, provided its boundary set is nonempty. Notice that a closed set can have an empty interior.

Fig. 11.8 Descriptive strongly near sets: $A_i \, \overset{\wedge}{\delta_\Phi} \, B, i \in \{1, 2, 3, 4, 5, 6, 7\}$

Fig. 11.9 NrvA, NrvB are
sample closure nerves

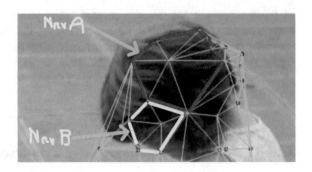

Example 11.5 Circle, triangles, and or any quadrilaterals are examples of closed sets
with either empty or nonempty interiors. Disks can be either closed or open sets with
nonempty interiors. ■

An extension of the notion of a nerve in a collection of nonempty sets $\mathscr{F}$ is a **closure
nerve** (denoted NCL$\mathscr{F}$), defined by

$$\text{NrvCL}\mathscr{F} = \left\{ X \in \mathscr{F} : \bigcap \text{cl}X \neq \varnothing \right\}.$$

Closure nerves are commonly found in triangulations of digital images.

Example 11.6 Examples of closure nerves are shown in Fig. 11.9. ■

From Example 11.6, we can arrive at a new form of closure nerve constructed from
filled triangles in a nerve, denoted by NrvCLt$\mathscr{F}$ defined by

$$\text{NrvCLt}\mathscr{F} = \left\{ \Delta \in \text{Nrv}F : \bigcap \text{cl}\Delta \neq \varnothing \right\}.$$

An easy next step is to consider nerve complexes that are descriptively near
and descriptively strongly near. Let NrvA, NrvB be a pair of nerves and let $\Delta_A \in$
NrvA, $\Delta_B \in$ NrvB. Then

$$\text{Nrv}A \; \delta_\Phi \; \text{Nrv}B, \text{ provided } \Delta_A \cap \Delta_B \neq \varnothing, \text{ i.e.,}$$

$\Delta_A \; \delta_\Phi \Delta_B$. Taking this a step further, whenever a region in interior of NrvA has a
description that matches the description of a region in the interior of NrvB, the pair
of nerves are descriptively strongly near. That is,

$$\text{Nrv}A \; \overset{\wedge}{\delta_\Phi} \; \text{Nrv}B, \text{ provided } \text{int}\Delta_A \underset{\Phi}{\cap} \text{int}\Delta_B \neq \varnothing.$$

Lemma 11.1 *Each closure nerve NrvCLt$\mathscr{F}$ has a nucleus.*

Proof Since all filled triangles in NrvCLt𝓕 have nonempty intersection, the triangles have a common vertex, the nucleus of the nerve.

Definition 11.2 A pair of filled triangles ΔA, ΔB in a triangulated region are **separated triangles**, provided $\Delta A \cap \Delta B = \varnothing$ (the triangles have no points in common) or ΔA, ΔB have an edge in common and do not have a common nucleus vertex.

Theorem 11.1 *A nonempty set of vertices V in a space X covered with filled triangles. If v, $v' \in V$ are vertices of separated filled triangles on X, then the space has more than one nerve.*

Proof Let $\Delta(v, p, q)$, $\Delta(v', p', q')$ be filled triangles on X. In a nerve, every filled triangle has a pair of edges in common with adjacent triangles and, from Lemma 11.1, the filled triangles in the nerve have a common vertex, namely, the nucleus. By definition, separated triangles have at most one edge in common and do not have a common nucleus. Hence, the separated triangles belong to separate nerves.

Theorem 11.2 *Filled triangles in nerves with a common edge are strongly near nerves.*

Proof Immediate from the definition of $\overset{\wedge\wedge}{\delta}$.

Algorithm 11.3: Construct Collection of Mesh Nerves

    Input   : Read set of generators S, $p, q \in S$.
    Output: Collection of mesh nerves NCL_{V_p}.
1 $S \longmapsto \mathcal{M}$/* Construct mesh using sites in S using Alg. 2 */;
2 **while** ($\mathcal{M} \neq \varnothing$ and $V_p \in \mathcal{M}$) **do**
3 | *Continue* ← *True*;
4 | *Select filled triangle* V_p;
5 | **while** ($\exists V_q \in \mathcal{M}$ & $V_q \neq V_p$ & *Continue*) **do**
6 | | *Select* $V_q \in \mathcal{M} \setminus V_p$;
7 | | **if** $V_p \cap V_q \neq \varnothing$ **then**
8 | | | $V_q \in NCL_{V_p}$;
9 | | | N.B.:$NCL_{V_p} := NCL_{V_p} \cup V_q$;
10 | | | $\mathcal{M} \leftarrow \mathcal{M} \setminus V_q$;
11 | | | /* V_q belongs to nerve NCL_{V_p} */
12 | | **else**
13 | | | *Continue* ← *False*;

Theorem 11.3 *Strongly near nerves are descriptively near nerves.*

Proof Let NrvA $\overset{\wedge\wedge}{\delta}$ NrvB be strongly near nerves. Then NrvA, NrvB have edge in common. Let A be the common edge. Then the description $\Phi(A)$ is common to both nerves. Hence, NrvA δ_Φ NrvB.

Theorem 11.4 *Nerves containing interior regions with matching descriptions are strongly descriptively near nerves.*

Proof Immediate from the definition of $\overset{\wedge}{\delta}_\Phi$.

We give a method for construction mesh nerves on an image in Algorithm 11.3. In the construction of a collection of nerve complexes, the step labelled N.B. is needed[1] to augment NCL_{V_p} as long as there are 2-simplexes that share a vertex p.

11.4 Features of Image Object Shapes

This section carries forward the notion of descriptively proximal images. The detection image object shapes is aided by detecting triangulation nerves containing the maximal number of filled triangles (denoted by maxNrv$K$).

Remark 11.2 (how to detect image object shapes with maximal nerve clusters) The following steps lead to the detection of image object shapes.

Triangulation: The triangulation of a digital image depends on the initial choice of vertices, used as generating points for filled triangles. In this work, keypoints have been chosen. A keypoint is defined by the gradient orientation (an angle) and gradient magnitudes (edge strength) of each image pixel. All selected image keypoints have different gradient orientations and edge strengths. Typically, in an image with a 100,000 pixels, we might find 1000 keypoints. Keypoints are ideally suited for shape detection, since each keypoint is also usually an edge pixel. For sample keypoints, see Figs. 11.10b, 11.12b, and in Fig. 11.14b.

(a)   **(b)**   **(c)**

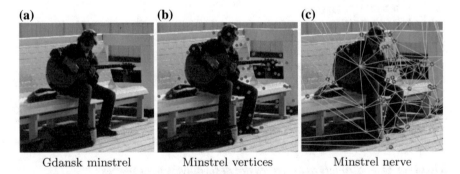

Gdansk minstrel   Minstrel vertices   Minstrel nerve

Fig. 11.10 Simplex vertices and musician nerve simplicial complex

---

[1]Many thanks to the reviewer for pointing this out.

Nerve Complexes: Keypoint-based nerve complexes have a nucleus that is a keypoint (e.g., on the approximate center of the hole covered by guitar strings the filled triangles in the nerve radiating out to edges of the guitar player in Fig. 11.11). For nerve complexes, see Figs. 11.10c, 11.13c, and in Fig. 11.14c.

Maximal Nerves: In the search for a principal image object shape, we start with maxNrv$K$ in a nerve with the maximum number of filled triangles that have a nucleus vertex in common. Experimentally, it has been found a large part of an image object shape will be covered by a maximal nerve (see, e.g., [33, Sects. 8.9 and 9.1]).

Maximal Nerve Clusters: Fourth, nerves strongly near maxNrv$K$ form a cluster useful in detecting an image object shape (denoted by maxNrvClu$K$).

Shape Contour: The outer perimeter of a maxNrvClu$K$ provides the contour of a shape that can be compared with other known shape contours, leading to the formation of shape classes. A maxNrvClu$K$ outer perimeter is called a shape edgelet [33, Sect. 7.6]. Shape contour comparisons can be accomplished by decomposing contour into nerve complexes and extracting geometric features of the nerves, e.g., nerve centroids, area (sum of the areas of the nerve triangles), number of nerve triangles, maximal nerve triangle area, which are easily compared across triangulations of different images.

Example 11.7 (*Minstrel closure nerve cluster shape*) The nerve complexes in Fig. 11.11 form a cluster with maxNrv$A$ doing most of the work in highlighting the shape of the guitar player's upper body.

Example 11.8 (*Girl closure nerve cluster head shape*) Based on a selection of keypoints shown in Fig. 11.12b, the triangulation of the girl image in Fig. 11.12a leads to a collection of filled triangle simplexes that cover the central region of the image as

Fig. 11.11 Minstrel nerve cluster shape

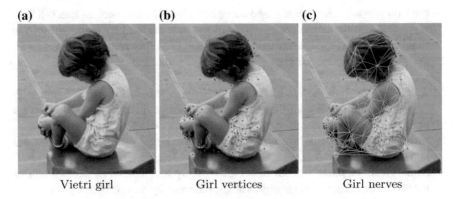

(a)   **(b)**   **(c)**

Vietri girl   Girl vertices   Girl nerves

Fig. 11.12 Simplex vertices and overlapping girl nerve simplicial complexes

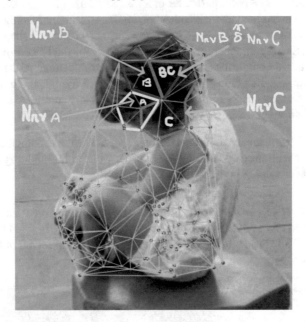

Fig. 11.13 Overlapping nerve complexes on girl image

shown in Fig. 11.12c. From this triangulation, maximal nucleus clusters can be identified. For example, we can begin to detect the shape of the head from the collection of overlapping nerve complexes in Fig. 11.13. The nerve complexes in Fig. 11.13 form a cluster with maxNrv A doing most of the work in highlighting the shape of a large part of the girl's head. Let the upper region of this space be endowed with what is known as proximal relator [31], which is a collection of proximity relations on the space. Let $(X, \mathscr{R}_\delta)$ be a proximal relator space with the upper region of Fig. 11.12a represented

(a)   **(b)**   **(c)**

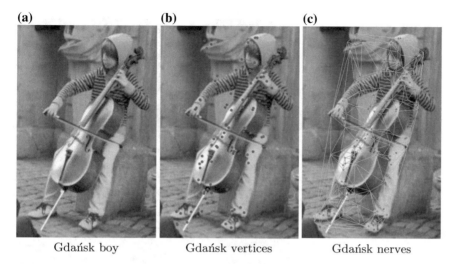

Gdańsk boy   Gdańsk vertices   Gdańsk nerves

Fig. 11.14 Simplex vertices and overlapping minstrel nerve simplicial complexes

by the set of points X and let $\mathscr{R}_\delta = \left\{ \overset{\wedge}{\delta}, \delta_\Phi, \overset{\wedge}{\delta_\Phi} \right\}$. Notice that the triangulation in Fig. 11.12c contains a number of separated triangles. Hence, from Theorem 11.1, we can expect to find more than one nerve. In this case, families of nerves can be found in this image, starting with the upper region of the image. Then observe the following things in the collections of nerve complexes in Fig. 11.13.

$$\text{Nrv}A \ \overset{\wedge}{\delta} \ \text{Nrv}B \ \text{(Nerves with a common edge are strongly near (Theorem 11.2))},$$

$$\text{Nrv}A \ \overset{\wedge}{\delta} \ \text{Nrv}C \ \text{(From Theorem 11.2, these nerves are strongly near)},$$

$$\text{Nrv}B \ \overset{\wedge}{\delta} \ \text{Nrv}C \ \text{(From Theorem 11.2, these nerves are strongly near)},$$

$$(\text{Nrv}A \ \cap \ \text{Nrv}B) \ \delta_\Phi \ \text{Nrv}C \ \text{(nerves with matching feature vectors, cf. Theorem 11.3)},$$

$$(Nrv A \ \cap \ \text{Nrv}C) \ \delta_\Phi \ \text{Nrv}B \ \text{(nerves with matching feature vectors, cf. Theorem 11.3)},$$

$$(\text{Nrv}A \ \cap \ \text{Nrv}C) \ \overset{\wedge}{\delta_\Phi} \ \text{Nrv}B \ \text{(nerve interiors with matching descriptions, cf. Theorem 11.4)},$$

$$(\text{Nrv}A \ \cap \ \text{Nrv}B) \ \overset{\wedge}{\delta_\Phi} \ \text{Nrv}C \ \text{(nerve interiors with matching descriptions, cf. Theorem 11.4)}.$$

From these proximities, we can derive the head shape from the contour formed by the sequence of connected line segments along the outer edges of the nerves.

Example 11.9 (Musician closure nerve cluster head shape) Based on a selection of keypoints shown in Fig. 11.14b, the triangulation of the Musician image in Fig. 11.14a leads to a collection of filled triangle simplexes that cover the central region of the image as shown in Fig. 11.14c. Notice, again, that from the triangulation of the musician image, maximal nucleus clusters can be identified. For example, we can

Fig. 11.15 Boy
tri-overlapping nerves

begin to detect the shape of the head and hand from the collection of overlapping
nerve complexes in Fig. 11.15. The nerve complexes in Fig. 11.15 form a cluster with
maxNrv B doing most of the work in highlighting the shape of a large part of the boy's
head. Let the upper region of this space be endowed with what is known as proximal
relator [31], which is a collection of proximity relations on the space. Let $(X, \mathscr{R}_\delta)$ be
a proximal relator space with the upper region of Fig. 11.14a represented by the set
of points X and let $\mathscr{R}_\delta = \left\{ \overset{\wedge}{\delta}, \delta_\Phi, \overset{\wedge}{\delta_\varphi} \right\}$. Notice again that a triangulation like the one
in Fig. 11.14c contains a number of separated triangles. Hence, from Theorem 11.1,
we can expect to find more than one nerve. Notice, again, that families of nerves can
be found in this image, starting with the upper region of the image. Then observe the
following things in the collections of nerve complexes in Fig. 11.15.

$\text{Nrv}A \stackrel{\wedge}{\delta} \text{Nrv}B$ (Nerves with a common edge are strongly near (Theorem 11.2)),

$\text{Nrv}A \stackrel{\wedge}{\delta} \text{Nrv}C$ (From Theorem 11.2, these nerves are strongly near),

$\text{Nrv}B \stackrel{\wedge}{\delta} \text{Nrv}C$ (From Theorem 11.2, these nerves are strongly near),

$(\text{Nrv}A \cap \text{Nrv}B)\, \delta_{\Phi}\, \text{Nrv}C$ (nerves with matching feature vectors, cf. Theorem 11.3),

$(Nrv A \cap \text{Nrv}C)\, \delta_{\Phi}\, \text{Nrv}B$ (nerves with matching feature vectors, cf. Theorem 11.3),

$(\text{Nrv}A \cap \text{Nrv}C)\, \stackrel{\wedge}{\delta_{\Phi}}\, \text{Nrv}B$ (nerve interiors with matching descriptions, cf. Theorem 11.4),

$(\text{Nrv}A \cap \text{Nrv}B)\, \stackrel{\wedge}{\delta_{\Phi}}\, \text{Nrv}C$ (nerve interiors with matching descriptions, cf. Theorem 11.4).

From these proximities, we can derive the head shape from the contour formed by the sequence of connected line segments along the outer edges of the nerves.

11.5 Concluding Remarks

A framework for the detection of image object shapes is given in this chapter. This framework is based on a proximal physical geometric approach in a descriptive proximity space setting that makes it possible to compare and classify image object shapes. Unlike the conventional Delaunay triangulation of spatial regions, the proposed triangulation introduced in this chapter results in simplexes that are filled triangles. By covering a part or all of a digital image with simplexes, the problem of describing object shapes is simplified. Both spatial and descriptive proximities are discussed at length. In addition, algorithms to generate meshes and to construct mesh nerve collections are given. Finally, illustrative examples for the framework have been presented.

References

1. Adamaszek, M., Adams, H., Frick, F., Peterson, C., Previte-Johnson, C.: Nerve complexes on circular arcs, 1–17. arXiv:1410.4336v1 (2014)
2. Ahmad, M., Peters, J.: Delta complexes in digital images. Approximating image object shapes, 1–18. arXiv:1706.04549v1 (2017)
3. Alexandroff, P.: Über den algemeinen dimensionsbegriff und seine beziehungen zur elementaren geometrischen anschauung. Math. Ann. **98**, 634 (1928)
4. Alexandroff, P.: Elementary concepts of topology. Dover Publications, Inc., New York (1965). 63 pp, translation of Einfachste Grundbegriffe der Topologie. Springer, Berlin (1932), translated by Alan E. Farley, Preface by D. Hilbert, MR0149463
5. Borsuk, K.: On the imbedding of systems of compacta in simplicial complexes. Fund. Math. **35**, 217–234 (1948)

6. Borsuk, K.: Theory of shape. Monografie Matematyczne, Tom 59. (Mathematical Monographs, vol. 59) PWN—Polish Scientific Publishers (1975). MR0418088, Based on K. Borsuk, Theory of shape. Lecture Notes Series, No. 28, Matematisk Institut, Aarhus Universitet, Aarhus (1971), MR0293602

7. Borsuk, K., Dydak, J.: What is the theory of shape? Bull. Austral. Math. Soc. **22**(2), 161–198 (1981). MR0598690

8. Bourbaki, N.: Elements of Mathematics. General Topology, Part 1. Hermann & Addison-Wesley, Paris (1966). I-vii, 437 pp

9. Čech, E.: Topological Spaces. John Wiley & Sons Ltd., London (1966). Fr seminar, Brno, 1936–1939; rev. ed. Z. Frolik, M. Katětov

10. Concilio, A.D., Guadagni, C., Peters, J., Ramanna, S.: Descriptive proximities I: properties and interplay between classical proximities and overlap, 1–12. arXiv:1609.06246 (2016)

11. Concilio, A.D., Guadagni, C., Peters, J., Ramanna, S.: Descriptive proximities. Properties and interplay between classical proximities and overlap, 1–12. arXiv:1609.06246v1 (2016)

12. Delaunay: Sur la sphère vide. Izvestia Akad. Nauk SSSR, Otdelenie Matematicheskii i Estestvennyka Nauk (7), 793–800 (1934)

13. Edelsbrunner, H.: Modeling with simplical complexes. In: Proceedings of the Canadian Conference on Computational Geometry, pp. 36–44, Canada (1994)

14. Edelsbrunner, H., Harer, J.: Computational Topology. An Introduction. American Mathematical Society, Providence, R.I. (2010). Xii+110 pp., MR2572029

15. Efremovič, V.: The geometry of proximity I (in Russian). Mat. Sb. (N.S.) **31(73)**(1), 189–200 (1952)

16. Fontelos, M., Lecaros, R., López-Rios, J., Ortega, J.: Stationary shapes for 2-d water-waves and hydraulic jumps. J. Math. Phys. **57**(8), 081520, 22 pp (2016). MR3541857

17. Gratus, J., Porter, T.: Spatial representation: discrete vs. continuous computational models: a spatial view of information. Theor. Comput. Sci. **365**(3), 206–215 (2016)

18. Guadagni, C.: Bornological convergences on local proximity spaces and ω_μ-metric spaces. Ph.D. thesis, Università degli Studi di Salerno, Salerno, Italy (2015). Supervisor: A. Di Concilio, 79 pp

19. Kokkinos, I., Yuille, A.: Learning an alphabet of shape and appearance for multi-class object detection. Int. J. Comput. Vis. **93**(2), 201–225 (2011). https://doi.org/10.1007/s11263-010-0398-7

20. Leray, J.: L'anneau d'homologie d'une reprësentation. Les Comptes rendus de l'Académie des sciences **222**, 1366–1368 (1946)

21. Lodato, M.: On topologically induced generalized proximity relations, Ph.D. thesis. Rutgers University (1962). Supervisor: S. Leader

22. Maggi, F., Mihaila, C.: On the shape of capillarity droplets in a container. Calc. Var. Partial Differ. Equ. **55**(5), 55:122 (2016). MR3551302

23. Naimpally, S., Peters, J.: Topology with Applications. Topological Spaces via Near and Far. World Scientific, Singapore (2013). Xv + 277 pp, American Mathematical Society. MR3075111

24. Opelt, A., Pinz, A., Zisserman, A.: Learning an alphabet of shape and appearance for multi-class object detection. Int. J. Comput. Vis. **80**(1), 16–44 (2008). https://doi.org/10.1007/s11263-008-0139-3

25. Peters, J.: Near sets. General theory about nearness of sets. Applied. Math. Sci. **1**(53), 2609–2629 (2007)

26. Peters, J.: Near sets. Special theory about nearness of objects. Fundamenta Informaticae **75**, 407–433 (2007). MR2293708

27. Peters, J.: Near sets: an introduction. Math. Comput. Sci. **7**(1), 3–9 (2013). http://doi.org/10.1007/s11786-013-0149-6. MR3043914

28. Peters, J.: Topology of Digital Images - Visual Pattern Discovery in Proximity Spaces, vol. 63. Intelligent Systems Reference Library. Springer (2014). Xv + 411 pp. Zentralblatt MATH Zbl 1295, 68010

29. Peters, J.: Visibility in proximal Delaunay meshes and strongly near Wallman proximity. Adv. Math. Sci. J. **4**(1), 41–47 (2015)

30. Peters, J.: Computational Proximity. Excursions in the Topology of Digital Images. Intelligent Systems Reference Library 102. Springer (2016). Viii + 445 pp. http://doi.org/10.1007/978-3-319-30262-1
31. Peters, J.: Proximal relator spaces. Filomat **30**(2), 469–472 (2016). MR3497927
32. Peters, J.: Two forms of proximal physical geometry. axioms, sewing regions together, classes of regions, duality, and parallel fibre bundles, 1–26. arXiv:1608.06208 (2016). To appear in Adv. Math. Sci. J. **5** (2016)
33. Peters, J.: Foundations of Computer Vision. Computational Geometry, Visual Image Structures and Object Shape Detection. Intelligent Systems Reference Library 124. Springer International Publishing, Switzerland (2017). I-xvii, 432 pp. http://doi.org/10.1007/978-3-319-52483-2
34. Peters, J.: Computational Topology of Digital Images. Visual Structures and Shapes. Intelligent Systems Reference Library. Springer International Publishing, Switzerland (2018)
35. Peters, J., Guadagni, C.: Strongly near proximity and hyperspace topology, 1–6. arXiv:1502.05913 (2015)
36. Peters, J., Ramanna, S.: Maximal nucleus clusters in Pawlak paintings. Nerves as approximating tools in visual arts. In: Proceedings of Federated Conference on Computer Science and Information Systems **8** (ISSN 2300-5963), 199–2022 (2016). http://doi.org/10.15439/2016F004
37. Peters, J., Tozzi, A., Ramanna, S.: Brain tissue tessellation shows absence of canonical microcircuits. Neurosci. Lett. **626**, 99–105 (2016)
38. Rényi, A.: On measures of entropy and information. In: Proceedings of the 4th Berkeley Symposium on Mathematical Statistics and Probability, vol. 1, pp. 547–547. University of California Press, Berkeley (2011). Math. Sci. Net. Review MR0132570
39. de Verdière, E., Ginot, G., Goaoc, X.: Multinerves and helly numbers of acylic families. In: Proceedings of 28th Annual Symposium on Computational Geometry, pp. 209–218 (2012)
40. Voronoï, G.: Nouvelles applications des paramètres continus à la théorie des formes quadratiques. premier mémoir. J. für die reine und angewandte math. **133**, 97–178 (1907). JFM 38.0261.01
41. Wallman, H.: Lattices and topological spaces. Ann. Math. **39**(1), 112–126 (1938)
42. Ziegler, G.: Lectures on polytopes. Graduate Texts in Mathematics, vol. 152. Springer, New York (1995). X+370 pp. ISBN: 0-387-94365-X, MR1311028
43. Ziegler, G.: Lectures on Polytopes. Springer, Berlin (2007). http://doi.org/10.1007/978-1-4613-8431-1

Chapter 12
Comparison of Classification Methods for EEG Signals of Real and Imaginary Motion

Piotr Szczuko, Michał Lech and Andrzej Czyżewski

Abstract The classification of EEG signals provides an important element of brain-computer interface (BCI) applications, underlying an efficient interaction between a human and a computer application. The BCI applications can be especially useful for people with disabilities. Numerous experiments aim at recognition of motion intent of left or right hand being useful for locked-in-state or paralyzed subjects in controlling computer applications. The chapter presents an experimental study of several methods for real motion and motion intent classification (rest/upper/lower limbs motion, and rest/left/right hand motion). First, our approach to EEG recordings segmentation and feature extraction is presented. Then, 5 classifiers (Naïve Bayes, Decision Trees, Random Forest, Nearest-Neighbors NNge, Rough Set classifier) are trained and tested using examples from an open database. Feature subsets are selected for consecutive classification experiments, reducing the number of required EEG electrodes. Methods comparison and obtained results are presented, and a study of features feeding the classifiers is provided. Differences among participating subjects and accuracies for real and imaginary motion are discussed. It is shown that though classification accuracy varies from person to person, it could exceed 80% for some classifiers.

Keywords Motion intent classification · EEG signal analysis · Rough sets

P. Szczuko (✉) · M. Lech · A. Czyżewski
Faculty of Electronics, Telecommunications and Informatics,
Gdańsk University of Technology, Gdańsk, Poland
e-mail: szczuko@sound.eti.pg.gda.pl

M. Lech
e-mail: mlech@sound.eti.pg.gda.pl

A. Czyżewski
e-mail: andcz@sound.eti.pg.gda.pl

© Springer International Publishing AG 2018
U. Stańczyk et al. (eds.), *Advances in Feature Selection for Data and Pattern Recognition*, Intelligent Systems Reference Library 138,
https://doi.org/10.1007/978-3-319-67588-6_12

227

12.1 Introduction

The classification of EEG signals is an important part of the brain-computer inter-face (BCI) application. It is required for the method to be highly accurate to maintain an efficient interaction between a human and a computer application [6, 15]. Applying a dedicated method of signal processing to EEG recordings allows for determining emotional states, mental conditions, and motion intents. Numerous experiments of imaginary motion recognition deal with unilateral, i.e. of left or right, hand motion. Such a classification is useful for locked-in-state or paralyzed subjects, thus it can be applied successfully to controlling computer applications [3, 11, 23–26, 31, 32, 52] or a wheelchair [7, 12] and communicating with locked-in patients and diagnosis of coma patients [8].

The motion intent classification can be performed in a synchronous or an asynchronous mode. The former method uses a visual cue, e.g. an icon on the screen flashing in timed intervals, and then verifies user's focus by means of the P300 potential induced in a reaction to this visual event [4, 5, 16, 33]. The latter approach is suited for self-paced interaction, but it requires a method of distinction between a resting and acting, in the latter case determining the type of the action [10, 40, 56]. The asynchronous approach is evaluated in our work, since the classification of left and right, and up and down motion intents and real motions is performed by various decision algorithms.

The main principle for detection and classification of imaginary motor activity in brain-computer interfaces is based on an observation that the real and imaginary motions involve similar neural activity of the brain [26]. It is indicated by an alpha wave signal power decrease in a motor cortex in a hemisphere contra-lateral to the movement side [25, 26, 31], usually registered by C_3 and C_4 channels [39, 43, 57]. It is related to a phenomena of event-related desynchronization (ERD) [20, 29, 58]. Such an activity can be detected and classified by various approaches.

Siuly et al. [42] employed a conjunction of an optimal allocation system and two-class Naïve Bayes classifier in the process of recognizing hand and foot movements. Data was partitioned in such a way that right hand movements were analyzed along with the right foot (first set) movements and left hand movements were analyzed also with right foot movements (second set). Left foot movements were not performed in the experiment. The global average accuracy over 10 folds, for the first and the second set, equalled to 96.36 and to 91.97%, respectively. The authors claimed to obtain the higher accuracy for the two-class Naïve Bayes classifier than for the Least Squares Support Vector Machine (LS-SVM), both cross-correlation (CC) and clustering technique (CT) based, examined in their earlier works [41, 59].

Schwarz et al. [38] aimed at developing BCI system that generates control signals for users with severe motor impairments, based on EEG signals processed using filter-bank common spatial patterns (fbCSP) and then classified with Random Forest which is a type of a random tree classifier, applied to experiments presented in their paper. In their experiments users were asked to perform right hand and feet motor imagination for 5 seconds according to the cue on the screen. For imagined right hand movement, each user was instructed to imagine sustained squeezing of a training ball. For motor imagery of both feet, the user was instructed to imagine repeated plantar flexion of both feet. The median accuracy of 81.4% over the feedback period (presenting information to the user about the motion intention) was achieved.

Kayikcioglu et al. [21] compared performance of k-NN , Multiple Layer Perceptron, which is a type of Artificial Neural Network tested herein, and SVM with RBF kernel. Training datasets were created based on one-channel EEG signal. The authors claim that the best accuracy was obtained for k-NN classifier but the presentation of the results is vague, thus not convincing.

Beside observing ERD occurrences, the motion intent classification is performed by: Linear Discriminant Analysis (LDA) [22, 31, 32, 58], k-means clustering and Principal Component Analysis (PCA) [48], or Regularised Fisher's Discriminant (RFD) [51]. The work presented in this chapter is inspired by previous results in applying Rough Set classifier of the real and imaginary motion activity over large database of 106 persons performing real and imaginary motion, resulting in accuracy exceeding 80%, and in some cases up to 100% [44, 45]. The main goal of this research is to determine the best method of signal classification, by applying selected classifiers, relatively simple and straightforward to use for practical applications. Another goal was to determine the impact of reducing the EEG signal representation on the accuracy: first by using a larger set of features (615), and then by limiting this amount of features (to 120 and to 50).

Despite the observed advancements in EEG classification there still remains a considerable group of users (15–30%) being "illiterate" in the Brain-Computer-Interfaces, thus unable to perform recognisable mental actions in a repeated manner. The exact reason is still unknown but the problem was formulated and studied [9, 53]. In this research there are subjects with relatively high and satisfactory results but the same methods yield poor results for other group of persons. The personal differences are discussed in Sect. 12.4.

The reminder of this chapter is structured as follows: Sect. 12.2 describes EEG signals preprocessing and feature extraction, Sect. 12.3 contains details of classifiers setup. Results are presented in Sects. 12.4, and 12.5 provides conclusions.

12.2 EEG Signal Parameterisation

EEG signals are parameterized in frequency bands associated experimentally with mental and physical conditions [55]. Following frequency ranges and their most popular interpretations are used: delta (2–4 Hz, consciousness and attention), theta (4–7 Hz, perceiving, remembering, navigation efforts) and alpha (8–15 Hz, thinking, focus, and attention), beta (conscious focus, memory, problem solving, information processing, 15–29 Hz), and gamma (learning, binding senses, critical thinking 30–59 Hz). Electrodes are positioned over crucial brain regions, and thus can be used for assessing activity of motor cortex, facilitating motion intent classification [1].

Recordings of EEG are polluted with various artifacts, originating from eye blinks, movement, and heartbeat. Dedicated methods were developed for detecting artifacts, filtering and improving signal quality. A Signal-Space Projection (SSP) [19, 50, 58], involving spatial decomposition of the EEG signals is used for determining contaminated samples. Such an artifact repeatedly originates from a given location, e.g. from eye muscles and is being recorded with distinct characteristics, amplitudes, and phases, thus the artifact pattern can be detected and filtered out. Signal quality improvements are also achieved by Independent Component Analysis (ICA) [19, 20, 50, 53].

The research approach presented in this chapter assumes an usage of Hilbert transform of the signal and of several parametrization methods based on envelope, power, and signal statistics, as well as a classification based on dedicated, carefully examined and trained classifiers. For those experiments a large EEG database was used: EEG Motor Movement/Imagery Dataset [14], collected with BCI2000 system [2, 37] and published by PhysioNet [14]. This database includes 106 persons and exceeds the amount of data collected by Authors themselves up to date, thus is more suitable for training and examining classification methods over a large population, facilitating also comparisons with research of others.

The dataset contains recordings of 4 tasks:

- A real movement of left-right hand,
- B real movement of upper-lower limbs,
- C imaginary movement of left-right hand,
- D imaginary movement of upper-lower limbs.

Sixty four electrodes were used located according to the 10–20 standard, with sampling rate 160 Sa/s, and timestamps denoting start and end of particular movement and one of 3 classes: rest, left/up, right/down. Among the available channels, only 21 were used, obtained from motor cortex: $FC_{Z,1,2,3,4,5,6}$, $C_{Z,1,2,3,4,5,6}$, $CP_{Z,1,2,3,4,5,6}$ (Fig. 12.1).

All 21 signals were processed in a similar manner, decomposed into the time-frequency domain (TF): delta (2–4 Hz), theta (4–7 Hz), alpha (8–15 Hz), beta (15–29 Hz), and gamma (30–59 Hz). Subsequently, each subband's envelope was obtained by Hilbert transform [27], reflecting activity in the given frequency band. This dataset was pre-processed employing the Brainstorm software, where segmentation and filtration of signals were performed [47]. Finally, 615 various features of envelopes were extracted. Authors of this chapter proposed a parametrization of envelopes of band-filtered signals. Consequently, 5 frequency subbands for each of 21 sensors, are parametrized as follows:

1. For a particular subband $j = \{delta, \ldots, gamma\}$ from a sensor $k = \{FC_1, \ldots, CP_6\}$, 5 activity features are extracted, reflecting the activity in the particular brain region: the sum of squared samples of the signal envelope (12.1), mean (12.2), variance (12.3), minimum (12.4), and maximum of signal envelope values (12.5),

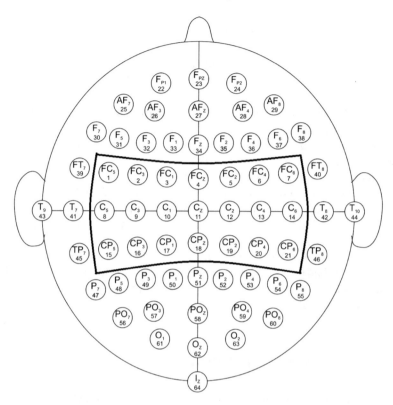

Fig. 12.1 A top view of a human head with electrodes in 10–20 setup, motor cortex channels in central region marked (*Source* [14])

2. For all 9 pairs of symmetrically positioned electrodes kL and kR (e.g. $kL = C_1$, and $kR = C_2$) the signal envelopes differences are calculated and summed up (12.6), to reflect asymmetry in hemispheres activity while performing unilateral motion:

$$SqSum_{j,k} = \sum_{i=1}^{N} \left(e_{j,k}[i] \right)^2,$$

(12.1)

$$Mean_{j,k} = \frac{1}{N} \sum_{i=1}^{N} \left(e_{j,k}[i] \right),$$

(12.2)

$$Var_{j,k} = \frac{1}{N} \sum_{i=1}^{N} \left(e_{j,k}[i] - Mean_{j,k} \right)^2,$$

(12.3)

$$Min_{j,k} = min(e_{j,k}[i]),$$

(12.4)

$$Max_{j,k} = max(e_{j,k}[i]), \tag{12.5}$$

$$SumDiff_{j,kL,kR} = \sum_{i=1}^{N} \left(e_{j,kL}[i] - e_{j,kR}[i] \right), \tag{12.6}$$

where, $e_{j,k}[i]$ is an envelope of the signal from particular subband j of electrode k and has length of N samples.

As a result there are 615 features extracted for every task. The result decision table includes also task number, person number and decision (T0, T1 or T2).

The multidimensional problem of classifying EEG signal is not straightforward, because personal biological and neurological features significantly influence values of registered signals and extracted features. In the following data classification (Sect. 12.3) every person is treated separately, thus for every task a new classifier is created with a different subset of useful and informative features.

EEG classification is hampered by personal biological and neurological differences, or other characteristics influencing EEG signal quality and features. Therefore each person is treated as individual classification case, and thus customized classifiers are created.

12.3   Data Classification Method

Data classification was performed in WEKA software package offering various data mining techniques [54], and in R programming environment [13] with RoughSets package [35].

All methods were applied in a 10 cross-validation runs, with a training and testing sets selected randomly in a 65/35 ratio split. These sets contain 1228 and 662 signals for a single person performing a particular task of 3 different action classes (rest, up/left motion, and down/right motion). The process is repeated for 106 persons, achieved average classification accuracy records are collected. In the described research three variants of features sets **P** were examined:

1. $\mathbf{P}_{615}$ with all 615 features.
2. $\mathbf{P}_{50}$ with features being the most frequently used in Rough Set classification rules from the first variant [44, 45]. Reducts from all iterations of given classification scenarios were analyzed for frequency of features and top 50 were used instead of 615 to repeat this experiment (Table 12.1). Other features appear in less than 3% of rules often matching only a single person, therefore are discarded to reduce overfitting. By this approach it is verified if a limited number of features is sufficient for accurate description of classes differences. Rough Set was used as a baseline, because of high accuracy achieved in previous experiments with this method [44, 45].
3. $\mathbf{P}_{C_3 C_4}$ with 120 features obtained only from signals from electrodes C_3 and C_4, as these were reported by other research to be the most significant for motion

Table 12.1 Top 50 features for classification rules in Rough Set method. A number of rules including the feature is provided. The set is used for other classifier in this chapter, denoted as $\mathbf{P}_{50}$

Attribute	No. of appear.	Attribute	No. of appear.	Attribute	No. of appear.
Var_{theta,FC_Z}	420	Max_{gamma,C_3}	279	Var_{theta,FC_6}	253
Min_{delta,C_1}	409	Min_{delta,C_5}	277	Max_{beta,C_4}	252
Min_{delta,FC_5}	389	Sum_{theta,FC_3}	277	Max_{gamma,FC_2}	250
$Mean_{gamma,C_6}$	388	Min_{delta,FC_3}	276	Min_{delta,CP_4}	248
Sum_{alpha,CP_4}	378	Var_{gamma,C_6}	275	Min_{delta,CP_Z}	248
Min_{delta,FC_Z}	367	Min_{beta,C_1}	274	Max_{theta,FC_1}	246
$Mean_{delta,FC_5}$	340	Min_{delta,FC_2}	273	Sum_{beta,FC_2}	246
Min_{delta,C_4}	337	Sum_{beta,FC_4}	272	Max_{gamma,C_1}	245
Max_{beta,C_1}	327	Sum_{gamma,FC_5}	269	Sum_{alpha,CP_2}	244
Min_{delta,CP_5}	326	Min_{delta,C_3}	268	Sum_{gamma,C_4}	239
Sum_{delta,FC_6}	316	Var_{beta,C_Z}	268	Max_{gamma,FC_5}	238
Var_{theta,CP_2}	310	Min_{gamma,C_4}	260	Min_{delta,CP_3}	238
Var_{alpha,FC_Z}	304	Sum_{theta,FC_Z}	259	Var_{theta,CP_1}	236
Sum_{gamma,FC_1}	299	Var_{alpha,FC_3}	259	$Mean_{theta,FC_3}$	231
Var_{theta,CP_6}	290	Max_{gamma,FC_Z}	258	Max_{alpha,FC_6}	229
Min_{delta,CP_2}	288	Var_{theta,C_4}	258	Var_{theta,C_Z}	229
Min_{delta,C_6}	284	Min_{delta,FC_4}	254		

classification [25, 31, 43], for verifying if limiting the region of interest to two regions on motor cortex decreases accuracy.

Five classification methods were chosen. Each have own parameters, and to determine the best setup a training-testing cycle with cross-validation was repeated with automatic changes of parameters from an arbitrary defined values sets (Table 12.2). As a result, for each classifier the best configuration was identified for $\mathbf{P}_{615}$, $\mathbf{P}_{50}$ and $\mathbf{P}_{C_3C_4}$ and then used for subsequent experiments. Following methods were used:

- Naïve Bayes (NB). Naïve Bayes method uses numeric estimator with precision values chosen based on analysis of the training data [18]. A supervised discretization was applied, converting numeric attributes to nominal ones.
- Classifier trees (J48). A pruned C4.5 decision tree was applied [34], with adjusted confidence factor used for pruning C, and a minimum number of instances for a leaf M. C was selected from a set $\{2^{-5}, 2^{-4}, \ldots, 2^{-1}\}$, M:$\{2^1, 2^2, \ldots, 2^5\}$.
- Random Forest (RF). This method constructs I random trees considering K randomly selected attributes at each node. Pruning is not performed. I and K were from a set $\{2^3, \ldots, 2^7\}$.
- Nearest-Neighbors (NNge). An algorithm of Nearest-neighbors using non-nested generalized exemplars (hyperrectangles, reflecting if-then rules) was used [28, 36]. The method uses G attempts for generalization, and a number of folder for mutual information I. G and I were from a set $\{2^0, \ldots, 2^6\}$.

Table 12.2 Classifiers parameters resulting with the highest classification accuracy for three used features sets

Classifier	Features set $\mathbf{P}_{615}$	Features set $\mathbf{P}_{50}$	Features set $\mathbf{P}_{C_3 C_4}$
NB	Not applicable	Not applicable	Not applicable
J48	C = 0.03125, M = 16	C = 0.03125, M = 16	C = 0.03125, M = 16
RF	I = 64, K = 64	I = 64, K = 32	I = 64, K = 16
NNge	G = 8, I = 2	G = 8, I = 8	G = 8, I = 4
RS	Not applicable	Not applicable	Not applicable

- Rough Set classifier (RS). A method applying Pawlak's Rough Set theory [30, 35] was employed to classification. It applies maximum discernibility method for data discretization and it selects a minimal set of attributes (a reduct) maintaining discernibility between different classes, by applying greedy heuristic algorithm [17, 45, 46]. A reduct is finally used to generate decision rules describing objects of the testing set, and applying these to the testing set.

12.4 Classification Results

Classification accuracies obtained for 106 persons by the best configuration of selected 5 classifiers are shown below as box-whiskers plots [49] (Fig. 12.2).

It can be observed that Rough Sets (RS) are significantly more accurate in classification than other methods. Random Forest (RF) is the second, but the advantage over Naïve Bayes (NB), J48 and Nearest-Neighbors (NNge) is not statistically significant. Nearest-Neighbors is usually the worst. There are a few cases of very high accuracy exceeding 90%, but also a few persons' actions were impossible to classify (observed accuracy lower than 33% is interpreted as random classification).

In each case the imaginary motion classification (task B and D) is not as accurate as classification of the real motion (task A and C). This can be justified by inability to perform a task restricted to only mental activity in a repeated manner, or subjects' fatigue, incorrect positioning of electrodes, or even BCI illiteracy. Classification of real upper/lower limbs movement (task C) is the easiest one for every method.

It can be observed that applying $\mathbf{P}_{615}$ to classification (Fig. 12.2) generally yields better results than limited features sets $\mathbf{P}_{50}$ or $\mathbf{P}_{C_3 C_4}$ (Fig. 12.3 and 12.4). The accuracy decrease of ca. 5%.

Personal differences must be taken into account in application of EEG classification, as our experiments show some individuals perform the best, and other the worst repeatedly. For example, the subject S004 from the database was the highest ranked in 103 cases of 192 classification attempts, followed by S072 being the top ranked in 26, and S022 in 19 cases. The worst performing subjects were: S031 in 15, S098 in 13, S047 in 12, S021 in 11, and S109 in 11 cases of 192 attempts. Subjects

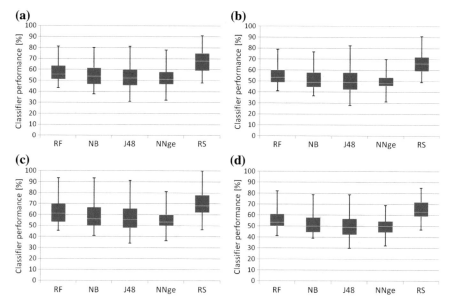

Fig. 12.2 Classification performance in 10 cross validation runs of selected classifiers for feature set P_{615}: (a)–(d) tasks A–D respectively

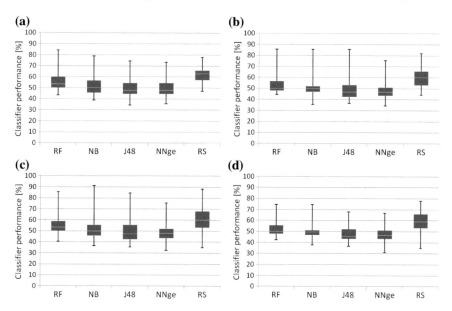

Fig. 12.3 Classification performance in 10 cross validation runs of selected classifiers for feature set P_{50}: (a)–(d) tasks A–D respectively

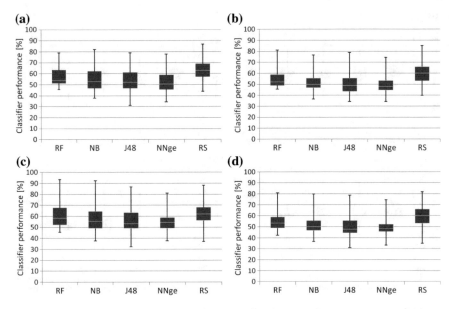

Fig. 12.4 Classification performance in 10 cross validation runs of selected classifiers for feature set $\mathbf{P}_{C_3 C_4}$: (a)–(d) tasks A–D respectively

are anonymous and no personal details are provided, so actual physical difference between them cannot be determined.

12.5 Conclusions

A method of EEG signal pre-processing, parametrization, and classification with selected 5 classifiers was presented. Among applied methods Rough Sets (RS) and Random Forest (RF) achieved the highest accuracy, with the Rough Set (RS) significantly outperforming other methods.

The presented procedure can be employed in a simple interface involving motion classification by EEG signals analysis. It opens a possibility to develop accurate and responsive computer applications to be interacted by intents of rest, left, right, up, and down motion. These five binary input controls are sufficient to perform complex actions such as navigating, confirming or rejecting options in a graphical user interface.

For each person the training and classification process must be repeated, because each case could differ, albeit slightly, with electrodes placements, signal registration conditions, hair and skin characteristics, varying level of stress and fatigue, varying manner of performing the imaginary motion, etc.

Subjects were anonymous, so their physical differences are unknown, but large discrepancy in classification accuracy was observed, probably impossible to be overcome. Still, it must be yet determined whether satisfactory accuracy can be achieved by applying processing and classification of signals from non-invasive registration of brain activity through the skull and the scalp.

The results presented in this chapter were achieved without a necessity to apply complex methods such as ICA or SSP described in literature, and blink and heartbeat artefacts elimination or signal improvements methods were not employed. Therefore main strength of the approach is its simplicity, and confirmed high accuracy, possible to achieve provided the person is able to perform defined actions in a repeated manner.

Acknowledgements The research is funded by the National Science Centre of Poland on the basis of the decision DEC-2014/15/B/ST7/04724.

References

1. Alotaiby, T., El-Samie, F.E., Alshebeili S.A.: A review of channel selection algorithms for eeg signal processing. EURASIP. J. Adv. Signal Process, **66** (2015)
2. BCI2000. Bci2000 instrumentation system project. http://www.bci2000.org, Accessed on 2017-03-01
3. Bek, J., Poliakoff, E., Marshall, H., Trueman, S., Gowen, E.: Enhancing voluntary imitation through attention and motor imagery. Exp. Brain Res. **234**, 1819–1828 (2016)
4. Bhattacharyya, S., Konar, A., Tibarewala, D.N.: Motor imagery, p300 and error-related eeg-based robot arm movement control for rehabilitation purpose. Med. Biol. Eng. Comput. **52**, 2014 (1007)
5. Chen, S., Lai, Y.A.: Sgnal-processing-based technique for p300 evoked potential detection with the applications into automated character recognition. EURASIP. J. Adv. Signal Process. **152** (2014)
6. Choi, K.: Electroencephalography (eeg)-based neurofeedback training for brain-computer interface (bci). Exp. Brain Res. **231**, 351–365 (2013)
7. Corralejo, R., Nicolas-Alonso, L.F., Alvarez, D., Hornero, R.: A p300-based brain-computer interface aimed at operating electronic devices at home for severely disabled people. Med. Biol. Eng. Comput. **52**, 861–872 (2014)
8. Czyżewski, A., Kostek, B., Kurowski, A., Szczuko, P., Lech, M., Odya, P., Kwiatkowska, A.: Assessment of hearing in coma patients employing auditory brainstem response, electroencephalography and eye-gaze-tracking. In: Proceedings of the 173rd Meeting of the Acoustical Society of America (2017)
9. Dickhaus, T., Sannelli, C., Muller, K.R., Curio, G., Blankertz, B.: Predicting bci performance to study bci illiteracy. BMC Neurosci. **10** (2009)
10. Diez, P.F., Mut, V.A., Avila Perona, E.M.: Asynchronous bci control using high-frequency. SSVEP. J. NeuroEngineering. Rehabil. **8**(39) (2011)
11. Doud, A.J., Lucas, J.P., Pisansky, M.T., He, B.: Continuous three-dimensional control of a virtual helicopter using a motor imagery based brain-computer interface. PLoS ONE. **6**(10) (2011)
12. Faller, J., Scherer, R., Friedrich, E., Costa, U., Opisso, E., Medina, J., Muller-Putz, G.R.: Non-motor tasks improve adaptive brain-computer interface performance in users with severe motor impairment. Front. Neurosci., 8 (2014)
13. Gardener, M., Beginning, R.: The statistical programming language, (2012). https://cran.r-project.org/manuals.html, Accessed on 2017-03-01

14. Goldberger, A.L., Amaral, L.A., Glass, L., Hausdorff, J.M., Ivanov, P.C., Mark, R.G., Mietus, J.E., Moody, G.B., Peng, C.K., Stanley, H.E.: Physiobank, physiotoolkit, and physionet: components of a new research resource for complex physiologic signals. Circulation **101**, 215–220 (2000)

15. He, B., Gao, S., Yuan, H., Wolpaw, JR.: Brain-computer interfaces, In: He, B. (ed.) Neural Engineering, pp. 87–151 (2012). https://doi.org/10.1007/978-1-4614-5227-0_2

16. Iscan, Z.: Detection of p300 wave from eeg data for brain-computer interface applications. Pattern Recognit. Image Anal. **21**(481) (2011)

17. Janusz, A., Stawicki, S.: Applications of approximate reducts to the feature selection problem. In: Proceedings of the International Conference on Rough Sets and Knowledge Technology (RSKT), number 6954 in Lecture Notes in Artificial Intelligence, pp. 45–50 (2011)

18. John, G.H., Langley, P.: Estimating continuous distributions in bayesian classifiers. In: Proceedings of the 11th Conference on Uncertainty in Artificial Intelligence, pp. 338–345 (1995)

19. Jung, T.P., Makeig, S., Humphries, C., Lee, T.W., McKeown, M.J., Iragui, V., Sejnowski, T.J.: Removing electroencephalographic artifacts by blind source separation. Psychophysiology **37**, 163–178 (2000)

20. Kasahara, T., Terasaki, K., Ogawa, Y.: The correlation between motor impairments and event-related desynchronization during motor imagery in als patients. BMC Neurosci. **13**(66) (2012)

21. Kayikcioglu, T., Aydemir, O.: A polynomial fitting and k-nn based approach for improving classification of motor imagery bci data. Pattern Recognit. Lett. **31**(11), 1207–1215 (2010)

22. Krepki, R., Blankertz, B., Curio, G., Muller, K.R.: The berlin brain-computer interface (bbci) - towards a new communication channel for online control in gaming applications. Multimed. Tools Appl. **33**, 73–90 (2007)

23. Kumar, S.U., Inbarani, H.: Pso-based feature selection and neighborhood rough set-based classification for bci multiclass motor imagery task. Neural Comput. Appl. **33**, 1–20 (2016)

24. LaFleur, K., Cassady, K., Doud, A.J., Shades, K., Rogin, E., He, B.: Quadcopter control in three-dimensional space using a noninvasive motor imagery based brain-computer interface. J. Neural. Eng. **10** (2013)

25. Leeb, R., Pfurtscheller, G.: Walking through a virtual city by thought. In: Proceedings of the 26th Annual International Conference of the IEEE EMBS, (2004)

26. Leeb, R., Scherer, R., Lee, F., Bischof, H., Pfurtscheller, G.: Navigation in virtual environments through motor imagery. In: Proceedings of the 9th Computer Vision Winter Workshop, pp. 99–108, (2004)

27. Marple, S.L.: Computing the discrete-time analytic signal via fft. IEEE Trans. Signal Proc. **47**, 2600–2603 (1999)

28. Martin, B.: Instance-based learning: nearest neighbour with generalization. Technical report, University of Waikato, Department of Computer Science, Hamilton, New Zealand (1995)

29. Nakayashiki, K., Saeki, M., Takata, Y.: Modulation of event-related desynchronization during kinematic and kinetic hand movements. J. NeuroEng. Rehabil. **11**(90) (2014)

30. Pawlak, Z.: Rough sets. Int. J. Comput. Inf. Sci. **11**, 341–356 (1982)

31. Pfurtscheller, G., Neuper, C.: Motor imagery and direct brain-computer communication. Proc. of IEEE **89**, 1123–1134 (2001)

32. Pfurtscheller, G., Brunner, C., Schlogl, A., Lopes, F.H.: Mu rhythm (de)synchronization and eeg single-trial classification of different motor imagery tasks. NeuroImage **31**, 153–159 (2006)

33. Postelnicu, C., Talaba, D.: P300-based brain-neuronal computer interaction for spelling applications. IEEE Trans. Biomed. Eng. **60**, 534–543 (2013)

34. Quinlan, R.: C4.5: Programs for Machine Learning. Morgan Kaufmann (1993)

35. Riza, S.L., Janusz, A., Slezak, D., Cornelis, C., Herrera, F., Benitez, J.M., Bergmeir, C., Stawicki, S.; Roughsets: data analysis using rough set and fuzzy rough set theories, (2015). https://github.com/janusza/RoughSets, Accessed on 2017-03-01

36. Roy, S.: Nearest neighbor with generalization. Christchurch, New Zealand (2002)

37. Schalk, G., McFarland, D.J., Hinterberger, T., Birbaumer, N., Wolpaw, J.R.: Bci 2000: A general-purpose brain-computer interface (bci) system. IEEE Trans. Biomed. Eng. **51**, 1034–1043 (2004)

38. Schwarz, A., Scherer, R., Steyrl, D., Faller, J., Muller-Putz, G.: Co-adaptive sensory motor rhythms brain-computer interface based on common spatial patterns and random forest. In: Proceedings of the 37th Annual International Conference of the Engineering in Medicine and Biology Society (EMBC), (2015)
39. Shan, H., Xu, H., Zhu, S., He, B.: A novel channel selection method for optimal classification in different motor imagery bci paradigms. BioMed. Eng. OnLine, **14** (2015)
40. Silva, J., Torres-Solis, J., Chau, T.: A novel asynchronous access method with binary interfaces. J. NeuroEng. Rehabil. **5**(24) (2008)
41. Siuly, S., Li, Y.: Improving the separability of motor imagery eeg signals using a cross correlation-based least square support vector machine for brain computer interface. IEEE Trans. Neural Syst. Rehabil. Eng. **20**(4), 526–538 (2012)
42. Siuly, S., Wang, H., Zhang, Y.: Detection of motor imagery eeg signals employing naive bayes based learning process. J. Measurement **86**, 148–158 (2016)
43. Suh, D., Sang Cho, H., Goo, J., Park, K.S., Hahn, M.: Virtual navigation system for the disabled by motor imagery. In: Advances in Computer, Information, and Systems Sciences, and Engineering, pp. 143–148 (2006). https://doi.org/10.1007/1-4020-5261-8_24
44. Szczuko, P., Lech, M., Czyżewski, A.: Comparison of methods for real and imaginary motion classification from eeg signals. In: Proceedings of ISMIS conference, (2017)
45. Szczuko, P.: Real and imagery motion classification based on rough set analysis of eeg signals for multimedia applications. Multimed. Tools Appl. (2017). https://doi.org/10.1007/s11042-017-4458-7
46. Szczuko, P.: Rough set-based classification of eeg signals related to real and imagery motion. In: Proceedings Signal Processing Algorithms, Architectures, Arrangements, and Applications, (2016)
47. Tadel, F., Baillet, S., Mosher, J.C., Pantazis, D., Leahy, R.M.: Brainstorm: A user-friendly application for meg/eeg analysis. Comput. Intell. Neurosci. vol. 2011, Article ID 879716 (2011). https://doi.org/10.1155/2011/879716
48. Tesche, C.D., Uusitalo, M.A., Ilmoniemi, R.J., Huotilainen, M., Kajola, M., Salonen, O.: Signal-space projections of meg data characterize both distributed and well-localized neuronal sources. Electroencephalogr. Clin. Neurophysiol. **95**, 189–200 (1995)
49. Tukey, J.W.: Exploratory Data Analysis. Addison-Wesley (1977)
50. Ungureanu, M., Bigan, C., Strungaru, R., Lazarescu, V.: Independent component analysis applied in biomedical signal processing. Measurement Sci. Rev. **4**, 1–8 (2004)
51. Uusitalo, M.A., Ilmoniemi, R.J.: Signal-space projection method for separating meg or eeg into components. Med. Biol. Eng. Comput. **35**, 135–140 (1997)
52. Velasco-Alvarez, F., Ron-Angevin, R., Lopez-Gordo, M.A.: Bci-based navigation in virtual and real environments. IWANN. LNCS **7903**, 404–412 (2013)
53. Vidaurre, C., Blankertz, B.: Towards a cure for bci illiteracy. Brain Topogr. **23**, 194–198 (2010)
54. Witten, I.H., Frank, E., Hall, M.A.: Data mining: Practical machine learning tools and techniques. In: Morgan Kaufmann Series in Data Management Systems. Morgan Kaufmann (2011). www.cs.waikato.ac.nz/ml/weka/, Accessed Mar 1st 2017
55. Wu, C.C., Hamm, J.P., Lim, V.K., Kirk, I.J.: Mu rhythm suppression demonstrates action representation in pianists during passive listening of piano melodies. Exp. Brain Res. **234**, 2133–2139 (2016)
56. Xia, B., Li, X., Xie, H.: Asynchronous brain-computer interface based on steady-state visual-evoked potential. Cogn. Comput. **5**(243) (2013)
57. Yang, J., Singh, H., Hines, E., Schlaghecken, F., Iliescu, D.: Channel selection and classification of electroencephalogram signals: an artificial neural network and genetic algorithm-based approach. Artif. Intell. Med. **55**, 117–126 (2012)
58. Yuan, H., He, B.: Brain-computer interfaces using sensorimotor rhythms: current state and future perspectives. IEEE Trans. Biomed. Eng. **61**, 1425–1435 (2014)
59. Zhang, R., Xu, P., Guo, L., Zhang, Y., Li, P., Yao, D.: Z-score linear discriminant analysis for EEG based brain-computer interfaces. PLoS ONE. **8**(9) (2013)

Chapter 13
Application of Tolerance Near Sets to Audio Signal Classification

Ashmeet Singh and Sheela Ramanna

Abstract This chapter is an extension of our work presented where the problem of classifying audio signals using a supervised tolerance class learning algorithm (TCL) based on tolerance near sets was first proposed. In the tolerance near set method(TNS), tolerance classes are directly induced from the data set using a tolerance level and a distance function. The TNS method lends itself to applications where features are real-valued such as image data, audio and video signal data. Extensive experimentation with different audio-video data sets were performed to provide insights into the strengths and weaknesses of the TCL algorithm compared to granular (fuzzy and rough) and classical machine learning algorithms.

Keywords Audio signal classification · Granular computing · Machine learning Rough sets · Tolerance near sets

13.1 Introduction

Automatic recognition (or classification) of speech and music content from acoustic features is a very popular application area of machine learning. Illustrative machine learning tasks include music information retrieval (MIR) [46, 49, 55], automatic music genre recognition [17], detection of commercial blocks in television news data, movie genre abstraction from audio cues [4].

This research has been supported by the Natural Sciences and Engineering Research Council of Canada (NSERC) Discovery grant. Special thanks to Dr. Rajen Bhatt, Robert Bosch Technology Research Center, US for sharing this data set.

A. Singh · S. Ramanna (✉)
Department of Applied Computer Science, University of Winnipeg, Winnipeg,
MB R3B 2E9, Canada
e-mail: s.ramanna@uwinnipeg.ca

A. Singh
e-mail: ashmeet908@gmail.com

© Springer International Publishing AG 2018   241
U. Stańczyk et al. (eds.), *Advances in Feature Selection for Data and Pattern Recognition*, Intelligent Systems Reference Library 138,
https://doi.org/10.1007/978-3-319-67588-6_13

This chapter is an extension of our work [38] on the problem of classifying audio signals with the near set-based method [30, 31] is addressed. Near set theory was influenced by rough set theory [26] and by the work of E. Orłowska on approximation spaces [24, 25]. The principal difference between rough sets and near sets is that near sets can be discovered without approximating sets with lower and upper approximation operators [12, 34]. Near sets are considered as a generalization of rough sets [50]. Application of near sets can be found in [11, 16, 33].

The basic structure which underlies near set theory is a perceptual system which consists of perceptual objects (i.e., objects that have their origin in the physical world) [12, 30, 50]. A great deal of human auditory perception is influenced by the structure of the ear, our limited dynamic range and physics [57]. Specifically, perception of music encompasses fields such as physics, psychoacoustics, mathematics, musicology, music theory, music psychology [17]. It deals with our ability to perceive such characteristics as pitch, musical scale, timbre, rhythm. And pitch is the perception of the frequency of sound. Hence, mathematical models aim to emulate the ear's ability to separate sound spectrally and temporally. Spectral separation is the ability to separate different frequencies in the sound signal. Temporal separation is the ability to separate sound events. The concept of tolerance plays a central role in distinguishing different sounds. The notion of tolerance is directly related to the idea of closeness between objects, such as image or audio segments that resemble each other with a tolerable level of difference. The term *tolerance space* was coined by Zeeman in modelling visual perception with tolerances [53]. Mathematically, a tolerance space $(X, \simeq)$ consists of a set X supplied with a binary relation $\simeq$ (i.e., a subset $\simeq \subset X \times X$) that is reflexive (for all $x \in X$, $x \simeq x$) and symmetric (for all $x, y \in X$, $x \simeq y$ and $y \simeq x$) but transitivity of $\simeq$ is not required. A modified form of a near sets algorithm based on tolerance spaces to classify audio signals is proposed in this chapter.

The supervised Tolerance Class Learner (TCL) algorithm is a modified form of the algorithm that was introduced in [37] to detect solar flare images. More recently, a tolerance form of rough sets (TRS) was successfully applied to text categorization [41, 42]. However, the key difference between these two approaches (TRS and TNS) is that in the TRS model, the data set is approximated into lower and upper approximations using a tolerance relation. In the TNS method, tolerance classes are directly induced from the data set using the tolerance level ε and a distance function. The TNS method lends itself to applications where features are real-valued such as image data [10], audio and video signals. However, the key difference between the TNS method used in Content-Based Image Retrieval (CBIR) and the proposed method is that our method is used in a supervised classification learning mode rather than a similarity-based image retrieval by ranking.

The representative set of features involve time and frequency representation. Our first dataset (referred to as *Speech vs. Non-Speech*) for experimentation consists of a collection of audio features extracted from videos with only two categories. The second dataset (referred to as *Music vs. Music + Speech*) is a further breakdown of one of the categories from the first data set. It is a part of a larger experiment for movie video abstraction with audio cues (speech, non-speech and music) from multi-genre movies [4].

Our third dataset (referred to as *Commercial Blocks in News Data*) [15, 47] includes features extracted from videos as well as audio. This is our first attempt at classifying video signals into two categories: commercial vs. non-commercial. This dataset was created out of footage from news channels to detect commercials which are shown in between the news broadcasts. This dataset includes audio features similar to the ones used in our first dataset. It also includes features extracted from the video. It is divided into 5 parts, each part representing a different news channel. The individual parts were combined and from those a random selection of objects was extracted.

The contribution of this work is a supervised learning algorithm Tolerant Class Learner (TCL) algorithm based on the TNS as well as extensive experimentation with different audio-video data sets to provide insights into the strengths and weaknesses of the TCL algorithm. TCL is compared with fuzzy, rough and classical machine learning algorithms. Our work has shown that TCL is able to demonstrate similar performance in terms of accuracy with the optimized Fuzzy IDT algorithm [4] and comparable performance with standard machine learning algorithms for the *Speech vs. Non-Speech* category. The results of this experiment were published in [38]. For the *Music vs. Music + Speech* category, the TCL algorithm is performing slightly worse. However, it should be noted that the machine learning algorithms available in WEKA[1] and RSES[2] has been optimized. TCL has not yet been optimized in terms of ideal tolerance classes and determination of best features since the features were already pre-determined as a part of the data set. Another important point to note, is that experiments with the Fuzzy IDT algorithm did not use a 10-fold cross-validation sampling. For the *Commercial Blocks in News Data*, TCL is comparable to the Naive Bayes algorithm, but not with Decision Trees and Support Vector Machines as reported in [47]. One important reason is that the authors performed post-and pre-processing on the data and used k-means to get a smaller more well divided dataset to optimize the Support Vector Machines algorithm which was not available for us.

The chapter is organized as follows: In Sect. 13.2, we discuss briefly research related to audio classification. In Sect. 13.5, we present some preliminaries for near sets and tolerance forms of near sets. We describe our proposed TCL algorithm in Sect. 13.6 followed by experiments in Sect. 13.7. We conclude the chapter in Sect. 13.8.

13.2 Related Work

Automatic music genre recognition (or classification) from acoustic features is a very popular music information retrieval (MIR) task where machine learning algorithms have found widespread use [3, 23, 46, 49, 55]. The broader field of MIR research

---

[1]http://www.cs.waikato.ac.nz/ml/weka/.

[2]http://www.mimuw.edu.pl/~szczuka/rses/start.html.

is a multidisciplinary study that encompasses audio engineering, cognitive science, computer science and perception (psychology). A survey of MIR systems can be found in [45]. Our ability to detect instruments stems from our ability to perceive the different features of musical instruments such as musical scale, dynamics, timbre, time envelope and sound radiation characteristics [17]. In this chapter, we restrict our discussion to machine learning methods applied to MIR and to audio/video features specific to this chapter.

Filters such as Principal Component Analysis are thus often applied to reduce the number of features. The most commonly used audio features include Mel-Frequency Cepstral Coefficients (MFCC), Zero Crossing Rate, Spectral Centroid [46]. MFCCs are perceptually motivated features commonly used in speech recognition research [13, 21, 46]. MFCCs initially described for speech analysis [21] were calculated over 25.6 ms windows applied on the audio file with a 10 ms overlap between each window.

Another important aspect of music classification is the size of the data sets. Datasets differ greatly in size, for example the MillionSong Dataset has a million files [22] or the GTZAN dataset contains 1000 files [44]. The MillionSong dataset uses HDF5 files which store tag data, pre-extracted features but not the audio signal itself. Others such as the GTZAN use actual audio files. The amount of pre-processing (or the lack thereof) is also a unique challenge. Datasets are often pre-processed to extract features [22] or are left as raw audio files [44] cut into clips.

In [3] the Adaboost algorithm was applied to the Magnatune dataset [3] consisting of 1005 training files and 510 testing files and the USPOP dataset[4] with 940 training and 474 testing files divided into 6 genres. The features included Fast Fourier Transform, Real Cepstral Features, Mel Frequency Cepstral Features, Zero Crossing Rate, Spectral Centroid, Rolloff, Auto-regression and Spread. In [23], a system that uses relevance feedback with Support Vector Machine (SVM) active learning was introduced. The audio features were based on Gaussian mixture models (GMMs) of MFCCs. The dataset included 1210 songs divided into 650 training, 451 testing and 103 validation songs. In [8], fuzzy-c means was used to classify music with the GTANZ and Magnatune dataset. Three time domain, five frequency domain and MFCC features for a total of 42 features from each of the 150 audio signals were used. By plotting the mean of a feature with its corresponding variance value for each signal, six optimal feature vectors were selected. Using Euclidean distance calculated on the feature vectors they performed clustering. Audio signals were classified into five broad categories: silence (SL), speech (SP), music (MS), speech with noise (SN), speech with music (SM).

In [14], music genre classification based on features involving timbral texture features, MFCC and wavelet features are discussed. The wavelet features were extracted using Discrete Wavelet Transform (DWT). The ensemble classifier uses a combination K-Nearest Neighbour (KNN), GMM and Artificial Neural Network (ANN)

---

[3]http://magnatune.com/.

[4]https://labrosa.ee.columbia.edu/millionsong/pages/additional-datasets.

classifiers using majority voting. PCA was used to reduce the dimensionality of the feature space. The Artificial Neural Network uses a feed-forward network with one input layer (300 input vectors of 38 dimensions), one output layer (for 10 music genres) with the pure linear transfer function and one hidden layer with sigmoid transfer function.

In [18] music genre classification was reported using MFCCs, Octave based spectral contrast (OSC and MPEG7 NASE) features on GTZAN dataset. Linear Discriminant Analysis was used for feature reduction and classification for classifying 10 music genres.

B. Kostek [17] presents a comprehensive discussion on perception-based data analysis in acoustics with applications in MIR and hearing. Granular computing methods (rough sets, fuzzy sets) [17, 35] and neural networks were applied to cognitive processing as well as to music genre recognition. In [56], several machine learning methods (Artificial Neural Networks, Support Vector Machines, Naive Bayes, Random Forest and J48) were applied to a database of 60 music composers/performers. 15–20 music pieces were collected for each of the musicians. All the music pieces were partitioned into 20 segments and then parameterized. The feature vector consisted of 171 parameters, including MPEG-7 low-level descriptors and Mel-Frequency Cepstral Coefficients (MFCC) complemented with time-related dedicated parameters. A more recent paper on the million song data set and the challenges of improving genre annotation can be found in [9]. Current developments in MIR can be found at.[5]

13.3 Automatic Detection of Video Blocks

Automatic detection of video blocks uses features and techniques that seem similar to those used for classification of audio. The video is divided into blocks, and from each block, video and audio data is extracted [15, 47]. The audio features extracted can be same as the ones used in MIR and audio classification [15, 47] such as MFCCs [13, 21, 46]. Previous techniques include detecting commercials using blank frames [40], product logos [1], repetitive video and audio cues [5] and analysis of on-screen text [20]. However the presence or absence of such features is less reliable especially now that broadcasts, channels and content have become digital and advertisements are targeted to consumers based on region, language, preference, gender, market demographics, age.

Commercial block classification methods can generally be divided into two categories - *knowledge-based* methods and *frequentist or repetition based* methods [15]. The knowledge-based approach uses information gathered from the video frames to make predictions. The frequentist-based method uses the fact that commercial content repeats with a higher frequency that news segments and other related types of

---

[5]http://www.ismir.net/.

content. The basic premise is that a commercial previously aired will repeat again, and then be identified.

The knowledge-based method extracts features or knowledge from the current frame, clip or segment like shot rate, presence or absence of logos, silent regions, blank frames, audio features, video features and so on [15]. Well-known machine learning methods such as SVM and ensemble classifiers such as Adaboost are then used for classification. Frequentist-based methods on the other hand operate on pre-saved content from commercials from which features have been extracted [15]. This data is then finger-printed or hashed. This data is then compared with data from the video feed to identify commercials. One shortfall of this technique is that only those exact commercials which have been previously identified, and then finger printed or hashed can be identified. They are most suitable for archived footage but are ill-suited for classifying commercials on the fly. What follows is a definition of features (which are used as probe functions). Probe functions are formally defined in Sect. 13.5.

13.4 Audio and Video Features

The following is an overview of the features used in each dataset. The video dataset [47] uses additional video features.

13.4.1 Audio Features

Definition 13.1 (*Spectral Centroid-SC_t*)

$$\frac{\sum_{n=1}^{N} M_t[n] \times n}{\sum_{n=1}^{N} M_t[n]},$$

where $M_t[n]$ is the magnitude of the Fourier transform at frame t and frequency bin n. The centroid is a measure of spectral shape and higher centroid values correspond to brighter textures with more high frequencies [46].

Definition 13.2 (*Spectral Flux-SF_t*)

$$\sum_{n=1}^{N} (N_t[n] - N_{t-1}[n])^2,$$

where $N_t[n]$ and $N_{t-1}[n]$ are normalized magnitude of the Fourier transform at the current frame t and the previous frame t-1. The spectral flux is a measure of the amount of local spectral change [46].

Definition 13.3 (*Spectral Roll Off-SRO_t*)

$$\sum_{n=1}^{R_t} M_t[n] = 0.85 \times \sum_{n=1}^{N} M_t[n],$$

where $M_t[n]$ is the magnitude of the Fourier transform at frame t and frequency bin n. Spectral Roll Off defines the frequency R_t below which 85% of the magnitude distribution of frequency lies.

Definition 13.4 (*Fundamental Frequency-$R(\tau)$*)

$$\frac{1}{N} \sum_{n=0}^{N} (x_r[n]x_r[n + \tau]).$$

It is a measure of the periodicity of the time-domain waveform. It may also be calculated from the audio signal as the frequency of the first order harmonic or the spacing between the different order harmonics in the periodic signal [39].

Definition 13.5 (*Zero Crossing Rate-Z_t*)

$$\frac{1}{2} \sum_{n=1}^{N} |sign(x[n] - sign(x[n - 1])|,$$

where the function returns 1 for a positive argument and 0 for a negative argument. It is the rate at which the signal changes sign. It is a measure of how noisy the signal is [46].

Definition 13.6 (*Short Time Energy-STE*)

$$\frac{1}{N} \sum_{m} [x(m)w(n - m)]^2,$$

where $x(m)$ is the audio signal, $w(m)$ is a rectangle window [54].

Definition 13.7 (*Low Short Time Energy Ratio-LSTER*)

$$\frac{1}{2N} \sum_{n=1}^{N} [(0.5 \times avSTE) - sign(STE(n)) + 1],$$

where

$$avSTE = \frac{1}{N} \sum_{n=1}^{N} [STE(n)],$$

and N is the number of frames, STE is the Short Time Energy of the nth frame and $avSTE$ is the average Short Time Energy in a window [19].

Definition 13.8 (*Root Mean Square Amplitude-E_{RMS}*)

$$\sqrt{\frac{1}{N} \sum_{n=1}^{N} [x^2(n)]}$$

Root Mean Square or RMS amplitude is the simple root mean square of the signal and is the same as that for any signal like a voltage signal. However, since a digital audio signal is non-continuous and is composed of n values we will use the above mentioned RMS function and not the regular RMS function which is described for continuous functions [6].

Definition 13.9 (*Log Energy-E_{log}*)

$$E_{bias} + E_0 \cdot log \sum_{n=0}^{N-1} x^2(n),$$

where log represents the natural logarithm and E_0 is a scaling factor to scale the energy of signal x to different scales. If $E_0 = 1$ then E_{log} is measured in 'neper'. If $E_0 = \frac{10}{log 10}$ which is approximately equal to 4.343, then E_{log} is measured in Decibels (dB) [6].

Mel-Frequency Cepstral Coefficient [21]

Mel-Frequency Cepstral Coefficient (or MFCC) is a combination of a cepstral representation of the spectrum using the Mel Frequency. It was initially described for use in speech analysis and has only recently been incorporated into music as well. MFCCs are also calculated in short windows on the audio file. The windows are generated using a Windowing function such as a Hamming or Hanning window. After the windows have been obtained, Discrete Fourier Transform is applied to them. Then Log Energy of the audio signal contained within the window is calculated, applying the Mel Scale to it, and then finally applying Discrete Cosine Transform on it, MFCC values of different orders are generated.

13.4.2 Video Features

The following is an overview of the video features. These are used only in the experiments involving the second dataset.

Definition 13.10 (*Shot Length*) Shot length [47] is simply the length of each shot. It is defined in terms of the frames contained within that shot and forms a 1D feature for the video segment. It was observed by the authors that commercials shots have a shorter duration than non-commercial shots.

Definition 13.11 (*Text Masked Edge Change Ratio*) Edge Change Ratio [47] (or ECR for short) detects the scene changes and the difference between the frames when the scene change occurs. It is a measure of the net motion at the object boundaries between successive frames. To calculate text Masked Edge Change Ratio, the text regions are masked to ignore the animation effects applied on the text. In the dataset, the authors used second order statistics to obtain a 2D feature for estimating motion content.

Definition 13.12 (*Text Masked Frame Difference*) Text Masked Frame Difference [47] estimates the amount of change between successive video frames. ECR may fail at times when motion occurs within the boundaries of the object, i.e., the object remains static (no translational motion) but some change occurs within it, like a color change, animation etc. In the dataset, the authors use second order statistics to obtain a 2D set of features for representing the net Frame Difference.

Definition 13.13 (*Overlaid Text Distribution*) Overlaid Text [47] are the text bands inside a frame which show information in a text based format within the frame. These are most often present during a news broadcast as a news ticker found most often in the lower third part of the screen and referred to as the chyron, captions, etc. During commercials, however, most of these text bands disappear to show the commercial content. Only a single news ticker may remain sometimes either at the top or at the bottom. To estimate the amount of overlaid text, the frame is divided into 5 horizontal strips of equal height. Within each strip the amount of text is analyzed. Second order statistics are then applied in each strip to generate a 10D feature set.

13.5 Theoretical Framework: Near Sets and Tolerance Near Sets

The TCL algorithm system uses a set-theoretic approach based on near set theory. The theory of near sets can be summarized in three simple concepts: a perceptual system, a perceptual indiscernibility relation and nearness relation and a near set [51]. Near set theory is characterized by a perceptual system, whose objects are associated by relations such as indiscernibility, weak indiscernibility, or tolerance. It is the perceptual tolerance relation that is used in this chapter. We now give formal defns for near sets and tolerance near sets. We start by giving a table of symbols.

Symbol	Interpretation
O	Set of perceptual objects
F	Set of probe functions defining a perceptual object
$\mathbb{R}$	Set of real numbers
ϕ	Probe function
$\mathscr{B}$	$\mathscr{B} \subseteq \mathbb{F}$, Subset of probe functions
x	$x \in O$, Sample perceptual object
$\phi_{\mathscr{B}}(x)$	$\phi_1(x), \ldots, \phi_i(x), \ldots, \phi_l(x)$ Description of a perceptual object
l	Length of object description
$\sim_{\mathscr{B}}$	Perceptual Indiscernibility Relation
$\cong_{\mathscr{B}}$	Weak Perceptual Indiscernibility Relation
$\bowtie_{\mathbb{F}}$	Nearness relation
$\langle X, \simeq \rangle$	Tolerance Space
ε	$\varepsilon \in \mathbb{R}$
$\cong_{\mathscr{B},\varepsilon}$	Perceptual Tolerance Relation
$N(x)$	Neighbourhood of x in tolerance space
TCL	Tolerance Class Learner

13.5.1 Preliminaries

Definition 13.14 (*Perceptual System* [30, 31]) A perceptual system is a pair $\langle O, F \rangle$, where O is a nonempty set of perceptual objects and F is a countable set of real-valued *probe functions* $\phi_i : O \to \mathbb{R}$.

Examples of probe functions include audio features such as MFCCs, Spectral Centroid defined in Sect. 13.4.

An object description is defined by means of a tuple of probe function values $\Phi(x)$ associated with an object $x \in X$, where $X \subseteq O$ as defined by Eq. 13.1.

Definition 13.15 (*Object Description* [30, 31])

$$\Phi(x) = (\phi_1(x), \phi_2(x), \ldots, \phi_n(x)), \tag{13.1}$$

where $\phi_i : O \to \mathbb{R}$ is a probe function of a single feature. Objects here are audio signals arising from a data set consisting of for example speech and non-speech elements.

Example 13.1 (Audio Signal Description)

$$\Phi(x) = \left(\phi_{SC_t}(x), \phi_{SF_t}(x), \phi_{SRO_t}(x), \phi_{R_\tau}(x), \phi_{Z_t}(x), \phi_{STE}(x), \ldots, \phi_{E_{log}}(x) \right) \tag{13.2}$$

A probe function is the equivalent of a sensor in the real world. Indeed the values for a description might come from an actual sensor like a sensor measuring the scale frequency and amplitude of sound and giving a value leading to a computation of the MFCC feature. Probe functions give rise to a number of perceptual relations between objects of a perceptual system. This approach is useful when decisions on nearness are made in the context of a perceptual system i.e., a system consisting of objects and our perceptions of what constitutes features that best describe these objects. In this chapter, the features of an audio signal are such characteristics as loudness, pitch, timbre.

Definition 13.16 (*Perceptual Indiscernibility Relation* [34]) Let $\langle O, F \rangle$ be a perceptual system and let $\mathscr{B} \subseteq F$,

$$\sim_{\mathscr{B}} = \{(x, y) \in O \times O : \text{for all } \phi_i \in \mathscr{B}, \ \phi_i(x) = \phi_i(y)\}. \tag{13.3}$$

Definition 13.16 is a refinement of the original idea of an indiscernibility relation [26] between objects to what is known as perceptual indiscernibility that is more in keeping with the perception of objects in the physical world such as perceiving music based on audio signals. The perceptual indiscernibility relation is reflexive, symmetric, and transitive. In other words, this relation aggregates objects with matching descriptions into equivalence classes. These descriptions are provided by the probe functions.

Definition 13.17 (*Weak Indiscernibility Relation* [34]) Let $\langle O, F \rangle$ be a perceptual system and let $\mathscr{B} \subseteq F$,

$$\cong_{\mathscr{B}} = \{(x, y) \in O \times O : \text{for some } \phi_i \in \mathscr{B}, \ \phi_i(x) = \phi_i(y)\}. \tag{13.4}$$

Definition 13.17 bears a resemblance to the human sound recognition system which compares similarity based on some features, but not all of them.

The study of near set focuses on the discovery of affinities between two disjoint perceptual objects. Disjoint sets containing objects with similar descriptions are near sets. In other words, pair of nonempty sets are considered near, if the intersection of sets is not empty, i.e., near sets have elements in common [32].

Definition 13.18 ([34] *Nearness Relation*) Let $\langle O, \mathbb{F} \rangle$ be a perceptual system and let $X, Y \subseteq O$. A set X is near to set Y within the perceptual system $\langle O, \mathbb{F} \rangle$ ($X \bowtie_{\mathbb{F}} Y$) iff there are $\mathscr{B}_1, \mathscr{B}_2 \subseteq \mathbb{F}$ and $\phi_i \in \mathbb{F}$ and there are $A \in O_{/\sim_{\mathscr{B}_1}}, B \in O_{/\sim_{\mathscr{B}_2}}, C \in O_{/\sim_{\phi_i}}$ such that $A \subseteq X$, $B \subseteq Y$, and $A, B \subseteq C$. If a perceptual system is understood, than a set X is near to set Y.

Following is the definition of perceptual near sets. X and Y are near sets if they satisfy the nearness relation.

Definition 13.19 ([34] *Perceptual Near Sets*) Let $\langle O, \mathbb{F} \rangle$ be a perceptual system and let $X, Y \subseteq O$ denote disjoint sets. Sets X, Y are near sets iff $X \bowtie_{\mathbb{F}} Y$.

Note 13.1 Near sets are also referred to as **descriptively near sets** [32] to distinguish them from **spatial near sets**. The description of sets of objects can be considered as points in an l-dimensional Euclidean space $\mathcal{R}^l$ called a feature space.

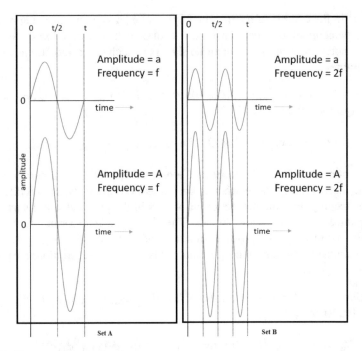

Fig. 13.1 Near sets of audio signals

Figures 13.1 and 13.2 are two examples of near sets in the context of audio signals. Notice that in Fig. 13.1, the audio signals from sets A and B are *near* if we consider the *probe function* that extract the amplitude feature. In Fig. 13.2, the two sets represented in ovals, are near based on the probe functions of musical notes emanating from instruments. These are examples of **descriptively near sets** where the descriptions are based on feature values of the audio signals.

13.5.2   Tolerance Near Sets

The proposed approach considered in this chapter in the context of tolerance spaces is directly related to work on sets of similar objects, starting with J.H. Poincaré [36]. Poincaré was inspired by the works of Ernst Weber in 1834 and Gustav Fechner in 1850. Using the laws and experiments of Weber and Fechner as a foundation, Poincaré defined tolerance in terms of an abstraction of sets. Zeeman noticed that a single eye could not distinguish between exact points in a 2D space (which is composed of infinite number of points) and could only distinguish within a certain tolerance [52].

Fig. 13.2 Near sets of
musical notes

Subsequently, tolerance relations were considered by E.C. Zeeman [52] in his study of the topology of the brain. Tolerance spaces as a framework for studying the concept of resemblance was presented in [43] and in [29, 48]. Recall, that a tolerance space $(X, \simeq)$ consists of a set X endowed with a binary relation $\simeq$ (i.e., a subset $\simeq \subset X \times X$) that is reflexive (for all $x \in X$, $x \simeq x$) and symmetric (for all $x, y \in X$, $x \simeq y$ and $y \sim x$) but transitivity of $\simeq$ is not required. Tolerance relations are considered generalizations of equivalence relations.

Definition 13.20 (*Perceptual Tolerance Relation* [27, 28]) Let $\langle O, F \rangle$ be a perceptual system and let $\mathscr{B} \subseteq F$,

$$\cong_{\mathscr{B},\epsilon} = \{(x, y) \in O \times O : \| \phi(x) - \phi(y) \|_2 \leq \varepsilon\}, \tag{13.5}$$

where $\| \cdot \|_2$ denotes the L^2 norm of a vector.

Definition 13.21 (*Neighbourhood* [48]) Let $\langle O, \mathbb{F} \rangle$ be a perceptual system, let $x \in O$, then for every set $\mathscr{B} \subseteq \mathbb{F}$ and $\varepsilon \in \mathbb{R}$, a *neighbourhood* is then defined by the following equation.
$$N(x) = \{y \in O : x \cong_{\mathscr{B},\varepsilon} y\}.$$

Definition 13.22 (*Pre-Class* [48]) Let $\langle O, \mathbb{F} \rangle$ be a perceptual system, then for $\mathscr{B} \subseteq \mathbb{F}$, $\varepsilon \in \mathbb{R}$, set $X \subseteq O$ is defined as a *Pre-Class* iff $x \cong_{\mathscr{B},\varepsilon} y$ for any pair $x, y \in X$.

Definition 13.23 (*Tolerance Class* [48]) A *Pre-Class* which is *Maximal* with respect to inclusion is called a *tolerance class*.

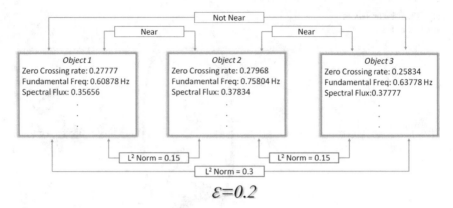

Fig. 13.3 Tolerance classes

The TCL learner algorithm uses tolerance class defined in Definition 13.23 for computing tolerance class for the feature vectors of audio signals. An example that demonstrates the determination of a tolerance class is shown in Fig. 13.3.

Example 13.2 (Audio Signal Features)

$$\Phi(x) = \left(\phi_{R_\tau}(x), \phi_{Z_t}(x), \phi_{SF_t}(x) \right) \tag{13.6}$$

Note, that in this example, the feature vector consists of 3 features. Based on the Euclidean distance measure and a $\varepsilon = 0.2$, one can see that object 1 and object 2 are *near* based on their feature values since their L^2 norm value is less than ε and object 2 and object 3 are *near* since their L^2 norm value is less than ε. However, the transitive property is violated since object 3 and object 1 are *not near* due their L^2 norm value being greater than ε. In other words, object 1 and 2 belong to one tolerance class and object 2 and object 3 belong to a second tolerance class.

13.6 Tolerance Class Learner - TCL

The Tolerance Class Learner [38] shown in Fig. 13.4 classifies audio and video data using a supervised approach.

In Module 1, the entire data set is loaded which includes all the feature values of the audio signal objects as well as the associated categories or labels. Each audio and video feature value is normalized between 0 and 1 using the Min- Max method shown in Module 2. Module 3 uses the normalized data and creates a distance matrix which calculates the Euclidean distance between each object in the training set. Module 4 (TCL Function) implements the *perceptual tolerance relation* defined in Definition 13.20 and computes the *tolerance classes* according to Definition 13.23. The best value for ε is chosen. The TCL function then creates tolerance classes by

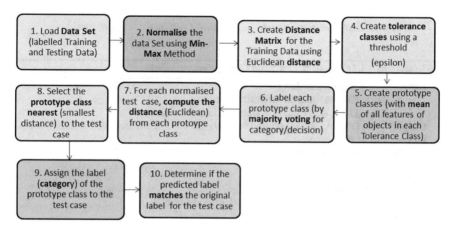

Fig. 13.4 Control flow diagram showing overview of TCL

collecting objects whose distances are less than ε into sets. Note that each set is maximal with respect to inclusion. A representative (prototype class) is created by finding the mean of all the objects inside a particular tolerance class in Module 5. This gives us a single-object-representation of every object in that tolerance class. The *feature values* for this prototype class object is simply the mean of feature values of all objects in that class. In Module 6, each prototype class object is then assigned a category (label) based on the most common label (majority category) of the objects in that tolerance class. In Module 7, for each normalized object in the test set (created in Module 2), the Euclidean distance between the object and the prototype class is computed. The category (label) for the prototype class that is closest (smallest distance) to the test object is chosen as the *category* for the test object in Modules 8 and 9. In Module 10, The *true category* of the test is compared with the *predicted category*. This process is repeated using a 10-fold cross validated (CV) method where the data is randomly divided into 10 pairs of training and testing sets. Classification is performed in two phases shown in Algorithms 13.1 and 13.2.

13.6.1 Algorithms

In the *learning phase* shown in Algorithm 13.1, given a tolerance level ε, tolerance classes are induced from the training set, and the representative of each tolerance class is computed as well as a category determination is performed based on majority voting. In the *testing phase*, the L^2 distance is computed for each element in test set with all of tolerance class representatives obtained in the first phase and assigned the category of the tolerance class representative based on the lowest distance value. The algorithms use the following notation:

$$procedure\_name(input_1, \ldots, input_n; output_1, \ldots, output_m)$$

Algorithm 13.1: Phase I: Learning Categories or labels

Input : $\varepsilon > 0$, // Tolerance level
         $\Phi(x)$, // Probe functions or Features vector
         $TR = \{TR_1, \ldots, TR_M\}$, // Training Data Set of size M
Output: $(NC, \{R_1, \ldots, R_{NC}\}, \{Cat_1, \ldots, Cat_{NC}\})$ //R_i is the representative class element
for $i \leftarrow 1$ **to** M **do**
  | computeMinMaxNorm(TR_i);
end
for $i \leftarrow 1$ **to** M **do**
    for $j \leftarrow i + 1$ **to** M **do**
      | computeNormL$_2$($TR_i, TR_j, ProbeValue_{ij}$; NormL2$_{ij}$);
    end
end
for $i \leftarrow 1$ **to** M **do**
    for $j \leftarrow i + 1$ **to** M **do**
      generateToleranceclass(NormL2$_{ij}$, ε; SetOfPairs);
      computeNeighbour(SetOfPairs, i, TR; N_i); // Compute the neighbourhood N_i of ith
      training data TR_i
      for all $x, y \in N_i$ **do**
        if $x, y \notin SetOfPairs$ **then**
          | $C_i \leftarrow N_i$; // Include y from class N_i into C_i
        end
      end
    end
    $H \leftarrow H \cup \{C_i\}$;
    // C_i is one tolerance class induced by the tolerance relation
    computeMajorityCat(C_i; Cat_i); // Determine Category by majority voting for each C_i
end
$NC \leftarrow |H|$; // Number of classes
// End of defineClass
defineClassRepresentative(NC, $\{R_1, \ldots, R_{NC}\}$, $\{Cat_1, \ldots, Cat_{NC}\}$);

The *computeMinMaxNorm* function normalizes each training set element TR_i. The L_2 norm distance between each pair of training set elements is computed using *computeNormL$_2$*. Next, by using the *generateToleranceclass* function, a set of training set pairs within the tolerance level ε is obtained.

For a training set element TR_i, its neighbourhood N_i is composed of all training set elements within the tolerance level ε of TR_i, including TR_i computed by the *computeNeighbour* function. If $N_i = \{TR_i\}$ then TR_i is a tolerance class with only one element. If a pair of neighbours in N_i does not satisfy the tolerance relation, the corresponding element is excluded from the tolerance class C_i. In addition, for each *representative* tolerance class C_i, the category information is obtained by a majority vote. Lastly, the *defineClassRepresentative* function computes the representatives (prototype) of each of the NC tolerance classes and assigns the majority class. The tolerance class prototype is a vector whose values are computed as the mean of the probe function values of those belonging to that class.

13.6.2 Phase II: Classification

Algorithm 13.2 in the classification phase uses the information obtained in the training phase, i.e., the set of representatives of training set and their associated category, inferred from the tolerance relation established between them. The *computeMin-MaxNorm* and *computeNormL$_2$* functions are the same as in phase I except that they require testing set data and the representative class data respectively. The *DetermineCat* function chooses the representative class that is closest to the test set element and assigns its category to that test set element.

Algorithm 13.2: Phase II: Assigning categories or labels

Input : $\varepsilon > 0$, // Tolerance level
$\Phi(x)$, // Probe functions or Features vector
$TS = \{TS_1, \ldots, TS_M\}$, // Testing Data Set
$\{R_1, \ldots, R_{NC}\}, \{Cat_1, \ldots, Cat_{NC}\}$ // Representative Class and their associated categories
Output: $(TS' = \{TS'_1, \ldots, TS'_M\})$ // Testing Data Set with assigned categories
for $i \leftarrow 1$ **to** M **do**
| computeMinMaxNorm(TS_i);
end
for $i \leftarrow 1$ **to** M **do**
| **for** $j \leftarrow i + 1$ **to** NC **do**
| | computeNormL$_2$(TS_i, RC_j, $ProbeValue_{ij}$; NormL2$_{ij}$);
| **end**
end
DetermineCat($NormL2_{ij}$; TS') // Computes min. distance and assigns category

---

The complexity of the algorithms is fairly straight forward to determine. In phase 1, the complexity of *computeMinMaxNorm* and *computeNormL$_2$* functions is $\mathcal{O}(n^2)$. The complexity of *generateToleranceclass* function is $\mathcal{O}(n^3)$. In phase 2, the complexity of *DetermineCat* function is $\mathcal{O}(n^2)$.

13.7 Experiments

13.7.1 Speech and Non-speech Dataset

The TCL algorithms were implemented in C++. Our dataset for training and testing has been restricted to a two-class problem with speech or non-speech categories. It consists of 3.5 h of audio data with a total of 12,928 segments (elements) with each segment marked as either speech or non-speech manually. The following nine (9) features were used in our experiments which were also used by Fuzzy Decision Tree Classifier [4]: Normalized zero-crossing rate, Variance of differential entropy computed from Mel-filter banks, Mean of Spectral flux, Variance of differential spectral centroid, Variance of first 5 Mel-Frequency cepstral coefficients (MFCC). MFCCs are perceptually motivated features commonly used in speech recognition research [13, 21, 46].

We have used 10-fold cross-validation for all our experiments except for FDT (whose results were reported in [4]). We have chosen a representative set of classical machine learning algorithms: Decision Trees(DT), Rule-based Systems (RBS-Ripper Algorithm), Support Vector Machines(SVM) and Naive Bayes(NB) implemented in WEKA [7] as well the classical rough sets model implemented in RSES [2]. For consistency, we created ten sets of training and testing pairs which were then used for experimentation across the three systems: RSES, WEKA and TCL. Table 13.1 gives accuracy calculated as percentage of correctly classified test set elements. The results reported for TCL are for the best value of ε though experiments with other values of ε starting from 0.1 to 0.9 were conducted. The average number of rules used by some of the appropriate classifiers is given in Table 13.2. The rough set classifier uses a form of genetic algorithm to generate the rules. Algorithmic details about the RSES system and the classifier can be found in [2]. The average execution time used by the classifiers is shown in Table 13.3. It should be pointed out that WEKA* indicates that time of execution for all classifiers implemented in WEKA is roughly the same. Table 13.4 gives accuracy values of our proposed TCL learner for different ε values for each FOLD.

Table 13.1 Accuracy (%) of TCL for all categories for each FOLD [38]

FOLD	RSES	DT(J48)	Bayes (NB)	RBS (JRIP)	SVM	TCL
1	90	90	87	89	90	86
2	90	93	87	91	90	88
3	92	92	89	91	90	87
4	91	92	89	92	91	86
5	91	92	89	91	90	86
6	92	91	89	91	91	86
7	90	91	88	91	90	86
8	90	88	88	90	90	87
9	90	91	90	91	90	88
10	89	91	87	90	90	86
Avg	90	91	88	91	90	87
FDT	87					

Table 13.2 Average number of rules

	RSES	WEKA-J48	WEKA-JRIP	FDT [4]
Num. of Rules	30,809	97 (equivalent)	16	2

Table 13.3 Execution Time for entire 10-Fold CV experiment

	RSES	WEKA*	FDT	TCL
Time in minutes	210	2	2 m sec	48

Table 13.4 Accuracy (%) of TCL for ε values for each FOLD

FOLD	$\varepsilon = 0.5$	$\varepsilon = 0.6$	$\varepsilon = 0.7$	$\varepsilon = 0.8$	$\varepsilon = 0.9$
1	86	82	86	80	85
2	85	77	85	88	87
3	86	74	87	86	83
4	86	79	82	74	82
5	85	71	71	84	72
6	85	79	86	86	86
7	87	79	85	87	84
8	88	79	87	87	85
9	86	79	86	87	97
10	85	84	86	86	84
Avg	86	78	84	84	84

13.7.2 Discussion

Overall, TCL is able to demonstrate similar performance in terms of accuracy with Fuzzy IDT algorithm and comparable performance with algorithms implemented in RSES and WEKA (as shown in Table 13.1). It should be noted that the Naive Bayes, SVM and TCL classifiers do not *discretize* the data set. In our experiments, it can be observed that the best accuracy that can be obtained with this data set is 91%. It is also interesting to note that the number of rules vary significantly across the rule-based classifiers (shown in Table 13.2). Clearly, this difference is reflected in their execution times. The fuzzy IDT algorithm requires pre-processing in terms of determination of membership functions (Gaussian) and clustering prior to classification. The fuzzy IDT algorithm was also designed for embedded platforms where execution time and compact classifiers are primary optimization criteria in addition to accuracy [4]. One of the key issues in TCL is determining the most optimal value of ε since this value is key to determining representative classes (see Table 13.4). This is similar to neighbourhood-based classifiers such as k-means. However, determining this parameter can be optimized, a process similar to the ones used for determining k-values.

13.7.3 Music and Music + Speech Dataset

The second dataset is an extension of the first dataset [4]. In this dataset the non-speech section is broken down further into music and music with speech. The data consists of 9348 objects labelled as music and music with speech. The objects are composed of 1 second segments collected form over 2.5 h of ground truth. The ground truth was movies from which audio was extracted as a WAV file. Each segment was marked as either music (purely instrumental) or music with speech (speech with background music or song). From each segment, the following features 10 were extracted - normalized zero-crossing rate (ZCR), variance of differential entropy computed from mel-filter banks, mean of spectral flux (SF), variance of spectral flux (SF), variance of differential spectral centroid (SC) and variance of first 5 Mel Frequency Cepstral Coefficients (MFCCs) all of which were used for our experiments. The MFCCs are the main driving features in this dataset too. As before the TCL algorithm was implemented in C++. The ten (10) features used in this dataset were: normalized Zero-Crossing Rate (ZCR), variance of differential entropy computed from mel-filter banks, mean of spectral flux (SF), variance of spectral flux (SF), variance of differential Spectral Centroid (SC) and variance of first 5 Mel Frequency Cepstral Coefficients (MFCCs) all of which were used for our experiments. The same features were used by the Fuzzy Decision Trees (FDT) [4]. Similar to the previous experiments, we have performed the 10-Fold CV for all our experiments except for FDT which gives an average accuracy. Table 13.5 gives an overview of the WEKA and RSES results with the 10-Fold CV data.

Similar to our previous experiment, we have used the same 10 pairs of training and testing data in our experiments for consistency. We use the same classical machine learning algorithms as used previously for speech non-speech data, Decision Trees (DT), Rule-based Systems (RBS), Support Vector Machines (SVM) and Naive Bayes

Table 13.5 Accuracy (%) of TCL for all categories for each FOLD

FOLD	RSES	DT(J48)	Bayes (NB)	RBS (JRIP)	SVM	TCL
1.	63	72	72	72	72	54
2.	59	68	68	68	68	68
3.	58	68	68	67	68	68
4.	56	68	69	69	69	69
5.	69	74	74	74	74	74
6.	63	69	69	69	69	68
7.	62	68	68	68	68	65
8.	56	69	69	69	69	68
9.	57	70	70	70	70	68
10.	62	71	71	71	71	67
Avg	54	70	70	70	70	68
FDT	70					

Table 13.6 Accuracy (%) of TCL for ε values for each FOLD

FOLD	$\varepsilon = 0.65$	$\varepsilon = 0.7$	$\varepsilon = 0.75$	$\varepsilon = 0.8$	$\varepsilon = 0.85$
1	66	66	58	54	56
2	70	64	68	68	65
3	67	68	68	68	68
4	65	69	66	70	69
5	62	56	74	74	74
6	57	61	69	69	68
7	57	61	59	68	65
8	59	65	65	69	68
9	63	64	70	70	68
10	69	66	64	71	67
Avg	63	64	66	68	67

(NB) implemented in WEKA [7] as well the classical rough sets model implemented in RSES [2] to validate our results. The results reported for TCL are for the best value of ε. Table 13.6 gives a detailed breakdown of accuracy vis-a-vis ε.

13.7.4   Detection of Commercial Blocks in News Data

The dataset was prepared by the authors in [15, 47] using footage from Indian and International News Channels. Over 30 h of footage was collected by the authors for a total of 129,676 shots. The footage was recorded at 720 X 576 resolution and 25 fps. It was collected from 5 television news channels CNN-IBN, TIMES NOW, NDTV 24X7, BBC WORLD and CNN with 33,117, 39,252, 17,052, 17,720, and 22,535 shots collected from each of channel respectively. The dataset is dominated by commercial shots with roughly 63% positives. From each shot, 13 features were extracted—4 video features and 7 audio features. The video features used were the mean and variance of Edge Change Ratio, Frame Difference Distribution over 32 bins, Text Area Distribution over 30 bins, Shot Length and Motion Distribution over 40 bins. Additionally the mean and variance of the 30 bins of Text Area Distribution and 32 bins of Frame Difference Distribution were also used. The video features looked for cues related to text on the screen, fast edits on the screen and fast movement of video content on the screen. The audio features look for cues like higher amplitude, music or jingles, sharp transitions between the commercials changes. The audio features used were the mean and variance of the following features: Zero Crossing Rate (ZCR), Short Time Energy (STE), Spectral Centroid (SC), Spectral Flux (SF) Spectral Roll-Off Frequency (SRF), Fundamental Frequency (FF). The final audio feature used were Mel-Frequency Cepstral Coefficients (MFCC) over 4000 bins [13, 21, 46]. The MFCCs were extracted from a bag-of-words and clustered around 4000 bins. The audio signal corresponding to each word was transformed into an MFCC value and these were then compiled into a 4000-dimensional histogram.

For our experiment we used about 17,000 shots due to memory limitations caused by the large sizes of certain feature vectors. The dataset is categorised as 1 and -1 for commercial and non-commercial respectively. It was distributed in LIBSVM format where each feature or feature value is associated with a key. Due to the large size of the dataset, it was shortened to 17,000 random objects from all channels. This was divided into 10 training and testing pairs to perform 10-fold CV with the TCL learner. The same 10 training and testing pairs were also used in WEKA where they were benchmarked with classical machine learning algorithms - SVM, Naive Bayes, BayesNet, Decision Trees(DT) and Rule-based classifiers(RBS). We were unable to run the rough-set based algorithm with RSES due to memory limitations.

The TCL algorithm performed best with a ε value of 0.5. We also discovered that the dataset published in the UCI Machine Learning repository[6] did not match the dataset that used in the paper [47] published by the creators of this dataset. In addition, the authors performed post-and pre-processing on the data and used k-means to get a smaller more well divided dataset. Hence the results with SVM from shown in Table 13.7 is an optimised result [47] using a well-balanced data set. We chose to benchmark with WEKA datasets instead. Due to the large size of the dataset, the run-time for the TCL algorithm on this dataset is one hour. Table 13.8 gives accuracy values of our proposed TCL learner for different ε values for each FOLD.

Table 13.7 Accuracy (%) of WEKA Algorithms and TCL for all categories for each FOLD

FOLD	BayesNet	SVM	DT	RBS	NB	TCL
1.	77	82	79	80	72	73
2.	76	80	80	80	72	72
3.	76	80	80	80	71	73
4.	76	82	80	80	71	75
5.	78	83	80	82	73	76
6.	77	82	78	80	73	74
7.	75	80	78	78	70	76
8.	75	80	80	80	71	74
9.	75	81	76	79	73	70
10.	76	81	80	81	69	74
Avg	76	81	79	80	72	74

---

[6]https://archive.ics.uci.edu/ml/datasets/TV+News+Channel+Commercial+Detection+Dataset.

Table 13.8 Accuracy (%) of TCL for ε values for each FOLD

FOLD	$\varepsilon = 0.5$	$\varepsilon = 0.6$	$\varepsilon = 0.7$	$\varepsilon = 0.8$	$\varepsilon = 0.9$
1.	73	50	47	46	52
2.	72	52	50	48	47
3.	73	51	52	46	50
4.	75	53	53	52	48
5.	76	48	56	54	49
6.	74	52	49	52	50
7.	76	48	47	54	51
8.	74	50	51	48	47
9.	70	53	50	48	53
10.	74	49	47	53	45
Avg	74	50	50	49	49

13.8 Conclusion and Future Work

We have proposed a supervised learning algorithm based on a tolerance form of near sets for classification learning. Extensive experimentation with different audio-video data sets were performed to provide insights into the strengths and weaknesses of the TCL algorithm compared to granular (fuzzy and rough) and classical machine learning algorithms. Based on the results, an obvious optimization is to remove outliers from tolerance classes since this has a bearing on the average feature vector value. The second optimization task that is also common to all learning problems is the determination of best features. Classification of TV programming is another interesting application for the TCL algorithm where a more fine grained categorization beyond news and commercials can be explored such as sports broadcasts, TV shows and other programming.

References

1. Banic, N.: Detection of commercials in video content based on logo presence without its prior knowledge. In: 2012 Proceedings of the 35th International Convention MIPRO, pp. 1713–1718 (2012)
2. Bazan, J.G., Szczuka, M.: The rough set exploration system. Springer Transactions on Rough Sets III. LNCS, vol. 3400, pp. 37–56. Springer, Berlin (2005)
3. Bergstra, J., Kegl, B.: Meta-features and adaboost for music classification. In: Machine Learning Journal: Special Issue on Machine Learning in Music (2006)
4. Bhatt, R., Krishnamoorthy, P., Kumar, S.: Efficient general genre video abstraction scheme for embedded devices using pure audio cues. In: 7th International Conference on ICT and Knowledge Engineering, pp. 63–67 (2009)

5. Covell, M., Baluja, S., Fink, M.: Advertisement detection and replacement using acoustic and visual repetition. In: 2006 IEEE Workshop on Multimedia Signal Processing, pp. 461–466 (2006)

6. Eyben, F.: Classification by Large Audio Feature Space Extraction, chap. Acoustic Features and Modelling, pp. 9–122. Springer International Publishing (2016)

7. Hall, M., Frank, E., Holmes, G., Pfahringer, B., Reutemann, P., Witten, I.H.: The WEKA data mining software: An update. SIGKDD Explorations 11(1), 10–18 (2009)

8. Haque, M.A., Kim, J.M.: An analysis of content-based classification of audio signals using a fuzzy c-means algorithm. Multimedia Tools Appl. 63(1), 77–92 (2013)

9. Hendrik, S.: Improving genre annotations for the million song dataset. In: 16th International Society for Music Information Retrieval Conference, pp. 63–70 (2015)

10. Henry, C., Ramanna, S.: Parallel computation in finding near neighbourhoods. In: Proceedings of the 6th International Conference on Rough Sets and Knowledge Technology (RSKT11). Lecture Notes in Computer Science, pp. 523–532. Springer, Berlin (2011)

11. Henry, C., Ramanna, S.: Signature-based perceptual nearness. Application of near sets to image retrieval. Springer J. Math. Comput. Sci. 7, 71–85 (2013)

12. Henry, C.J.: Near sets: theory and applications. Ph.D. thesis, University of Manitoba (2011)

13. Hunt, M., Lennig, M., Mermelstein, P.: Experiments in syllable-based recognition of continuous speech. In: Proceedings of International Conference on Acoustics, Speech and Signal Processing, pp. 880–883 (1996)

14. Jinsong, Z., Oussalah, M.: Automatic system for music genre classification (2006) (PGM-Net2006)

15. Kannao, R., Guha, P.: Tv commercial detection using success based locally weighted kernel combination. In: Proceedings on the the International Conference on MultiMedia Modeling, pp. 793–805. Springer International Publishing (2016)

16. Kaur, K., Ramanna, S., Henry, C.: Measuring the nearness of layered flow graphs: application to content based image retrieval. Intell. Decis. Technol. J. 10, 165–181 (2016)

17. Kostek, B.: Perception-based data processing in acoustics, applications to music information retrieval and psychophysiology of hearing. Series on Cognitive Technologies. Springer, Berlin (2005)

18. Lee, C.H., Shih, J.L., Yu, K.M., Lin, H.S.: Automatic music genre classification based on modulation spectral analysis of spectral and cepstral features. IEEE Trans. Multimed. 11(4), 670–682 (2009)

19. Lie, L., Hao, J., HongJiang, Z.: A robust audio classification and segmentation method. In: Proceedings of the Ninth ACM International Conference on Multimedia, MULTIMEDIA '01, pp. 203–211 (2001)

20. Liu, N., Zhao, Y., Zhu, Z., Lu, H.: Exploiting visual-audio-textual characteristics for automatic tv commercial block detection and segmentation. IEEE Trans. Multimed. 13(5), 961–973 (2011)

21. Logan, B.: Mel frequency cepstral coefficients for music modeling. In: Proceedings of 1st International Conference on Music Information Retrieval, Plymouth, MA (2000)

22. Bertin-Mahieux, T., Ellis, D.P., Whitman, B., Lamere, P.: The million song dataset. In: Proceedings of the 12th International Conference on Music Information Retrieval (ISMIR 2011), pp. 7–12 (2011)

23. Mandel, M., Poliner, G., Ellis, D.: Support vector machine active learning for music retrieval. Multimed. Syst. 12(1), 3–13 (2006)

24. Orłowska, E.: Semantics of Vague Concepts. Applications of rough sets. In: Technical Report 469, Institute for Computer Science. Polish Academy of Sciences (1982)

25. Orłowska, E.: Semantics of vague concepts. In: Dorn, G., Weingartner, P. (eds.) Foundations of Logic and Linguistics. Problems and Solutions, pp. 465–482. Plenum Press, London (1985)

26. Pawlak, Z.: Rough sets. Int. J. Comput. Inf. Sci. 11(5), 341–356 (1982)

27. Peters, J.: Tolerance near sets and image correspondence. Int. J. Bio-Inspired Comput. 1(4), 239–245 (2009)

28. Peters, J.: Corrigenda and addenda: tolerance near sets and image correspondence. Int. J. Bio-Inspired Comput. **2**(5), 310–318 (2010)
29. Peters, J., Wasilewski, P.: Tolerance spaces: origins, theoretical aspects and applications. Inf. Sci.: An Int. J. **195**(5), 211–225 (2012)
30. Peters, J.F.: Near sets. General theory about nearness of objects. Appl. Math. Sci. **1**(53), 2609–2629 (2007)
31. Peters, J.F.: Near sets. Special theory about nearness of objects. Fundam. Inf. **75**(1–4), 407–433 (2007)
32. Peters, J.F.: Near sets: an introduction. Math. Comput. Sci. **7**(1), 3–9 (2013)
33. Peters, J.F., Naimpally, S.: Applications of near sets. Am. Math. Soc. Not. **59**(4), 536–542 (2012)
34. Peters, J.F., Wasilewski, P.: Foundations of near sets. Inf. Sci. **179**(18), 3091–3109 (2009)
35. Hoffmann, P., Kostek, B.: Music genre recognition in the rough set-based environment. In: Proceedings of 6th International Conference, PReMI 2015, pp. 377–386 (2015)
36. Poincaré, J.: Sur certaines surfaces algébriques; troisième complément 'a l'analysis situs. Bulletin de la Société de France **30**, 49–70 (1902)
37. Poli, G., Llapa, E., Cecatto, J., Saito, J., Peters, J., Ramanna, S., Nicoletti, M.: Solar flare detection system based on tolerance near sets in a GPU-CUDA framework. Knowl.-Based Syst. J. Elsevier **70**, 345–360 (2014)
38. Ramanna, S., Singh, A.: Tolerance-based approach to audio signal classification. In: Proceedings of 29th Canadian AI Conference, LNAI 9673, pp. 83–88 (2016)
39. Rao, P.: Audio signal processing. In: S.P. B. Prasad (ed.) Speech, Audio, Image and Biomedical Signal Processing using Neural Networks, pp. 169–190. Springer International Publishing (2008). https://doi.org/10.1007/978-3-319-27671-7_66
40. Sadlier, D.A., Marlow, S., O'Connor, N.E., Murphy, N.: Automatic TV advertisement detection from MPEG bitstream. In: Proceedings of the 1st International Workshop on Pattern Recognition in Information Systems: In Conjunction with ICEIS 2001, PRIS '01, pp. 14–25 (2001)
41. Sengoz, C., Ramanna, S.: A semi-supervised learning algorithm for web information extraction with tolerance rough sets. In: Proceedings of Active Media Technology 2014. LNCS, vol. 8610, pp. 1–10 (2014)
42. Sengoz, C., Ramanna, S.: Learning relational facts from the web: a tolerance rough set approach. Pattern Recognit. Lett. Elsevier **67**(2), 130–137 (2015)
43. Sossinsky, A.B.: Tolerance space theory and some applications. Acta Appl. Math.: Int. Survey J. Appl. Math. Math. Appl. **5**(2), 137–167 (1986)
44. Sturm, B.L.: An analysis of the GTZAN music genre dataset. In: Proceedings of the Second International ACM Workshop on Music Information Retrieval with User-centered and Multimodal Strategies, pp. 7–12 (2012)
45. Typke, R., Wiering, F., Veltkamp, R.C.: A survey of music information retrieval systems. In: ISMIR 2005, 6th International Conference on Music Information Retrieval, London, UK, 11–15 September 2005, Proceedings, pp. 153–160 (2005)
46. Tzanetakis, G., Cook, P.: Musical genre classification of audio signals. IEEE Trans. Speech Audio Process. **10**(5), 293–302 (2002)
47. Vyas, A., Kannao, R., Bhargava, V., Guha, P.: Commercial block detection in broadcast news videos. In: Proceedings of the 2014 Indian Conference on Computer Vision Graphics and Image Processing, No. 63 in ICVGIP '14, pp. 63:1–63:7 (2014)
48. Wasilewski, P., Peters, J.F., Ramanna, S.: Perceptual tolerance intersection. Trans. Rough Sets XIII Springer LNCS **6499**, 159–174 (2011)
49. Wold, E., Blum, T., Keislar, D., Wheaton, J.: Content-based classification, search, and retrieval of audio. IEEE Multimed. **3**(2), 27–36 (1996)
50. Wolski, M.: Perception and classification. A note on near sets and rough sets. Fundam. Inf. **101**, 143–155 (2010)
51. Wolski, M.: Toward foundations of near sets: (pre-)sheaf theoretic approach. Math. Comput. Sci. **7**(1), 125–136 (2013)

52. Zeeman, E.: The topology of the brain and visual perception. University of Georgia Institute Conference Proceedings, pp. 240–256 (1962); Fort, M.K., Jr. (ed.) Topology of 3-Manifolds and Related Topics. Prentice-Hall, Inc

53. Zeeman, E., Buneman, O.: Tolerance spaces and the brain. In: Towards a Theoretical Biology, vol. 1, pp. 140–151; Published in Waddington, C.H. (ed.) Towards a Theoretical Biology. The Origin of Life, Aldine Pub, Co. (1968)

54. Zhang, T., Kuo, C.: Content-Based Audio Classification and Retrieval for Audiovisual Data Parsing. Kluwer Academic Publishers, Norwell (2001)

55. Zhouyu, F., Guojun, L., Kai, M., Dengsheng, Z.: A survey of audio-based music classification and annotation. IEEE Trans. Multimed. **13**(2), 303–319 (2011)

56. Zwan, P., Kostek, B., Kupryjanow, A.: Automatic classification of musical audio signals employing machine learning approach. Audio Engineering Society Convention, vol. 130, pp. 2–11 (2011)

57. Zwicker, E., Fastl, H.: Psychoacoustics, facts and models. Springer Series in Information Sciences. Springer, Berlin (1990)

Part IV
Decision Support Systems

Chapter 14
Visual Analysis of Relevant Features in Customer Loyalty Improvement Recommendation

Katarzyna A. Tarnowska, Zbigniew W. Raś, Lynn Daniel and Doug Fowler

Abstract This chapter describes a practical application of decision reducts to a real-life business problem. It presents a feature selection (attribute reduction) methodology based on the decision reducts theory, which is supported by a designed and developed visualization system. The chapter overviews an application area - Customer Loyalty Improvement Recommendation, which has become a very popular and important topic area in today's business decision problems. The chapter describes a real-world dataset, which consists of about 400,000 surveys on customer satisfaction collected in years 2011–2016. Major machine learning techniques used to develop knowledge-based recommender system, such as decision reducts, classification, clustering, action rules, are described. Next, visualization techniques used for the implemented interactive system are presented. The experimental results on the customer dataset illustrate the correlation between classification features and the decision feature called "Promoter Score" and how these help to understand changes in customer sentiment.

Keywords NPS · Recommender system · Feature selection · Action rules · Meta actions · Semantic similarity · Sentiment analysis · Visualization

K. A. Tarnowska (✉) · Z. W. Raś
University of North Carolina, Charlotte, NC, USA
e-mail: ktarnows@uncc.edu

Z. W. Raś
Polish-Japanese Academy of IT, Warsaw University of Technology, Warsaw, Poland
e-mail: ras@uncc.edu

L. Daniel · D. Fowler
The Daniel Group, Charlotte, NC, USA
e-mail: LynnDaniel@theDanielGroup.com

D. Fowler
e-mail: DougFowler@theDanielGroup.com

© Springer International Publishing AG 2018
U. Stańczyk et al. (eds.), *Advances in Feature Selection for Data and Pattern Recognition*, Intelligent Systems Reference Library 138,
https://doi.org/10.1007/978-3-319-67588-6_14

14.1 Introduction

Nowadays most businesses, whether small-, medium-sized or enterprise-level orga-
nizations with hundreds or thousands of locations collect their customers feedback
on products or services. A popular industry standard for measuring customer satisfaction is so called "Net Promoter Score"[1] [15] based on the percentage of customers
classified as "detractors", "passives" and "promoters". Promoters are loyal enthusi-
asts who are buying from a company and urge their friends to do so. Passives are
satisfied but unenthusiastic customers who can be easily taken by competitors, while
detractors are the least loyal customers who may urge their friends to avoid that
company. The total Net Promoter Score is computed as %Promoters -%Detractors.
The goal here is to maximize NPS, which in practice, as it turns out, is a difficult
task to achieve, especially when the company has already quite high NPS.

Most executives would like to know not only the changes of that score, but also
why the score moved up or down. More helpful and insightful would be to look
beyond the surface level and dive into the entire anatomy of feedback.

The main problem we tried to solve is to understand the difference in data patterns
of customer sentiment on a single company personalization level, in years 2011–
2015. The same we should be able to explain changes, as well as predict sentiment
changes in the future.

The second section presents work related to customer satisfaction software tools.
The third section describes dataset and application area - decision problem. The
fourth presents approach proposed to solve the problem, including machine learning
techniques and visualization techniques. Lastly, the fifth section provides evaluation
results and sixth - final conclusions.

14.2 Related Applications

Horst Schulz, former president of the Ritz-Carlton Hotel Company, was famously
quoted as saying: "Unless you have 100% customer satisfaction...you must improve".
Customer satisfaction software helps to measure customers' satisfaction, as well as
gain insight into ways to achieve higher satisfaction. *SurveyMonkey* [3] is the industry leading online survey tool, used by millions of businesses across the world. It
helps to create any type of survey, but it also lacks features with regard to measuring satisfaction and getting actionable feedback. *Client Heartbeat* [2] is another
tool built specifically to measure customer satisfaction, track changes in satisfaction
levels and identify customers 'at risk'. *SurveyGizmo* [16] is another professional
tool for gathering customer feedback. It offers customizable customer satisfaction
surveys, but it also lacks features that would help to intelligently analyze the data.
Customer Sure [4] is a tool that focuses on customer feedback: facilitates distribu-

---

[1]NPS®, Net Promoter®and Net Promoter®Score are registered trademarks of Satmetrix Systems,
Inc., Bain and Company and Fred Reichheld.

tion of customer surveys, gathering the results. It allows to act intelligently on the feedback by tracing customer satisfaction scores over time and observe trends. *Floqapp* [8] is a tool that offers customer satisfaction survey templates, collects the data and puts it into reports. *Temper* [18] is better at gauging satisfaction as opposed to just being a survey tool. Similar to *Client Heartbeat*, Temper measures and tracks customer satisfaction over a period of time. The *Qualtrics* Insight Platform [19] is the leading platform for actionable customer, market and employee insights. Besides customers' feedback collection, analysis and sharing it offers extensive insight capabilities, including tools for end-to-end customer experience management programs, customer and market research and employee engagement.

These types of tools mostly facilitate design of surveys, however, offer very limited analytics and insight into customer feedback. It is mostly confined to simple trend analysis (tracing if the score has increased or decreased over time).

14.3 Dataset and Application

The following section describes the dataset we worked on and the corresponding application area. The problem we deal with includes attribute analysis and relevant feature selection (reduction) and ultimately an analysis of customer satisfaction and providing recommendations to improve it.

14.3.1 Dataset

The data was provided by a consulting company based in Charlotte within a research project conducted in the KDD Lab at UNC-Charlotte. The company collects data from telephone surveys on customer's satisfaction from repair service done by heavy equipment repair companies (called clients). There are different types of surveys, depending on which area of customer satisfaction they focus on: service, parts, rentals, etc. The consulting company provides advisory for improving their clients' Net Promoter Score rating and growth performance in general. Advised companies are scattered among all the states in the US (as well as south Canada) and can have many subsidiaries. There are above 400,000 records in the dataset, and the data is kept being continuously collected. The dataset consists of features related to:

1. Companies' details (repair company's name, division, etc.);
2. Type of service done, repair costs, location and time;
3. Customer's details (name, contact, address);
4. Survey details (timestamp, localization) and customers' answers to the questions in a survey;

5. Each answer is scored with 1–10 (optionally textual comment) and based on the total average score (*PromoterScore*) a customer is labeled as either promoter, passive or detractor of a given company.

The data is high-dimensional with many features related to particular assessment (survey questions') areas, their scores and textual comments. The consulted companies as well as surveyed customers are spread geographically across United States. Records are described with some temporal features, such as *DateInterviewed*, *InvoiceDate* and *WorkOrderCloseDate*.

14.4 Decision Problem

The goal is to find characteristics (features) which most strongly correlate with *Promoter/Detractor* label (*PromoterScore*), so that we can identify areas, where improvement can lead to changing a customer's status from "Detractor" to "Promoter" (improvement of customer's satisfaction and company's performance). Identifying these variables (areas) helps in removing redundancy.

Analysis should not only consider global statistics, as global statistics can hide potentially important differentiating local variation (on a company level). The goal is to explore the geography of the issue and use interactive visualization to identify interdependencies in a multivariate dataset. It should support geographically informed multidimensional analysis and discover local patterns in customers' experience and service assessment. Finally, classification on NPS should be performed on semantically similar customers (similarity can be also based on geography). A subset of most relevant features should be chosen to build a classification model.

14.4.1 Attribute Analysis

The first problem we need to solve is to find out which benchmarks are the most relevant for Promoter Status. There is also a need to analyze how the importance of benchmarks changed over years for different companies (locally) and in general (globally), and additionally how these changes affected changes in Net Promoter Score, especially if this score is deteriorated (which means customer satisfaction worsened). We need to identify what triggered the highest NPS decreases and the highest NPS growths.

The business questions to answer here are:

- What (which aspects) triggered changes in our Net Promoter Score?
- Where did we go wrong? What could be improved?
- What are the trends in customer sentiment towards our services? Did more of them become promoters? passives? petractors? Did promoters become passives? promoters become detractors? If yes, why?

The problem with the data is that the set of benchmarks asked is not consistent and varies for customers, companies and years. Customer expectations change as well. Therefore, we have to deal with a highly incomplete and multidimensional data problem.

14.4.2 Attribute Reduction

The consulting company developed over 200 such benchmarks in total, but taking into considerations time constraints for conducting a survey on one customer it is impossible to ask all of them. Usually only some small subsets of them are asked. There is a need for benchmarks (survey) reduction, but it is not obvious which of them should be asked to obtain the most insightful knowledge. For example, consulting aims to reduce the number of questions to the three most important, such as "Overall Satisfaction", "Referral Behavior" and "Promoter Score", but it has to be checked if this will not lead to a significant knowledge loss in the customer satisfaction problem. We need to know which benchmarks can/cannot be dropped in order to control/minimize the knowledge loss. For example, in years 2014–2015 some questions were asked less frequently since questionnaire structure changed and survey shortened for some companies. There is a need for analysis regarding how these changes in the dataset affect the previously built classification and model in the system.

14.4.3 Customer Satisfaction Analysis and Recognition

The second application area is tracking the quality of the data/knowledge being collected year by year, especially in terms of its ability to discern between different types of customers defined as *Promoters*, *Passives* and *Detractors*. The main questions business would like to know the answers to are:

- What (which aspect of service provided) makes their customers being promoters or detractors?
- Which area of the service needs improvement so that we can maximize customer satisfaction?

For every client company we should identify the minimal set of features (benchmarks) needed to classify correctly if a customer is a promoter, passive or detractor. We need to determine the strength of these features and how important a feature is in the recognition process. However, answering these questions is not an easy task, as the problem is multidimensional, varies in space and time and is highly dependent on the data structure used to model the problem. Sufficient number of customer feedback on various aspects must be collected and analyzed in order to answer these questions. Often human abilities are not sufficient to analyze such huge volume of data in terms of so many aspects. There is a need for some kind of automation of the task or visual analytics support.

14.4.4 Providing Recommendations

The ultimate goal of our research project is to support business with recommendations (recommendable sets of actions) to companies, so that we can improve their NPS. The items must be evaluated in terms of some objective metrics.

Besides, we need to make the recommendation process more transparent, valid and trustworthy. Therefore, we need to visualize the process that leads to generating a recommendation output. The end user must be able to understand how recommendation model works in order to be able to explain and defend the model validity. Visual techniques should facilitate this process.

14.5 Proposed Approach

Addressing the problem of feature analysis, there are two approaches to the feature evaluation and its importance towards the classification problem.

The first one is based on the discrimination power of a set of features and how the classification problem (in terms of accuracy) is affected if we discard one or more of them in a dataset. It always starts with the set of all features used in classification. This approach is logic-based and it can be called top-down approach.

The second one is a statistic-based and it can be called bottom-up approach. It talks about the discrimination power of a single feature or a small set of features. It does not make any reference to discrimination power of combined effect of features together. To compare them, we can say that the first approach is focused more on minimizing knowledge loss, the second one more on maximizing knowledge gain.

14.5.1 Machine Learning Techniques

In this subsection we present major machine learning and data mining techniques we used to solve the problem stated in the previous section. These include: decision reducts and action rules (stemming from Pawlak theory on Information Systems, Decision Tables and Reducts), classification and clustering.

14.5.1.1 Decision Reducts

To solve decision problems as stated in the previous sections we propose applying attribute reduction techniques. The one we propose is based on decision reducts and stems from rough set theory (logic-based).

Rough set theory is a mathematical tool for dealing with ambiguous and imprecise knowledge, which was introduced by Polish mathematician Professor Pawlak in 1982 [10]. The rough set theory handles data analysis organized in the form of tables. The data may come from experts, measurements or tests. The main goals of the data analysis is a retrieval of interesting and novel patterns, associations, precise problem analysis, as well as designing a tool for automatic data classification.

Attribute reduction is an important concept of rough set for data analysis. The main idea is to obtain decisions or classifications of problems on the conditions of maintaining the classification ability of the knowledge base. We introduce basic concepts of the theory below.

Information Systems

A concept of *Information System* stems from the theory of rough sets.

Definition 14.1 An *Information System* is defined as a pair $S = (U, A)$, where U is a nonempty, finite set, called *the universe*, and A is a nonempty, finite set of attributes i.e. $a : U \rightarrow V_a$ for $a \in A$, where V_a is called the domain of a [12].

Elements of U are called *objects*. A special case of *Information Systems* is called a *Decision Table* [9].

Decision Tables

In a decision table, some attributes are called *conditions* and the others are called *decisions*. In many practical applications, decision is a singleton set. For example, in table in Fig. 14.1 decision is an attribute specifying *Promoter Status*. The conditions would be all the attributes that determine *Promoter Status*, that is, question benchmarks and also other attributes (geographical, temporal, etc.).

Based on knowledge represented in a form of a decision table, it is possible to model and simulate decision-making processes. The knowledge in a decision table is represented by associating or identifying decision values with some values of conditional attributes.

Client attributes			Customer attributes			Service attributes		Survey attributes and questions (customer experience on client's service)					NP S	Stat us
ID	Name	Adress, ...	Name	Location	...	Time	Cost	Q1 (score)	Q2 (score)	Q... (score)	QN (score)			Promoter
1														Passive
2														Detractor

Fig. 14.1 Illustration of NPS dataset structure - features and decision attribute

For extracting action rules, it is also relevant to differentiate between so-called *flexible* attributes, which can be changed, and *stable* attributes [12], which cannot be changed. $A = A_{St} \cup A_{Fl}$, where A_{St} and A_{Fl} denote *stable* attributes and *flexible* attributes respectively. Example of stable attributes in customer data would be company's and survey's characteristics, while flexible would be assessment areas (benchmarks), which can be changed by undertaking certain actions (for example, staff training).

Reducts

In decision systems often not every attribute is necessary for the decision-making process. The goal is to choose some subset of attributes essential for this. It leads to the definition of *reducts*, that is, minimal subsets of attributes that keep the characteristics of the full dataset. In the context of action rule discovery an *action reduct* is a minimal set of attribute values distinguishing a favorable object from another. In our application area, it is of interest to find unique characteristics of the satisfied customers that can be used by the company to improve the customer satisfaction of 'Detractors'. We need to find a set of distinct values or unique patterns from the 'Promoter' group that does not exist in the 'Detractor' group in order to propose an actionable knowledge. Some of attribute values describing the customers can be controlled or changed, which is defined as an action. Before defining formally a *reduct*, it is necessary to introduce a *discernibility relation*.

Definition 14.2 Let objects $x, y \in U$ and set of attributes $B \subset A$. We say that x, y are *discernible* by B when there exists $a \in B$ such that $a(x) \neq a(y)$. x, y are *indiscernible* by B when they are identical on B, that is, $a(x) = a(y)$ for each $a \in B$. $[x]_B$ denotes a set of objects indiscernible with x by B.

Furthermore, following statements are true:

- for each objects x, y either $[x]_B = [y]_B$ or $[x]_B \cap [y]_B = \emptyset$,
- *indiscernibility relation* is an equivalence relation,
- each set of attributes $B \subset A$ determines a partition of a set of objects into disjoint subsets.

Definition 14.3 A set of attributes $B \subset A$ is called *reduct of the decision table* if and only if:

- B keeps the discernibility of A, that is, for each $x, y \in U$, if x, y are discernible by A, then they are also discernible by B,
- B is irreducible, that is, none of its proper subset keeps discernibility properties of A (that is, B is minimal in terms of discernibility).

The set of attributes A appearing in every reduct of information system (decision table DT) is called *the core*.

In order to allow for local and temporal analysis we divide the yearly global NPS data (2011, 2012, 2013, 2014, 2015) into separate company data (38 datasets for each year in total). For each such extracted dataset we perform a corresponding feature selection, transformation and then run an attribute reduction algorithm (in the RSES system [14]).

14.5.1.2 Classification

To guarantee mining high quality action rules, we need to construct the best classifiers first. Also, the results of classification provide an overview of the consistency of knowledge hidden in the dataset. The better the classification results, the more consistent and accurate knowledge stored in the data. Also, better ability of the system to recognize promoters/passives/detractor correctly is a foundation for the system to give accurate results.

To track the accuracy of the models built on the yearly company data we performed classification experiments for each company's dataset and for each year. Evaluation was performed with 10-fold cross-validation on decomposition tree classifiers.

Decomposition trees are used to split dataset into fragments not larger than a predefined size. These fragments, represented as leaves in decomposition tree, are supposed to be more uniform. The subsets of data in the leaves of decomposition tree are used for calculation of decision rules.

We saved results from each classification task: accuracy, coverage, confusion matrix.

Decision rule

The decision rule, for a given decision table, is a rule in the form: $(\phi \rightarrow \delta)$, where ϕ is called *antecedent* (or *assumption*) and δ is called *descendant* (or *thesis*) of the rule. The antecedent for an atomic rule can be a single term or a conjunction of k elementary conditions: $\phi = p_1 \wedge p_2 \wedge \ldots \wedge p_n$, and δ is a decision attribute. Decision rule describing a class K_j means that objects, which satisfy (match) the rule's antecedent, belong to K_j.

In the context of prediction problem, decision rules generated from training dataset, are used for classifying new objects (for example classifying a new customer for NPS category). New objects are understood as objects that were not used for the rules induction (new customers surveyed). The new objects are described by attribute values (for instance a customer with survey's responses). The goal of classification is to assign a new object to one of the decision classes.

14.5.1.3 Clustering

It is believed that companies can collaborate with each other by exchanging knowledge hidden in datasets and they can benefit from others whose hidden knowledge is similar. In order to recommend items (actions to improve in the service, products), we need to consider not only historical feedback of customers for this company, but we also propose looking at companies who are similar in some way, but perform better. We use concept of semantic similarity to compare companies, which is defined as similarity of their knowledge concerning the meaning of three concepts: promoter, passive, and detractor. Clients who are semantically close to each other can have their datasets merged and the same considered as a single company from the business perspective (customers have similar opinion about them).

We use hierarchical clustering algorithm to generate the nearest neighbors of each company. Given the definition of semantic similarity, the distance between any pairs of companies are quantified in a semantic way and the smaller the distance is, the more similar the companies are. A semantic similarity-based distance matrix is built on top of the definition. With the distance matrix, a hierarchical clustering structure (dendrogram) is generated by applying an agglomerative clustering algorithm.

The sample dendrogram (for year 2015) is shown in Fig. 14.2 as a part of visualization system. The figure shows the hierarchical clustering of 38 companies with respect to their semantic similarity. In the following analysis, companies' names are replaced by numbers based on their alphabetical order, rather than using exact actual names due to the confidentiality.

A dendrogram is a node-link diagram that places leaf nodes of the tree at the same depth. If describing it using tree-structure-based terminology, every leaf node in the dendrogram represents the corresponding company as the number shown, and the depth of one node is the length of the path from it to the root, so the lower difference of the depth between two leaf nodes, the more semantically similar they are to each other. With the dendrogram it is easy to find out the groups of companies which are relatively close to each other in semantic similarity. In this example, the companies (leaf nodes) are aligned on the right edge, with the clusters (internal nodes) to the left. Data shows the hierarchy of companies clusters, with the root node being "All" companies.

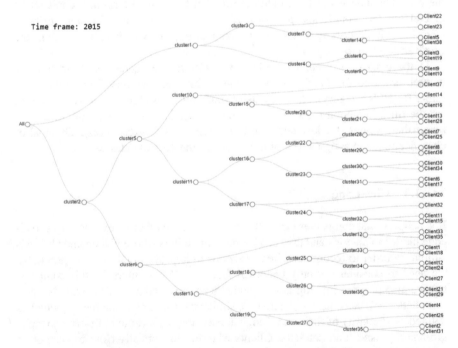

Fig. 14.2 Sample hierarchical clustering dendrogram for company clients based on 2015 Service data (part of the visualization system)

14.5.1.4 Action Rules

An *action* is understood as a way of controlling or changing some attribute values in an information system to achieve desired results [6]. An *action rule* is defined [12] as a rule extracted from an information system, that describes a transition that may occur within objects from one state to another, with respect to decision attribute, as defined by the user. In nomenclature, action rule is defined as a term: $[(\omega) \wedge (\alpha \rightarrow \beta) \rightarrow (\Phi \rightarrow \Psi)]$, where ω denotes conjunction of fixed condition attributes, $(\alpha \rightarrow \beta)$ are proposed changes in values of flexible features, and $(\Phi \rightarrow \Psi)$ is a desired change of decision attribute (action effect).

Let us assume that Φ means 'Detractors' and Ψ means 'Promoters'. The discovered knowledge would indicate how the values of flexible attributes need to be changed under the condition specified by stable attributes so the customers classified as detractors will become promoters. So, action rule discovery applied to customer data would suggest a change in values of flexible attributes, such as benchmarks, to help "reclassify" or "transit" an object (customer) to a different category ("Passive" or "Promoter") and consequently, attain better overall customer satisfaction.

An action rule is built from *atomic action sets*.

Definition 14.4 *Atomic action term* is an expression $(a, a_1 \rightarrow a_2)$, where a is attribute, and $a_1, a_2 \in V_a$, where V_a is a domain of attribute a.

If $a_1 = a_2$ then a is called stable on a_1.

Definition 14.5 By *action sets* we mean the smallest collection of sets such that:

1. If t is an atomic action term, then t is an action set.
2. If t_1, t_2 are action sets, then $t_1 \wedge t_2$ is a candidate action set.
3. If t is a candidate action set and for any two atomic actions $(a, a_1 \rightarrow a_2), (b, b_1 \rightarrow b_2)$ contained in t we have $a \neq b$, then t is an action set. Here b is another attribute $(b \in A)$, and $b_1, b_2 \in V_b$.

Definition 14.6 By an *action rule* we mean any expression $r = [t_1 \Rightarrow t_2]$, where t_1 and t_2 are action sets.

The interpretation of the action rule r is, that by applying the action set t_1, we would get, as a result, the changes of states in action set t_2.

The ultimate goal of building an efficient recommender system is to provide actionable suggestions for improving a company's performance (improving its NPS efficiency rating). Extracting action rules is one of the most operative methods here and it has been applied to various application areas like medicine - developing medical treatment methods [17, 22, 23] sound processing [11, 13] or business [12].

In our application, we first extend the single-company dataset (by adding to it datasets of its semantic neighbors). Next, we mine the extended datasets to find action rules, that is, rules which indicate action to be taken in order to increase Net Promoter Score. The first step of extracting action rules from the dataset is to complete the initialization of mining program by setting up all the variables.

The process of initialization consists of selecting stable and flexible attributes and a decision attribute, determining the favorable and unfavorable state for a decision attribute, and defining a minimum confidence and support for the resulting rules. The results of action rule mining are used in the process of generating recommendations in our approach.

14.5.1.5 Meta Actions and Triggering Mechanism

Recommender system model proposed by us is driven by action rules and meta-actions to provide proper suggestions to improve the revenue of companies. Action rules, described in the previous subsection, show minimum changes needed for a client to be made in order to improve its ratings so it can move to the promoter's group. Action rules are extracted from the client's dataset.

Meta-actions are the triggers used for activating action rules and making them effective. The concept of *meta-action* was initially proposed in Wang et al. [21] and later defined in Raś et al. [20]. Meta-actions are understood as higher-level actions. While an action rule is understood as a set of atomic actions that need to be made for achieving the expected result, meta-actions need to be executed in order to trigger corresponding atomic actions.

For example, the temperature of a patient cannot be lowered if he does not take a drug used for this purpose—taking the drug would be an example of a higher-level action which should trigger such a change. The relations between meta-actions and changes of the attribute values they trigger can be modeled using either an influence matrix or ontology. An example of an influence matrix is shown in Table 14.1. It describes the relations between the meta-actions and atomic actions associated with them. Attribute a denotes stable attribute, b - flexible attribute, and d - decision attribute. $\{M_1, M_2, M_3, M_4, M_5, M_6\}$ is a set of meta-actions which hypothetically triggers action rules. Each row denotes atomic actions that can be invoked by the set of meta-actions listed in the first column. For example, in the first row, atomic actions $(b_1 \rightarrow b_2)$ and $(d_1 \rightarrow d_2)$ can be activated by executing meta-actions M_1, M_2 and M_3 together.

In our domain, we assume that one atomic action can be invoked by more than one meta-action. A set of meta-actions (can be only one) triggers an action rule that consists of atomic actions covered by these meta-actions. Also, some action rules can be invoked by more than one set of meta-actions.

If the action rule $r = [\{(a, a_2), (b, b_1 \rightarrow b_2)\} \implies \{(d, d_1 \rightarrow d_2)\}]$ is to be triggered, we consider the rule r to be the composition of two association rules r_1 and

Table 14.1 Sample meta-actions influence matrix

	a	b	d
$\{M_1, M_2, M_3\}$		$b_1 \rightarrow b_2$	$d_1 \rightarrow d_2$
$\{M_1, M_3, M_4\}$	a_2	$b_2 \rightarrow b_3$	
$\{M_5\}$	a_1	$b_2 \rightarrow b_1$	$d_2 \rightarrow d_1$
$\{M_2, M_4\}$		$b_2 \rightarrow b_3$	$d_1 \rightarrow d_2$
$\{M_1, M_5, M_6\}$		$b_1 \rightarrow b_3$	$d_1 \rightarrow d_2$

r_2, where $r_1 = [\{(a, a_2), (b, b_1)\} \implies \{(d, d_1)\}]$ and $r_2 = [\{(a, a_2), (b, b_2)\} \implies \{(d, d_2)\}]$. The rule r can be triggered by the combination of meta-actions listed in the first and second row in Table 14.1, as meta-actions $\{M_1, M_2, M_3, M_4\}$ cover all required atomic actions: (a, a_2), $(b, b_1 \rightarrow b_2)$, and $(d, d_1 \rightarrow d_2)$ in r. Also, one set of meta-actions can potentially trigger multiple action rules. For example, the mentioned meta-action set $\{M_1, M_2, M_3, M_4\}$ triggers not only rule r, but also another rule, such as $[\{(a, a_2), (b, b_2 \rightarrow b_3)\} \implies \{(d, d_1 \rightarrow d_2)\}]$, according to the second and fourth row in Table 14.1, if such rule was extracted.

The goal is to select such a set of meta-actions which would trigger a larger number of actions and the same bring greater effect in terms of NPS improvement. The effect is quantified as following: supposing a set of meta-actions $M = \{M_1, M_2, \ldots, M_n : n > 0\}$ triggers a set of action rules $\{r_1, r_2, \ldots, r_m : m > 0\}$ that covers objects in a dataset with no overlap. We defined the coverage (support) of M as the summation of the support of all covered action rules. That is, the total number of objects that are affected by M in a dataset. The confidence of M is calculated by averaging the confidence of all covered action rules:

$$sup(M) = \sum_{i=1}^{m} sup(r_i),$$

$$conf(M) = \frac{\sum_{i=1}^{m} sup(r_i) \cdot conf(r_i)}{\sum_{i=1}^{m} sup(r_i)},$$

The effect of applying M is defined as the product of its support and confidence: $(sup(M) \cdot conf(M))$, which is a base for calculating the increment of NPS rating. The increment in NPS associated with different combinations of meta-actions (recommendations) are visualized in an interactive recommender system (Figs. 14.3, 14.4).

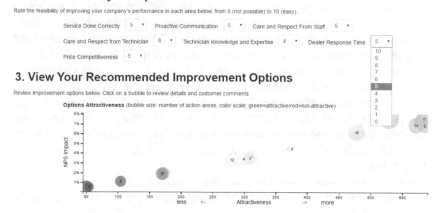

Fig. 14.3 Extracted meta actions are visualized as recommendations with different NPS impact scores. Each meta action can be assigned different feasibility

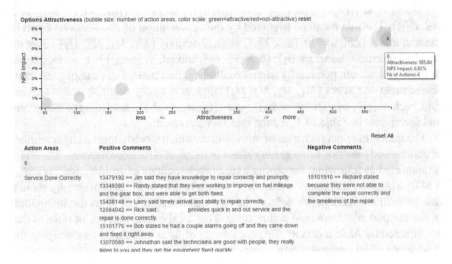

Fig. 14.4 Each recommendable item (set of meta-actions) can be further analyzed by means of its attractiveness, feasibility and associated raw text comments

14.5.2 Visualization Techniques

This subsection is related to the visualization system developed within this research and some visualization work related to solving similar problems.

14.5.2.1 Related Work

The work in "Visualizing Multiple Variables Across Scale and Geography" [5] from the 2015 IEEE VIS conference attempted a multidimensional attribute analysis varying across time and geography. Especially interesting is approach for analyzing attributes in terms of their correlation to the decision attribute, which involves both global and local statistics. The sequence of panels (Fig. 14.5) allows for a more fine-grained, sequential analysis by discovering strongly correlated variables at a global level, and then investigating it through geographical variation at the local level. The paper's presented approach supports a correlation analysis in many dimensions, including geography and time, as well as, in our case - particular company, which bears similarity to the problems we tried to solve. The visualization proposed in the paper helps in finding more discriminative profiles when creating geo-demographic classifiers.

Other papers are more related to visualizing recommendations, as the visual system developed by us is mainly to support recommender system and its output (a list of recommendable items). The first such paper was "Interactive Visual Profiling of Musicians" [7], from the 2015 VIS conference. The visualization system supports interactive profiling and allows for multidimensional analysis. It introduces a concept

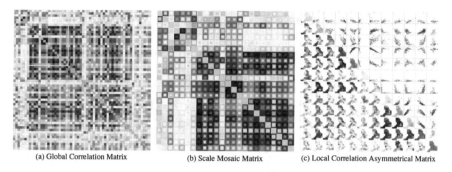

(a) Global Correlation Matrix   (b) Scale Mosaic Matrix   (c) Local Correlation Asymmetrical Matrix

Fig. 14.5 Multivariate comparison across scale and geography showing correlation in [5]

of similarity and how similar items (musicians) can be presented on visualization. We are dealing with analogous problem, as we want to compare companies (clients) based on similarity as well. The related work presented in that paper led us to the other work on recommender systems visualizations. For example, Choo et al. [1] present a recommendation system for a vast collection of academic papers.

The developed VisRR is an interactive visual recommendation system which shows the results (recommended documents) in the form of a scatterplot. Problem area is similar to our project as it combines visualizing recommendations and information retrieval results.

14.5.2.2   Heatmap

The implemented visual system supports a feature analysis of 38 companies for each year between 2011–2015 in the area of customer service (divided into shop service/field service) and parts.

It serves as a feature selection method showing the relative importance of each feature in terms of its relevance to the decision attribute - Promoter Status, and it is based on the algorithm that finds minimal decision reducts. The reducts were discovered using RSES (Rough Set Exploration System [14]). The results were saved to files from which the data-driven web-based visualization was built. Only these attributes were used that occurred in reducts and therefore, have occurrence percentage assigned as a metric displayed by the cell's color, were displayed on the heatmap (as columns).

The visualization allows the user to interactively assess the changes and implications onto predictive characteristics of the knowledge-based model. It is supported by visual additions in form of charts showing accuracy, coverage and confusion matrix of the model built on the corresponding, user-chosen dataset.

For a basic attribute analysis we propose a heatmap (Fig. 14.6) which bears similarity to a correlation matrix. However, attribute relevance to the decision attribute (Promoter Score) is defined here by means of reducts' strength (a percentage occurrence of an attribute in reducts).

Customer Sentiment Analysis Net Promoter Score

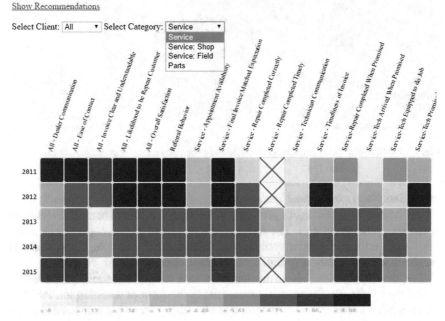

Fig. 14.6 Visual design of a correlation matrix based on a heatmap

The analyst can choose the Company category (analysis supported for 38 companies and "All" the companies) and Survey Type category ("Service"/"Service:Field"/ "Service:Shop"/"Parts"). The columns correspond to the benchmarks (that is surveys' attributes) found in reducts. The rows represent years, in which customer satisfaction assessment was performed (current version supports datasets from years 2011–2015).

The cells represent benchmark strength in a given year—the color linear scale corresponds to an occurrence percentage. The darker cells indicate benchmarks that belong to more reducts than benchmarks represented by the lighter cells - the darker the cell, the stronger the impact of the associated benchmark on promoter score (color denotes the strength of what 'statement of action rules are saying'). The average benchmark score is visible after hovering over a cell. Additionally, we use red-crossed cells to denote benchmarks that were not asked as questions in a given year for a chosen company so that we do not confuse the benchmarks' importance with the benchmarks' frequency of asking.

The design allows to track which new benchmarks were found in reducts in relation to the previous year, which disappeared, which gained/lost in strength.

The benchmark matrix should be analyzed together with the NPS row chart to track how the changes in benchmarks' importance affected NPS (decision attribute) changes (Fig. 14.7). The row chart showing yearly NPS changes per company is

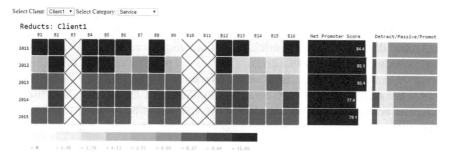

Fig. 14.7 Attribute heatmap with NPS row chart and NPS category distribution chart

complementary to the reducts' matrix. Further, we added "Stacked area row chart" to visualize distribution of different categories of customers (detractors, passives, promoters) per year. It is complementary to the two previous charts, especially to the row NPS chart and helps to understand changes in Net Promoter Score. The distribution is shown based on percentage values, which are visible after hovering over the corresponding area on the chart. We used colors for differentiating categories of customers as analogy to traffic lights: red means detractors ("angry" customers), yellow for passives, and green for promoters.

14.5.2.3 Dual Scale Bar Chart and Confusion Matrix

For the purpose of visualizing classification results we have used a dual scale bar chart (Fig. 14.8). It allows to additionally track knowledge losses by means of both: "Accuracy" and "Coverage" of the classifier model trained on single-company datasets. It is interactive and updates after the user chooses a "Client" from the drop-down menu.

Additionally, we use the confusion matrix to visualize the classifier's quality for different categories (Fig. 14.8). From the recommender system point of view it is important that the classifier can recognize customers (covered by the model), because

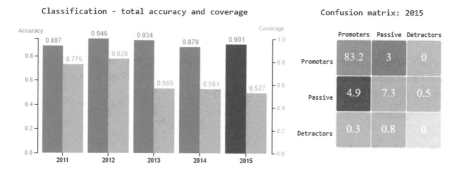

Fig. 14.8 Visualizations for the classifier's results - Accuracy, Coverage and Confusion Matrix

it means we can change their sentiment. The chart updates after interacting with the dual scale bar classification chart (hovering over the bar for the corresponding year) and shows detailed information on classifier's accuracy per each year. The rows in this matrix correspond to the actual decision classes (all possible values of the decision), while columns represent decision values as returned by the classifier in discourse. The values on diagonal represent correctly classified cases. Green colors were used to represent correctly classified cases, while red scale to represent misclassified cases. The color intensity corresponds to the extent of the confusion or the right predictions.

14.5.2.4   Multiple Views

We use multiple views (see Fig. 14.9) to enable a multidimensional problem analysis. However, this was quite challenging to design, as these views require both a sophisticated coordination mechanism and layout. We have tried to balance the benefits of multiple views and the corresponding complexity that arises. The multiple view was designed to support deep understanding of the dataset and the methodology for recommendations to improve NPS. The basic idea behind it is based on hierarchical structure of the system and a dendrogram was also used as an interface to navigate to more detailed views on a specific company. All other charts present data related to one company (client) or aggregated view ("All"):

- A reducts matrix is a way to present the most relevant attributes (benchmarks) in the NPS recognition process in a temporal aspect (year by year),
- NPS and Detractor/Passives/Promoters charts are complementary to reducts matrix and help track changes in Net Promoter Score in relation to changes in the benchmark importance,
- A Detractor/Passives/Promoter distribution chart helps understand the nature behind the changes in the NPS and customer sentiment,
- A classification and confusion matrix chart help in tracking the quality of gathered knowledge of NPS in datasets within different years and in understanding the reasons for the knowledge loss (seen as decreases in classification model accuracy) in terms of benchmark changes (survey structure changes),
- A confusion matrix is a more detailed view of a classification accuracy chart and helps in understanding the reasons behind accuracy yearly changes along with the customer's distribution chart.

In conclusion, the problem of changing datasets per year and per company (client) and therefore the model and the algorithm can be fully understood after considering all the different aspects presented on each chart.

14.6   Evaluation Results

In this subsection, we present evaluation of the system on the chosen use-case scenarios.

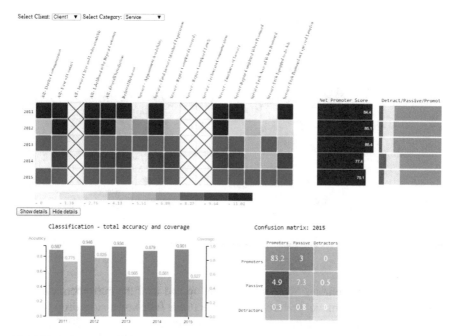

Fig. 14.9 Visual analysis for a single client ("Client1")

14.6.1   Single Client Data (Local) Analysis

Let us present a case study of analysis process on the example of the chosen Client1 (see Fig. 14.9). In the first row of the reduct matrix only dark colors can be observed, which means that all the colored benchmarks are equally high important. Customers' sentiment is pretty certain and defined on benchmarks.

Next year, importance of benchmarks dropped a lot, the values decreased. For some reason, customers changed their opinion. Average score of these benchmarks says that customers are little less satisfied. New benchmarks are showing up, but they are not that important. Some benchmarks lost importance changing the color from the darkest to lighter. We can observe movement of people from Detractors to Promoters, in connection with benchmark analysis it provides much wider view. Checking confusion matrix allows to go deeper into what customers are thinking. In the third year all benchmark lost in their way. Customers do not have strong preference on which benchmarks are the most important, but NPS is still going up. In the fourth year two benchmarks disappeared, in the fifth year all of benchmarks are again getting equal strength. We could observe a huge movement into the "Detractor" group in 2014, but at the same time customers got more confidence in which benchmarks are the most important. In 2015 all of the benchmarks have the same strength.

The number of "Passives" increases from 2014 to 2015, and the move to this group is from the "Detractor" group (we could observe a huge movement from detractors

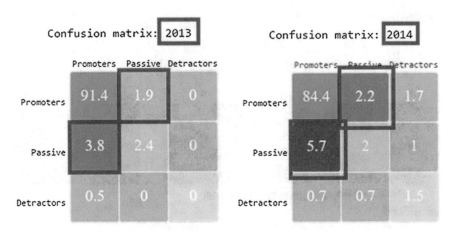

Fig. 14.10 Classification results analysis - confusion matrices

to passives). Net Promoter Score was moving up and later the customers were not certain (about benchmarks preference) and then there followed a huge drop in NPS.

Looking at the classification model, coverage jumped down from 2013 to 2014 incredibly that is, the number of customers which system can classify. Accuracy went down, and by looking at the confusion matrix (see Fig. 14.10) we can conclude that more people who are similar in giving their rates are confused by the system between Promoters and Passives. Some of such customers end up in Promoter category, while the others in Passives.

14.6.2 Global Customer Sentiment Analysis and Prediction

By looking at the "Reducts" heatmap, we can draw the conclusion that the customers' sentiment towards the importance of benchmarks is apparently changing year by year. Some benchmarks are getting more important, some less. By looking at the benchmarks for "All" companies (see Fig. 14.6), "Overall Satisfaction" and "Likelihood to be Repeat Customer" seem to be the winners. Clearly, it will differ when the personalization goes to the companies' level. It might be interesting to personalize further - going to different seasons (Fall, Spring,...) and seeing if the customers' sentiment towards the importance of benchmarks is changing. Our recommendations for improving NPS are based on the customers' comments using the services of semantically similar companies (clients) which are doing better than the company we want to help. So, it is quite important to check which benchmarks they favor. It is like looking ahead how our customers most probably will change their sentiment if the company serving these customers will follow the recommendation of the system we build. In other words, we can somehow control the future sentiment of our customers and choose the one which is the most promising for keeping NPS improving instead of improving it and next deteriorating.

From the proposed and developed charts, we can see how customers' sentiment towards the importance of certain benchmarks is changing year by year. Also, we can do the analysis trying to understand the reasons behind it with a goal to prevent them in the future if they trigger the decline in NPS.

The advantage of using recommender system the way it is proposed—based on hierarchical structure (dendrogram), assuming that companies use its recommendations, will be the ability to predict how customers' sentiment towards the importance of certain benchmarks will probably change with every proposed recommendation. So, we can not only pick up a set of benchmark triggers based on their expected impact on NPS but also on the basis of how these changes most probably will impact the changes in customers' sentiment towards the importance of benchmarks. This way we can build a tool which will give quite powerful guidance to companies. It will show which triggers are the best, when taking into account not only the current company's performance but also on the basis of the predicted company's performance in the future assuming they follow the system's recommendations.

14.6.3 Recommendations for a Single Client

The first step is to find a desired client on the map (Fig. 14.11) and click the associated dot. One can see, Client16 was found semantically similar to 2 other clients (called here extensions): (1) - most similar, (2) - secondly similar.

Since these two clients were performing better than Client16 in 2015 by means of NPS, also the customers' feedback from these two clients is used to suggest recommendations for improving NPS as it shows how customers would evaluate Client16

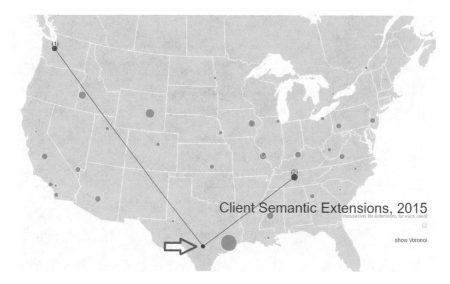

Fig. 14.11 Visualization of semantic extensions of Client16

if its performance was better but similar to its extensions. This way the base for discovering recommendations for Client16 is richer and the same should guarantee its better improvement in NPS. By similarity we mean similarity in customers' understanding of three concepts: promoter, detractor, passive.

After clicking the point for Client16, the recommendation chart for this client displays (Fig. 14.12). We can see that in total 18 combinations of different changes recommended to implement were generated, from which recommendation "nr 17" seems to be the most attractive (top right)-assuming all actions have the same default feasibility of 10. The generated actions are related to 8 different areas:

- service done correctly,
- price competitiveness,
- proactive communication,
- technician's knowledge,
- invoice accuracy and clarity,
- need for more technicians,
- care and respect from technician,
- staff care and respect to customers.

We can tweak with different feasibilities for improving each of this areas after consulting with the client to help choose the optimal combination of changes so that to both: maximize NPS Impact and facilitate implementing the changes. The bubble chart adjusts according to the chosen feasibility; choosing the feasibility of 0 means we do not want to consider a set containing this area.

When we hover the mouse over the bubble we can see the quantified details of the changes (Fig. 14.13):

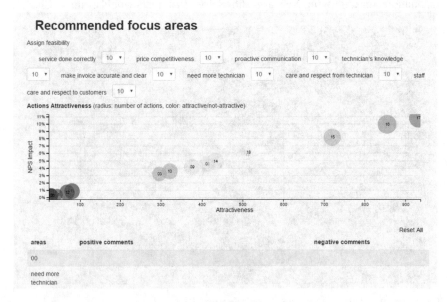

Fig. 14.12 Visualized recommendations for the Client16

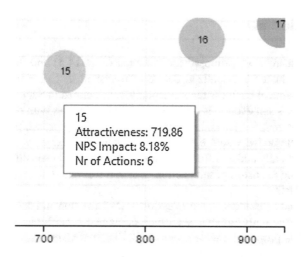

Fig. 14.13 Details of the chosen recommendation

- Ordering number (ID) - 15, 16, etc.,
- Number of changes (actions)-the greater the bubble the more changes it involves,
- NPS Impact-the predicted impact on NPS,
- Attractiveness-computed as Impact multiplied by the total Feasibility.

When we click on the bubble we can explore details in the data table (Fig. 14.14):

- Areas of focus,
- Positive comments about this client associated with this area,
- Negative comments about this client associated with this area.

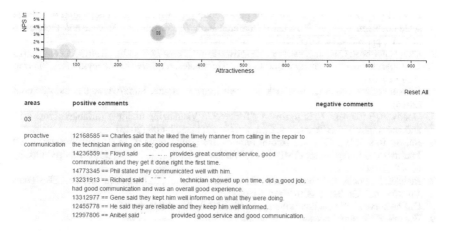

Fig. 14.14 Raw text comments associated with the recommendation "nr 13" for the Client16

14.7 Conclusions

Within this research, we proposed an approach (Sect. 14.5) and designed, implemented and evaluated a visualized data-driven system that supports a real-world problem in the area of business recommendation, as defined in Sect. 14.2. The problem of customer's satisfaction is multidimensional and varies with different characteristics of customer, service and repairing company. The approach presented in this chapter uses interactive visual exploration to identify variables whose correlation varies geographically and at different scales. It also allows for sensitivity analysis of variables relating to customer satisfaction, which is relevant to the repair company's consulting industry.

We performed a number of preprocessing, transformation and data mining experiments on the originally raw dataset which contained superfluous and incomplete information. We proposed a wide spectrum of different visualization techniques to support analytics in the given problem area looking at different analytical aspects: locality, temporal and spatial. We attempted to balance the cognitive load of multiple views with the amount of information we needed to convey to enable effective problem insights. The developed system can also serve businesses for further customer feedback methodology's enhancement. The results in the form of NPS improvement can be explored in an interactive web-based recommender system. The results have been validated with the future users and they are convincing and satisfying from the business point of view.

References

1. Choo, J., Lee, C., Kim, H., Lee, H., Liu, Z., Kannan, R., Stolper, C.D., Stasko, J., Drake, B.L., Park, H.: VisIRR: visual analytics for information retrieval and recommendation with large-scale document data. In: 2014 IEEE Conference on Visual Analytics Science and Technology (VAST), pp. 243–244. IEEE (2014)
2. Client heartbeat: Customer satisfaction software tool. https://www.clientheartbeat.com/ (2016)
3. Customer satisfaction surveys: Questions and templates. https://www.surveymonkey.com/mp/csat/ (2016)
4. Customer sure customer feedback software | customersure. http://www.customersure.com/ (2016)
5. Goodwin, S., Dykes, J., Slingsby, A., Turkay, C.: Visualizing multiple variables across scale and geography. IEEE Trans. Vis. Comput. Gr. **22**(1), 599–608 (2016)
6. Im, S., Raś, Z.W., Tsay, L.: Action reducts. In: Foundations of Intelligent Systems - 19th International Symposium, ISMIS 2011, Warsaw, Poland, June 28–30, 2011. Proceedings, pp. 62–69 (2011)
7. Janicke, S., Focht, J., Scheuermann, G.: Interactive visual profiling of musicians. IEEE Trans. Vis. Comput. Gr. **22**(1), 200–209 (2016)
8. Online survey and benchmarking application | floq. http://floqapp.com/
9. Pawlak, Z.: Rough Sets and Decision Tables. Springer, Berlin (1985)
10. Pawlak, Z., Marek, W.: Rough sets and information systems. ICS. PAS. Reports **441**, 481–485 (1981)
11. Raś, Z.W., Dardzinska, A.: From data to classification rules and actions. Int. J. Intell. Syst. **26**(6), 572–590 (2011)

12. Raś, Z.W., Wieczorkowska, A.: Action-rules: how to increase profit of a company. In: Principles of Data Mining and Knowledge Discovery, 4th European Conference, PKDD 2000, Lyon, France, September 13–16, 2000, Proceedings, pp. 587–592 (2000)
13. Raś, Z.W., Wieczorkowska, A. (eds.): Advances in Music Information Retrieval. Studies in Computational Intelligence, vol. 274. Springer, Berlin (2010)
14. Rses 2.2 user's guide. http://logic.mimuw.edu.pl/~rses
15. SATMETRIX: Improving your net promoter scores through strategic account management (2012)
16. Surveygizmo | professional online survey software and tools. https://www.surveygizmo.com/
17. Tarnowska, K.A., Raś, Z.W., Jastreboff, P.J.: Decision Support System for Diagnosis and Treatment of Hearing Disorders - The Case of Tinnitus. Studies in Computational Intelligence, vol. 685. Springer, Berlin (2017)
18. Temper - find out how your customers feel about every aspect of your business. https://www.temper.io/
19. The world's leading research and insights platform | qualtrics. https://www.qualtrics.com/
20. Tzacheva, A.A., Raś, Z.W.: Association action rules and action paths triggered by meta-actions. In: 2010 IEEE International Conference on Granular Computing, GrC 2010, San Jose, California, USA, 14–16 August 2010, pp. 772–776 (2010)
21. Wang, K., Jiang, Y., Tuzhilin, A.: Mining actionable patterns by role models. In: Liu, L., Reuter, A., Whang, K.Y., Zhang, J. (eds.) ICDE, p. 16. IEEE Computer Society (2006)
22. Wasyluk, H.A., Raś, Z.W., Wyrzykowska, E.: Application of action rules to hepar clinical decision support system. Exp. Clin. Hepatol. **4**(2), 46–48 (2008)
23. Zhang, C.X., Raś, Z.W., Jastreboff, P.J., Thompson, P.L.: From tinnitus data to action rules and tinnitus treatment. In: 2010 IEEE International Conference on Granular Computing, GrC 2010, San Jose, California, USA, 14–16 August 2010, pp. 620–625 (2010)

Chapter 15
Evolutionary and Aggressive Sampling for Pattern Revelation and Precognition in Building Energy Managing System with Nature-Based Methods for Energy Optimization

Jarosław Utracki and Mariusz Boryczka

Abstract This chapter presents a discussion on an alternative attempt to manage the grids that are in intelligent buildings such as central heating, heat recovery ventilation or air conditioning for energy cost minimization. It includes a review and explanation of the existing methodology and smart management system. A suggested matrix-like grid that includes methods for achieving the expected minimization goals is also presented. Common techniques are limited to central management using fuzzy-logic drivers, but referred redefining of the model is used to achieve the best possible solution with a surplus of extra energy. Ordinary grids do not permit significant development in the present state. A modified structure enhanced with a matrix-like grid is one way to eliminate basic faults of ordinary grids model, but such an intricate grid can result in sub-optimal resource usage and excessive costs. The expected solution is a challenge for different Ant Colony Optimization (*ACO*) techniques with an evolutionary or aggressive approach taken into consideration. Different opportunities create many latent patterns to recover, evaluate and rate. Increasing building structure can surpass a point of complexity, which would limit the creation of an optimal grid pattern in real time using the conventional methods. It is extremely important to formulate more aggressive ways to find an approximation of the optimal pattern within an acceptable time frame.

Keywords ACS · ACO · BEMS · Energy optimization · TSP · Matrix-like grid · Central heating system

J. Utracki (✉) · M. Boryczka
Institute of Computer Science, University of Silesia in Katowice, Będzińska 39,
41-200 Sosnowiec, Poland
e-mail: jaroslaw.utracki@us.edu.pl

M. Boryczka
e-mail: mariusz.boryczka@us.edu.pl

© Springer International Publishing AG 2018
U. Stańczyk et al. (eds.), *Advances in Feature Selection for Data and Pattern Recognition*, Intelligent Systems Reference Library 138,
https://doi.org/10.1007/978-3-319-67588-6_15

15.1 Introduction

Nature-based methods such as swarm optimization are very promising in simplifying function, and some of these methods have been considered and advanced to further processing with actual existing technical installations in modern buildings. Ant Colony Optimization, which is based on behavior of ants, appear to be one promising method. Ants scour all of the possible paths in a grid in order to determine the optimal path. A variety of *ACO* approaches have been tested in order to prove that some of them cannot be processed within a feasible time period. A building's grids are similar to a 3D matrix and therefore are too complex to be fully processed and 2D pattern samples are insufficient and inadequate. The quantum part of this assignment is similar to the asymmetric traveling salesman problem (*aTSP*). However, the textbook example model of the TSP has to be modified before being applied in an infinitely looped time period and an environment that users can change. Pattern modifications that are made by a user or system scheduler create new maps that must be re-evaluated. Each map shows a disjunctive part of the matrix-like grid that is used in evaluating the snapshot. Temporary maps are a specific TSP environment. Algorithms in individual maps work in parallel aggregate together in the entire grid. All of these processes and the usage of *ACO* algorithms could be used to ensure the optimization of energy costs in future intelligent buildings. This chapter consist of three main sections: – introduction to Ant Colony Optimization and an Ant Colony System; – introduction to an *Extended Buildings' Energy Concept* with matrix-like grids; and – implementation in a simulation as the proof-of-concept followed by a discussion and conclusion.

15.2 Ant Colony Optimization

Ant Colony Optimization (*ACO*) is a probabilistic technique for solving computational problems in computer science and technical research, which can be narrowed in order to find the best path in graph (in relation to the problem being considered). *ACO* Algorithms are part of a large family of different ant algorithms that have evolved from the ancestor Ant System [5].

15.2.1 Ant System

Ants, which belong to social species live in colonies and due to the interaction and cooperation them, they are capable of complex behaviors that permit them to solve difficult problems from the point of view of a single individual. An interesting characteristic of several ant species is their ability to establish the shortest path between the anthill and the food source that we found [11] by using chemical trails that are deposited as pheromone paths [30]. A pheromone path is the way that the

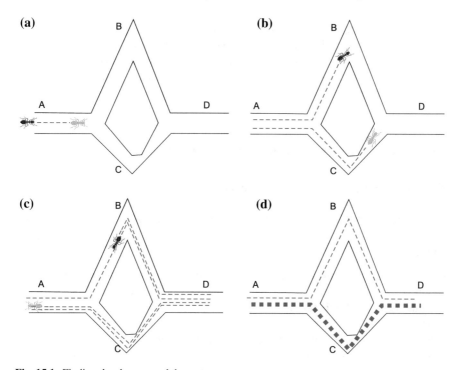

Fig. 15.1 Finding the shortest path by ants

ants in an anthill communicate in order to solve the problem of navigation. The communication methods and the use of pheromones that are deposited on the path by individual ants that is used to find the shortest path between two points is presented in Fig. 15.1.

Figure 15.1a presents a situation in which the ants that are traveling from point A to point D (or in the opposite direction) have to make a decision about whether to turn left or right in order to choose the way through point B or point C. This is a random decision and the ants choose to continue along (in this move) one of the two possible edges, which each have an approximately fifty percent probability. Figure 15.1b, c present the process of depositing the pheromone trail in the paths that are selected. Due to the length of each path on the path that is adjacent to point C, more pheromone will be deposited on the trail because of the shorter length and the greater number of ants that have used it. The same number of ants traveling at the same speed on the shorter path means that the accumulation of pheromones is quicker. The differences in the pheromone concentration on the paths are significant (Fig. 15.1d), which is then noticed by the others ants. As a result, ants (in a probability meaning) take path ACD more often (or in the opposite direction: – DCA) due to the stronger pheromone trail. This is the reason for individual ants to choose this path and to thus strengthen the decision that was made. This process is called auto-catalysis (positive feedback). Finally, each ant will choose path ACD (DCA) as the shortest path.

This description of behavior of ants inspired Marco Dorigo, Vittorio Maniezzo and Alberto Colorni [5] to create an Ant System in which artificial ants exchange information via pheromone trails and cooperate with each other. This was the dawn of the paradigm for the metaheuristic algorithms that are used to solve various optimization problems.

15.2.2 Ant Colony Optimization Metaheuristics

The Ant Colony Algorithm is a constructional algorithm because a partial solution is built by each ant during as it travels through the graph in each iteration. The edges of the graph, which represent the possible paths than can be chosen by an ant, have two types of information that help the ants to make their decision o which way to go [23]:

- Heuristic information of attractiveness. Heuristic preference of the movement from node r to s through $a = (r, s)$ edge. It is denoted as η_{rs} and it cannot be modified while the algorithm is running;
- Artificial pheromone trail concentration level information about the movement of the ants from node r to s rate (how willingly the ants choose this edge). This is denoted as τ_{rs} and it cannot be modified while the algorithm is running and is dependent on the solutions that have been explored by the ants, which indicates the amount of experience that has been gained by the ants.

Heuristic information is the minimum amount of knowledge that is required to solve a specific problem and is gained in increments in contrast to the pheromone trail information. All of the ants in a colony have the following characteristics [12]:

- It searches for the solution with the lowest possible cost;
- It is equipped with memory $\mathcal{M}$ and stored information about the traversed path;
- It has an initial state and at least one stop circumstance;
- It starts at the initial state and explores the possible nodes to obtain a solution incrementally (the problem presents several states, which are defined as sequences of the nodes that have been visited by the ant);
- If the ant is in state $\delta_r = \langle \delta_{r-1}, r \rangle$ – it means that the ant is in node r and that the previously visited node was the last node in the sequence δ_{r-1} – it can move to any node s in its feasible neighbourhood $\mathcal{N}_r = \{s \mid (a_{rs} \in A) \wedge \langle \delta_r, s \rangle \in D\}$, where A is the set of edges and D is the set of feasible states;
- The move from r node to s node is made by applying the transition rule, whose characteristics are:

(a) $\mathcal{A}_{rs}$, i.e. locally stored information about the heuristics and pheromone trail for the $a = (r, s)$ edge;
(b) private ant memory $\mathcal{M}$;
(c) restrictions that have been defined by the problem being solved (Ω);

- While completing the solution, the ant that is on the way from node r to node s can update the pheromone trail τ_{rs} that is associated with $a = (r, s)$ edge. This is called: *an online step-by-step pheromone trail update*;
- After a solution is generated, the ant can recall the traversed path and deposit a pheromone trail on the visited edges. This is called: *an online delayed pheromone trail update*;
- The process of finding a solution ends when stop circumstances have been reached, after the expected state has been fulfilled.

15.2.3 Ant Colony System

The Ant Colony System (*ACS*) is one of the first successors of the Ant System (AS) and was presented in 1997 by Marco Dorigo and Luca Gambardella [9]. This system, like its ancestors, is based on simulating the reality of an actual ant colony.
  It is:

- multi-agent,
- decomposed,
- parallel,
- stochastic, or in other words – probabilistic,
- a system with positive feedback.

  As a follow-up of the previous ant systems, three major modifications were made to the Ant Colony System [2, 12] (Fig. 15.2) and implemented:

- different, so-called: *pseudo-random proportional transition rule*,
- local deposition of the pheromone trail (by ants),
- global deposition of the pheromone trail (by daemon).

15.2.3.1 Transition Rule

An ant k in r node choose s node with the probability specified by the following formula [12].
  If $q \leq q_0$ then the best available node is selected (exploitation):

$$s = \arg\max_{u \in \mathcal{N}_k(r)}\{[\tau_{ru}]^\alpha \cdot [\eta_{ru}]^\beta\}, \tag{15.1}$$

else if $q > q_0$ then s node is chosen randomly with probability usage (exploration):

$$p_{rs}^k(t) = \begin{cases} \dfrac{[\tau_{rs}(t)]^\alpha \cdot [\eta_{rs}]^\beta}{\sum\limits_{u \in \mathcal{N}_k(r)} [\tau_{rs}(t)]^\alpha \cdot [\eta_{ru}]^\beta}, & \text{if } s \in \mathcal{N}_k(r) \\[4mm] 0, & \text{otherwise,} \end{cases} \tag{15.2}$$

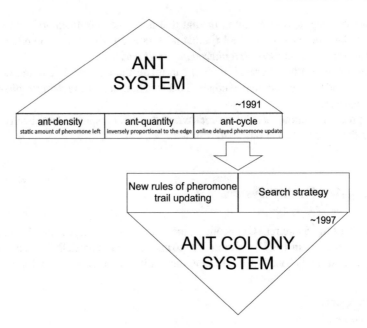

Fig. 15.2 AS evolution to ACS

where:

q_0 — parameter, $q_0 \in \langle 0, 1 \rangle$,

q — random number from the interval $\langle 0, 1 \rangle$,

$\tau_{rs}(t)$ — density of the pheromone trail on edge (r, s) at time t,

η — heuristic information,

α — parameter effecting pheromone to decision, due to the Marco Dorigo suggestion [23] α equals 1 and can be omitted,

β — relative importance of the pheromone trail and the heuristic information factor,

$\mathcal{N}_k(r)$ — feasible neighbourhood of k ant at r node.

u — one feasible edge from node r, where $u \in \mathcal{N}_k(r)$.

The transition rule in *ACO* is non-deterministic but also not a stochastic guarantee of an algorithm's performance because the probability of an ant transitioning to the next node not only depends on a randomly generated q number, but also on the density of the pheromone trail that has been deposited on the edge and heuristic information.

15.2.3.2 Pheromone Trail Local Updating Rule

After each step, an ant deposits pheromone on the traversed a_{rs} edge according to the formula:

$$\tau_{rs}(t+1) = (1-\varphi) \cdot \tau_{rs}(t) + \varphi \cdot \tau_0, \tag{15.3}$$

where:

φ — pheromone vaporization factor, $\varphi \in (0, 1)$,
τ_0 — value equal to the initial pheromone trail.

During an ant's travels through the edge, *pheromone trail local updating rule* the density of the pheromone assigned to this edge diminishes and therefore the attractiveness of the edge decreases for the other ants. This is an impulse to try unknown and unexplored edges and the probability of choosing the visited edges decreases while the algorithm is running. It is worth noting that the density of the pheromone trail cannot decrease below its initial value.

15.2.3.3   Pheromone Trail Global Updating Rule

After an ant's cycle of iteration ends, only the daemon (and not an individual ant) is responsible for updating the pheromone trail in the *Ant Colony System*. In order for this to be done *ACS* processing the ant's best generated solution $S_{global\text{-}best}$ from the entire set, which means that the best elaborated solution since calculations began is taken.

Updating the pheromone trail is preceded by the evaporation of the pheromone trail on every edge that belongs to the global-best solution. It should be noted that pheromone evaporation is only applied to those edges that should be updated – used for pheromone deposition. The rule is:

$$\tau_{rs}(t+n) = (1-\rho) \cdot \tau_{rs}(t), \qquad \forall a_{rs} \in S_{global\text{-}best}, \tag{15.4}$$

where:

ρ — evaporation pheromone the second factor, where $\rho \in (0, 1)$,
a_{rs} — edge connecting r node and s node.

After that, the daemon deposit pheromones according to the following rule:

$$\tau_{rs}(t+n) = \tau_{rs}(t) + \rho \cdot f(C(S_{global\text{-}best})), \qquad \forall a_{rs} \in S_{global\text{-}best}, \tag{15.5}$$

These rules are thoroughly explained in [1, 22, 23].

As an additional activity before *pheromone trail global updating stage* the daemon can explore the local search algorithm in order to improve the ant's solution.

15.2.4 Type of Issues That Are Able To Be Solved Using ACO

In the combinatorial theory, the optimization problem is a computational problem where it is necessary to find the maximal or minimal value of a specified parameter that remains in a certain property. The parameter that is used to estimate the quality of the solution is called the cost function. The optimization problem is called a maximization problem when the maximal value of the cost function is to be determined; and on the other hand, it is called a minimization problem when the minimal value of this function is to be determined.

The optimization problem that is to be resolved by the ant colony optimization algorithm is defined as a finite set of elements $N \in \{n_1, n_2, \ldots, n_m\}$ with set of constraints Ω that is adequate for the specified issue [13]. Let $G \subseteq 2^N$ be a subset of feasible solutions, so the S solution of the problem is admissible if and only if $S \in G$. In other words, if the problem is defined as several states that represent ordered sequences $\delta = < n_r, n_s, \ldots, n_u, \ldots >$ of elements belongs to the N set, then the G is a set of all of the states that fulfill the constraints that are specified in Ω, and the S solution is an G element that fulfills the problem requirements. Additionally, the structure of neighbourhood is defined, which means that the δ_2 state is a neighbor of δ_1, if δ_1 and δ_2 belongs to G and the δ_2 state is feasible for δ_1 in logical one step. The purpose of the algorithm, when assignments of every S solution with its cost $C(S)$ have been done, is to find the lowest cost solution. Over the years, many of ant colony systems have been presented and successfully used to explore various computational problems that are based on the above-mentioned model of optimization. Several examples are listed in Table 15.1 [1].

Table 15.1 Application of ant colony algorithms

Type of Problem	Algorithm	Reference
Traveling Salesman Problem (TSP)	AS	[5]
Quadratic Assignment Problem (QAP)	AS-QAP	[8]
Job-Shop Scheduling Problem (JSP)	AS-JSP	[14]
Vehicle Routing Problem (VRP)	AS-VRP	[21]
Load Balancing Problem (comm-nets)	ABC	[4]
Load Balancing and Routing (telecom-nets)	CAF	[20]
Sequential Ordering Problem (SOP)	HAS-SOP	[10]
Graph Coloring Problem (GCP)	ANTCOL	[7]
Shortest Common Super-sequence Problem (SCSP)	AS-SCS	[16]
Frequency Assignment Problem (FAP)	ANTS-FAP	[6]
Multiple Knapsack Problem (MKP)	AS-MKP	[15]
Wave-Length-Path Routing and Allocation Problem	ACO-VWP	[18]
Redundancy Allocation Problem	ACO-RAP	[19]

15.3   Building's Installation as 3D Matrix-Like Grids

After a short introduction of the methods for the optimization of the Ant Colony System, an attempt can be made to use these techniques in the modern buildings that are equipped with any type of Energy Management System. Intelligent buildings have a large number of various installations such as electrical, ventilation, drainage, water-heating etc. One of the most promising kind of grids for energy optimization is water-heating.

One alternative to the classical conception was introduced in [25] where a water-heating installation is represented as the 3D matrix-like grid that is illustrated in Fig. 15.3. This representation leads to the creation of multi-way paths that reach every node in the matrix.

As can be noticed, this notation in connection with the building's plans creates a non-regular structure with undefined ways to reach the nodes in an energy optimized manner. A visual comparison of both of the attempts: classical non-Matrix and Matrix ones that were thoroughly explained in [26, 27] is presented in Fig. 15.4. In short, it can be observed that the non-matrix grid only has a vertical heat factor supply with one horizontal source line for all of the vertical ones, which is usually placed in the

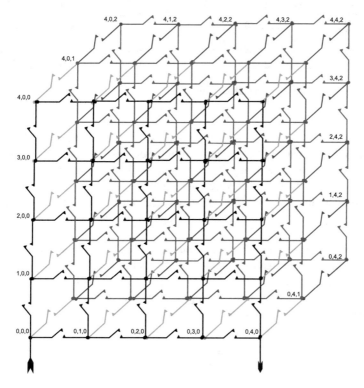

Fig. 15.3 3D matrix-like grid with regular spread nodes

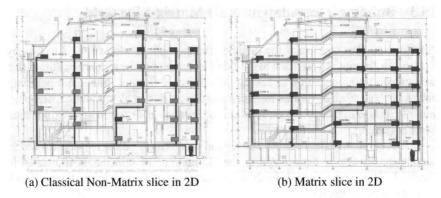

(a) Classical Non-Matrix slice in 2D   (b) Matrix slice in 2D

Fig. 15.4 Two different models of water-heating grid

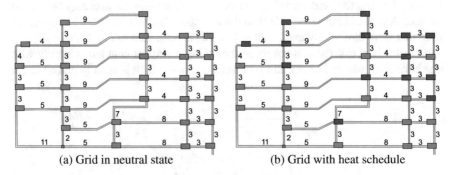

(a) Grid in neutral state   (b) Grid with heat schedule

Fig. 15.5 Matrix-like grid as a weighted graph

basement of a building. The nodes are placed in the line of the vertical supply. In contrast to this, a matrix-like grid has multiple vertical and horizontal connections in each node and can offer many different configurations for heating water.

A grid can be represented as a weighted graph similar to Fig. 15.5a in which the vertices (nodes) are connected by edges (existing connection of the grid) of a specified weight (energy transfer cost). Example settings for heat requisition are presented in Fig. 15.5b.

The energy spread costs that are disbursed by the nodes (heaters placed in the nodes) are the amount that is programmed by a scheduler for thermal comfort and that can tuned by system sensors due to existing conditions [28]. The equation for the required heat power and the mean average temperature between the heating medium and surroundings [24] in the node is described below.

$$\Delta T_\alpha = \frac{1}{2} \left(T_{iz} + T_{ip} \right) - T_i, \tag{15.6}$$

where:

Δt_g — mean average temperature,
T_{iz} — temperature of the heater supply,
T_{ip} — temperature of the heater return,
T_i — temperature of the surroundings.

The heater power is described by:

$$\Phi_g = (\Phi_o - \Phi_p) \cdot \beta_{1..5},$$ (15.7)

where:

Φ_g — heater power,
Φ_o — counted requirement for heat power,
Φ_p — heat power surplus spread by connections (edges),
$\beta_{1..5}$ — specified heater factors.

And finally heating power is described as:

$$\Phi = a \cdot \Delta T_\alpha^{1+b} \cdot A_g \cdot \varepsilon_{\Delta T},$$ (15.8)

where:

a, b — factors dedicated to the type of heater,
A_g — surface of the heater,
$\varepsilon_{\Delta T}$ — correction factor for different non-standard modes of heater work,
ΔT_α — mean average temperature.

This is required in order to estimate the energy costs to ensure the proper adjustments to the scheduled thermal parameters (and that are used in the simulator described in Sects. 15.5.2 and 15.6). It has to be mentioned that the transport capacity of the grid is not fully scalable and each connection in the grid has a maximum capacity that cannot be exceeded. The weight of the edge (connection between two nodes) is changeable from a nominal static cost to a dynamically cost that can be increased even to infinity to force the use of another edge for the heating medium transfer. This factor means that entire grid must be divided into smaller grids with an unknown shape and size. A relict graph, which needs to be revealed, must be placed inside each of the smaller grid. Every revealed relict graph needs to be reconnected into one matrix graph in order to ensure the operability of the system.

ACO optimization can lead to finding the shortest energy path for the heating medium as is presented in Fig. 15.6a, which can reduce the entire graph into the relict graph presented in Fig. 15.6b by ignoring the nodes that are not used in a particular iteration. This process is discussed further in Sect. 15.5.2.

The capacity (heating medium transport ability) of each edge is the limiter for one path and can be the incentive to seek a second path (Fig. 15.7).

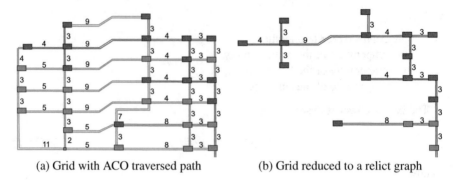

(a) Grid with ACO traversed path   (b) Grid reduced to a relict graph

Fig. 15.6 Matrix-like grid transposed to relict graph

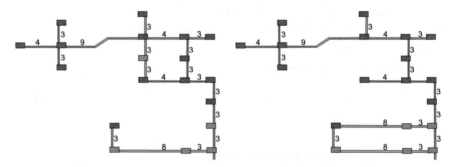

Fig. 15.7 Different but equivalent relict graphs for the same configuration

15.4   Classical *ACO* Approach in Revealing a Pattern

The classical approach is appropriate in grids that have a varied number of vertices and edges. However, typical TSP examples are based on a set of up to 1,000 towns that are to be visited by a virtual traveler. This amount can be too small for a matrix-like grid in which more than this number of nodes can be found in a typical building. Revealing a pattern using a classical algorithmic approach is far from perfect, which can be observed in the historical example algorithmic experiment TSP instance att532 [22] presented in Fig. 15.8 or other examples published in [3, 17]. The "long shots" edges are noticeable in the attached figure and prove the imperfection of this algorithm.

15.4.1   Capability of Revealing with ACO

The Ant Colony seems to be perfect solution to approximate the TSP path function, when there is no need to find the exact solution in polynomial time. It must be stated that the classical *ACO* methods are capable of finding "the Best Optimal Path"[1] (the

---

[1]Mathematically: an optimal path is only one or an issue has equivalent solutions, but within the meaning of programming and algorithms can be more than one and slightly different, non congruent

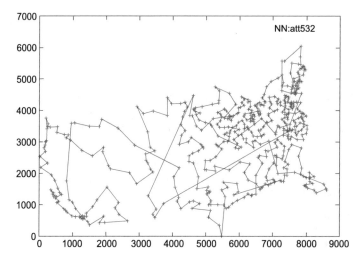

Fig. 15.8 Tour optimized by the nearest neighbor - instance att532 from TSPLIB as an example of classical algorithmic approach

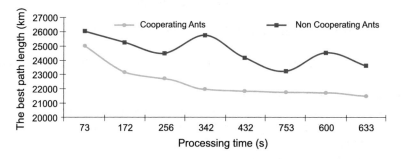

Fig. 15.9 Tour optimized by the cooperating and non-cooperating agents in ACS

abbreviation *BOP* instead of the full notation will be used further) in almost every optimization. One significant feature of *ACO* is that the Ant Colony defines a synergy effect as the cooperation of individual ants (agents) together, which leads to next level of results that surpass the sum of the result of particular ants [3], which as a result of another experiment is presented in Fig. 15.9.

It should be noted that the cooperating agents provide an improved approximation of the function.

The ant system that was considered as the starting point with the classical approach is a successor of the ant-cycle [5, 12], which was extended to *ACS* and modified in order to be used for a matrix-like grid. A weighted graph defines the possible paths for agents. Each agent has a further duty to transport its additional load [25] which

---

solutions that can be distinguished by i.e. the time cost to be revealed. "The Best Optimal Path" (*BOP*) means the best solution than can be achieved under given assumptions.

also determines any possible paths due to the path's throughput. Each edge can be traversed only in one fixed direction during one period of the iteration, which is set by the result of *BOP* search.

15.5 Aggressive *ACO* Sampling in Revealing a Pattern

There are several possible solutions to prevent a system failure due to inefficiency in the *ACO* time process. One of these is to use parallel threads in *ACO* processing and/or a more efficient central computer system with greater computing power.

Other, but not fulfilling presented list, that are more adequate and hardware independent prevention methods are:

- Ants decomposition and specialization,
- Precognition paths,
- Ants spreading.

15.5.1 Ant Decomposition and Specialization

Ants are separated into specialized divisions that are aimed at testing the hypothesis that two independent group of ants, path-seekers and carriers, are more efficient. In the first stage, the path-seekers traverse the edges of the weighted graph to pre-optimize the path.

In the second stage, the *carriers* try to deliver enough of the cargo (heating agent), which is a success when no clogging occurs on the path. In the event that some edges are blocked due to their throughput ability, the third stage is fired up with local path-seekers exploring to find another sub-path for the carriers and processing the return to the second stage. This principle is presented in Fig. 15.10.

Stage I is shown in Fig. 15.10 I and Stage II is presented in Fig. 15.10 II in which clogging occurs. Stage III in Fig. 15.10 III shows the way to eliminate clogging by adding a sub-path to *BOP* that were evaluated in Stage I. Stage IV is a representation of the return to Stage II with no clogging. Stages II and III are deployed repeatedly in the event of demand. The most significant point is that the carriers can move through an edge in only one direction, while the path-seekers can move in both directions.

The actual use of this model is based on a pre-programmed grid shape that is used as a central heating grid in a building with the matrix-like style shown in Fig. 15.11. In this figure, all of the nodes are hidden in order to emphasize the edges. The action is presented in the simulation shown in Fig. 15.12.

As can be observed, there are two kind of ants the path-seekers which are the violet dots and the carriers, which are the yellow dots. The path-seekers are much faster than the carriers and their role is to prepare the pre-route for the carriers. The path-seekers deposit the pheromone trails that will be used by the carriers. The carriers do not deposit the pheromone trails; they have a different role, which is to deliver their cargo to the nodes and to distribute the cargo during passage. The algorithm that was used for the implementation is presented in Sect. 15.6.

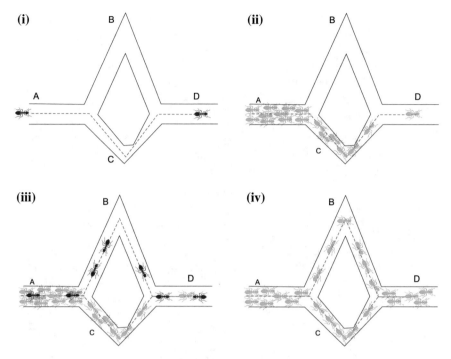

Fig. 15.10 The principle of overcoming the overload of the path

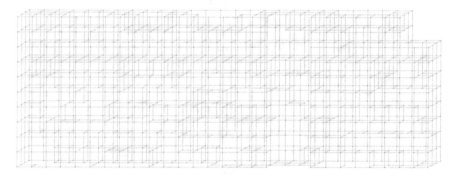

Fig. 15.11 Central heating grid in a simulated building

15.5.2   Precognition Paths

The concept of using precognition paths is based on the scheduler role in which the required human-comfort parameters are pre-ordered. Each node with that is not in a "neutral" state is left and each node that is in a "neutral" state is removed from grid. Thus, the "active" nodes are grouped in one graph, which is connected with

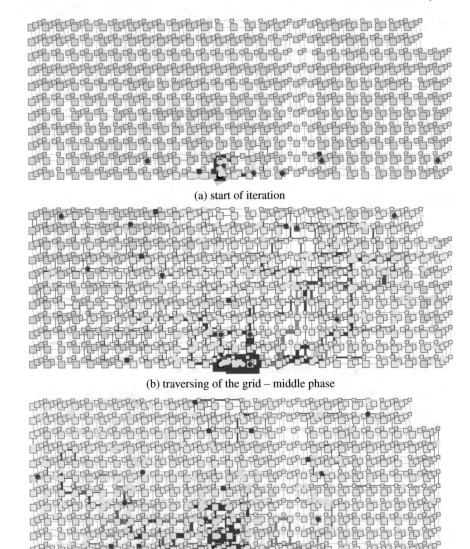

(a) start of iteration

(b) traversing of the grid – middle phase

(c) traversing of the grid – near end phase

Fig. 15.12 Snapshots of one system iteration where the ants are traversing the grid to achieve all of the requirements pre-programmed in the scheduler

tour optimization, by the nearest neighbor. After that, the graphs are transposed into the physical entity that existed from the initial state with "neutral" nodes and edges only needed to be a coupler between the "active" nodes.

The process of creating paths is presented in Figs. 15.13 and 15.14. The starting point is presented in Fig. 15.13a – pre-ordered scheduled parameters overlaid on

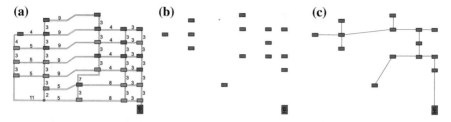

Fig. 15.13 Precognition paths generating - part 1

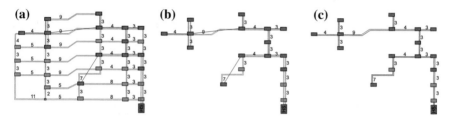

Fig. 15.14 Precognition paths generating - part 2

a matrix-like grid. The "active" nodes are illustrated in Fig. 15.13b. The result of linking the "active" nodes in a grid by the nearest neighbor acquisition is shown in Fig. 15.13c.

That result is again overlaid on a pre-ordered matrix-like grid as in Fig. 15.14a, and the shaped grid is reduced to a pre-relict graph by adjusting the graph presented in Fig. 15.13c into the actual existing routes; as is shown in Fig. 15.14b. All of the unexplored or explored with lower than expected frequencies are eliminated from the graph, which leads to a simplification of the graph. Finally, the created graph is reduced to the relict graph as in Fig. 15.14c.

15.5.3 Ants Spreading

The last presented (but not last existing) technique for overcoming time execution limit deadlock is the ants spreading to the nodes of the grid and beginning to evaluate the paths separately for each node with a mutual pheromone trail information exchange with a common TSP path search in the first stage and a an modified second stage with examining path found for heat agent delivery purposes with one starting point. This technique is mentioned as an entry point for future examination and testing. At this time, it is a concept that must be confirmed or rejected.

15.6 Algorithm and Procedures

The system that is evaluated is an *ACS* with the aggressive sampling described in previous section Ant decomposition and specialization. The main part of the algorithm implementing the described achievement is presented in the listing 15.1.

Algorithm 15.1 Main process

1: **procedure** ACO_GRIDMETAPROCESSING
2:　　initialize_parameters()
3:　　**while** *unheated_node_exists* **do** ▷ We have the answer if final state is achieved
4:　　　　create_new_ants()
5:　　　　evaporate_pheromone_from_every_edge()
6:　　　　daemon_actions()
7:　　**end while**
8:　　**return** *exitcode* ▷ The exitcode of the terminal state
9: **end procedure**

All of the procedures that are used in the main algorithm box are described below.

The procedure *initialize_parameters*() looks like Algorithm 15.2.

Algorithm 15.2 Initializaton of the parameters

1: **procedure** INITIALIZE_PARAMETERS
2:　　**for** every edge **do** *quantity_of_ants_using_edge* $= 0$
3:　　**end for**
4: **end procedure**

The procedure *create_new_ants*() looks like Algorithm 15.3.

Algorithm 15.3 Creation of new ants

1: **procedure** CREATE_NEW_ANTS
2:　　**repeat** *inparallel*
3:　　　　**for** $k = 1$ **to** m **do** ▷ **where m is a quantity of ants**
4:　　　　　　*new_ant*(k) ▷ **ants acting like carriers**
5:　　　　　　*new_ant_pathfinder*(k) ▷ **ants acting like path seekers**
6:　　　　**end for**
7:　　**until** *doneinparallel*
8: **end procedure**

The procedure *new_ant*() looks like Algorithm 15.4.

Algorithm 15.4 New Ant procedure

1: **procedure** NEW_ANT(*ant_id*)
2:   *initialize_ant*(*ant_id*)
3:   *initialize_ant_memory*()
4:   **while** *unheated_node_exists* **and** *cargo* > 0 **do**
5:     *P* = *compute_transition_probability*(*L*)
6:     *next_node* = *choose_next_node*(*P*)
7:     *move_ant*(*next_node*)
8:     *L* = *update_internal_state*()
9:   **end while**
10:   *release_ant_resources*(*ant_id*)
11: **end procedure**

The procedure *initialize_ant*() looks like Algorithm 15.5.

Algorithm 15.5 Initialization of the ant

1: **procedure** INITIALIZE_ANT(ant_id)
2:   *move_ant_to_boiler*()
3:   *cargo* = *initialize_cargo*()
4:   *unheated_nodes_on_path* = 0
5: **end procedure**

The procedure *initialize_ant_memory*() looks like Algorithm 15.6.

Algorithm 15.6 Initialization of the ant's memory

1: **procedure** INITIALIZE_ANT_MEMORY()
2:   $L+ = boiler$
3:   *distance* = 0
4: **end procedure**

The procedure *compute_transition_probability*() looks like Algorithm 15.7.

Algorithm 15.7 Computing of ant's transition probability

1: **procedure** COMPUTE_TRANSITION_PROBABILITY(L)
2:   **for** every_available_edge **do**
3:     **if** (*quantity_of_ants_using_edge* < *max_edge_capacity*) **and** (*pheromone_on_edge* > *min_pheromone_amount*) **then**
4:       *compute_probability*()
5:     **end if**
6:   **end for**
7: **end procedure**

The procedure *move_ant*() looks like Algorithm 15.8.

Algorithm 15.8 Moving of the ant

1: **procedure** MOVE_ANT(next_node)
2:     $cargo- = transmission\_energy\_loss$
3: **end procedure**

The procedure *update_internal_state*() looks like Algorithm 15.9.

Algorithm 15.9 Internal state updating

1: **procedure** UPDATE_INTERNAL_STATE
2:     $L+ = next\_node$
3:     $quantity\_of\_ants\_using\_edge(L, L - 1)+ = 1$
4:     $distance+ = distance\_between\_nodes(L, L - 1)$
5:     **if** $L is unheated$ **then**
6:         $unheated\_nodes\_on\_path+ = 1$
7:         $heat\_node(L)$
8:         $cargo- = heating\_energy\_loss$
9:     **end if**
10: **end procedure**

The procedure *release_ant_resources*() looks like Algorithm 15.10.

Algorithm 15.10 Releasing ant resources

1: **procedure** RELEASE_ANT_RESOURCES(ant_id)
2:     **for** every visited edge **do**
3:         $quantity\_of\_ants\_using\_edge- = 1$
4:     **end for**
5: **end procedure**

The procedure *initialize_cargo*() looks like Algorithm 15.11.

Algorithm 15.11 Initializing cargo

1: **procedure** INITIALIZE_CARGO
2:     $cargo = x$ ▷ where x is heat load
3:   ▷ or *time-to-live* for path seekers
4: **end procedure**

The procedure *daemon_actions*() look like Algorithm 15.12.

Algorithm 15.12 Daemon actions

1: **procedure** DAEMON_ACTION
2:   *Reset_iteration_best_path*()
3: **end procedure**

The procedure *search_node_with_max_pheromone(L)* looks like Algorithm 15.13

Algorithm 15.13 Search node with a MAX pheromone concentration

1: **procedure** SEARCH_NODE_WITH_MAX_PHEROMONE(L)
2:   **for** *every available edge* **do**
3:     **if** *quantity_of_ants_using_edge* < *max_edge_capacity* **then**
4:       *get_node_with_max_pheromone*()
5:     **end if**
6:   **end for**
7: **end procedure**

The procedure *best_iteration_path_found*() looks like Algorithm 15.14.

Algorithm 15.14 Best Iteration Path Found

1: **procedure** BEST_ITERATION_PATH_FOUND()
2:   **if** *unheated_nodes_on_path* > *iteration_max_unheated_nodes_on_path* **then**
3:     *iteration_max_unheated_nodes_on_path* = *unheated_nodes_on_path*
4:     return TRUE;
5:   **else**
6:     **if** (*path_distance* < *iteration_min_path_distance*) **and** (*unheated_nodes_on_
path* = *iteration_max_unheated_nodes_on_path*) **then**
7:       *iteration_min_path_distance* = *path_distance*
8:       **return TRUE;**
9:     **end if**
10: **end if**
11: **return FALSE**
12: **end procedure**

The procedure *reset_iteration_best_path*() looks like Algorithm 15.15.

Algorithm 15.15 reset_iteration_best_path()

1: **procedure** RESET_ITERATION_BEST_PATH()
2:   *iteration_max_unheated_nodes_on_path* = 0
3:   *iteration_min_path_distance* = *MAX_AVAILABLE_VALUE*
4: **end procedure**

The procedure *new_ant_pathfinder*() looks like Algorithm 15.16.

Algorithm 15.16 New Ant Pathfinder procedure

1: **procedure** NEW_ANT_PATHFINDER()(*ant_id*)
2:     *initialize_ant*(*ant_id*)
3:     *initialize_ant_memory*()
4:     *unheated_nodes_on_path* = 0
5:     *path_distance* = 0
6:     **while** *unheated_node_exists* **and** *cargo* > 0 **do**   ▷ **Pathfinder do not carry a cargo**
7:   ▷ **but here it is a factor**
8:   ▷ **which determine ant** *time-to-live* **limit**
9:         *draw_a_number_q*   ▷ **random value where** $q \in \langle 0, 1 \rangle$
10:        **if** $q \leq q_0$ **then**   ▷ **where** q_0 **- exploration percent parameter - exploatation**
11:            $P = search\_node\_with\_max\_pheromone(L)$
12:        **else**   ▷ **exploration**
13:            $P = compute\_pathfinder\_transition\_probability(L)$
14:        **end if**
15:        *next_node* = *choose_next_node*(*P*)
16:        **move_ant(next_node)**
17:        $L = update\_internal\_state()$
18:        *path_distance*+ = 1
19:    **end while**
20:    **if** *unheated_nodes_on_path* > 0 **then**
21:        **for** *every visited edge* **do**
22:            *unheated_nodes_on_path*+ = 1
23:            *put_pheromone_on_edge*()
24:        **end for**
25:        **if** *best_iteration_path_found*() **then**
26:            **for** *every visited edge* **do**
27:                *put_extra_pheromone_on_edge*()
28:            **end for**
29:        **end if**
30:    **end if**
31:    *release_ant_resources*(*ant_id*)
32: **end procedure**

The procedures:

- *evaporate_pheromone_from_every_edge*(),
- *compute_probability*(),
- *compute_pathfinder_transition_probability*(),
- *put_pheromone_on_edge*(),
- *put_extra_pheromone_on_edge*(),
- *get_node_with_max_pheromone*,

are based on mathematical formulas according to the formulas presented and described in the Sect. 15.2.3.

The most significant issue in revealing a pattern and achieving the best optimal solution is time required to evaluate the algorithm. *ACO* needs time to recover the paths and to test them. Time restrictions between the phase iterations (scheduler service) that are too narrow can cause some paths to remain unprocessed and therefore the optimal pattern will not be revealed. Due to *Extended Buildings' Energy Concept* [29], this problem can lead to insufficiency in energy distribution and total system failure. A state in which there unprocessed nodes (vertices) in a building is unacceptable.

15.7 Conclusion

The presented techniques and modified *ACO* algorithms appear to be adequate for energy saving in modern buildings that are equipped with a centrally managed artificial intelligence system. The simulations that were performed delivered promising results, which are presented in [25], especially with *Extended Buildings' Energy Concept* [28, 29]. Simulations of energy distribution in a heating grid (one room under continuous acquisition) based on a common central fuzzy logic driven control as the best known system that is implemented around the world and its results (C_{fl}), which were used to compare with an alternative approach that is based on an Ant Colony Optimized heat factor transport in a matrix-like grid (A) revealed that were 2.9% fewer energy requests in the *ACO*-driven grid. Some of the other factors that are presented in the figure are: the standard common thermostatic regulation without central driving (C_{ni}), external temperature (Ex) and the preprogrammed temperature (P) that is expected to be achieve for the human-comfort environment parameters. The sample 24 h characteristics are combined in Fig. 15.15.

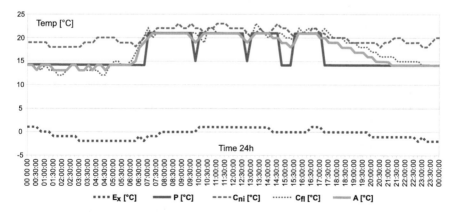

Fig. 15.15 The 24 h temperature acquisition

One technical problem that must be overcome is the infrastructure that needs to be built from scratch. An essential element of the executive system is e.g. a non-stop working multi-way valve that can switch up to five or six ways. The algorithmic model that was considered revealed some benefits but still needs further experiments and testing. It was created in a classical and matrix-like grid simulator, which is based on the experience collected, in order to estimate the expected profits, find imperfections or habitual faults and to prepare further experiments. An indispensable component of an efficiently working BEMS[2] is accordant and precise feedback system collecting environmental parameters of the internal and external milieu. This kind of subsystem is designed [28] as a part of *Extended Buildings' Energy Concept* and has been implemented. It is undeniable that not only energy saving but also human comfort are a priorities, and that is possible to achieve this state only with hybrid and holistic management.

References

1. Boryczka, M.: The Ant Colony Programming in the process of the automatic approximation of function. University of Silesia, Katowice (2006). The title of the original: "Programowanie mrowiskowe w procesie aproksymacji funkcji", In Polish
2. Boryczka, U.: Algorithms of the ant colony optimization. University of Silesia, Katowice (2006). The title of the original: "Algorytmy optymalizacji mrowiskowej", In Polish
3. Boryczka, U.: Study of synergistic effect in ant colony optimization. http://prac.us.edu.pl/~uboryczk/strony/prace/efektsyn.html. In Polish (2017)
4. Bruten, J., Rothkrantz, L., Schoonderwoerd, R., Holland, O.: Ant-based load balancing in telecommunications networks. Adapt. Behav. **5**(2), 169–207 (1996)
5. Colorni, A., Dorigo, M., Maniezzo, V.: Positive feedback as a search strategy. Technical report 91-016, Dipartimento di Elettronica, Politectico di Milano (1991)
6. Colorni, A., Maniezzo, V.: An ants heuristic for the frequency assignment problem. Future Gener. Comput. Syst. **1**(16), 927–935 (2000)
7. Costa, D., Hertz, A.: Ants can colour graphs. J. Oper. Res. Soc. **48**(3), 295–305 (1997)
8. Dorigo, M., Maniezzo, V., Colorni, A.: The ant system applied to the quadratic assignment problem. Technical report 94-28, Dipartimento di Elettronica e Informazione, Politecnico di Milano (1994)
9. Dorigo, M., Gambardella, L.M.: Ant colony system: a cooperative learning approach to the traveling salesman problem. IEEE Transactions on Evolutionary Computation **1**(1), 53–66 (1997)
10. Dorigo, M., Gambardella, L.M.: Has-sop: hybrid ant system for the sequential ordering problem. Technical report 11, Istituto Dalle Molle Di Studi Sull Intelligenza Artificiale (1997)
11. Goss, S., Beckers, R., Deneubourg, J.L.: Trails and u-turns in the selection of shortest path by ant lasius niger. Theor. Biol. **159**, 397–415 (1992)
12. Herrera, T.F., Stützle, T., Cordón, G.O.: A review on the ant colony optimization metaheuristic: basis, models and new trends. Mathware Soft Comput. [en línia] **9**(2) (2002). Accessed 2017
13. Maniezzo, V.: Exact and approximate nondeterministic three-search procedures for quadratic assignment problem. INFORMS J. Comput. **11**(4), 358–369 (1999)
14. Maniezzo, V., Trubian, M., Colorni, A., Dorigo, M.: Ant system for job-shop scheduling. Belgian J. Oper. Res., Stat. Comput. Sci. (JORBEL) **34**, 39–53 (1994)

---

[2]Abbreviation BEMS means: Building Energy Managing System.

15. Michalewicz, Z., Leguizamon, G.: A new version of Ant System for subset problems. In: Proceedings of the 1999 Congress on Evolutionary Computation. IEEE Press, Piscataway (1999)
16. Middendorf, M., Michel, R.: An aco algorithm for shortest common super-sequence problem. In: New Methods in Optimisation, pp. 51–61 (1999)
17. Optimal solutions for symmetric tsps. http://comopt.ifi.uni-heidelberg.de. In German (2017)
18. Sinclair, M.C., Navarro Varela, G.: Ant Colony Optimisation for virtual-wave-length-path routing and wavelenght allocation. In: Proceedings of the 1999 Congress on Evolutionary Computation, pp. 1809–1816. IEEE Press, Piscataway (1999)
19. Smith, A.E., Liang, Y.C.: An Ant System approach to redundancy allocation. In: Proceedings of the 1999 Congress on Evolutionary Computation, pp. 1478–1484. IEEE Press, Piscataway (1999)
20. Snyers, D., Kuntz, P., Heusse, M., Guerin, S.: Adaptive agent-driven routing and load balancing in communication networks. Technical report RR-98001-IASC, ENST de Bretagne, BP 832, Brest Cedex (1998)
21. Strauss, C., Bullnheimer, B., Hartl, R.F.: An improved ant system algorithm for the vehicle routing problem. Technical report POM-10/97, Institute of Management Science, University of Vienna (1997)
22. Sttzle, T., Dorigo, M.: The ant colony optimization metaheuristic: algorithms, applications, and advances. In: Handbook of Metaheuristics, pp. 250–285. Springer US, Boston (2003)
23. Stützle, T., Dorigo, M.: The ant colony optimization. In: A Bradford Book. MIT Press, London (2004)
24. Szymański, W., Babiarz, B.: The heating systems. Oficyna Wydawnicza Politechniki Rzeszowskiej, Rzeszów (2015). The title of the original: "Ogrzewnictwo", In Polish
25. Utracki, J.: Building management system–artificial intelligence elements in ambient living driving and ant programming for energy saving–alternative approach. Proceedings: 5th International Conference – Information Technologies in Medicine. ITIB 2016, vol. 2, pp. 109–120. Springer International Publishing, Kamień Śląski (2016)
26. Utracki, J.: Intelligent building systems: the swarm optimization of a grid-based systems – scheme of action, vol. 10, pp. 229–233. Creativetime, Kraków, Poland (2017). The title of the original: Systemy Domów Inteligentnych: Schemat Funkcjonowania Systemów Sieciowych Wykorzystujących Optymalizację Rojową. Zagadnienia aktualne poruszane przez młodych naukowców In Polish
27. Utracki, J.: Intelligent building systems: ants in energy optimization management duty, vol. 10, pp. 226–229. Creativetime, Kraków, Poland (2017). The title of the original: Systemy Domów Inteligentnych: Mrówki W Służbie Optymalizacji Zarządzania Energią. Zagadnienia aktualne poruszane przez młodych naukowców In Polish
28. Utracki, J.: The human-comfort parameters acquisition system as a precise control feedback role in an intelligent building energy management systems. Technical Issues (3) (2017)
29. Utracki, J.: Intelligent building systems: passive or reactive buildings - new solutions. Agorithmic support., vol. 10, pp. 220–225 (2017). Creativetime, Kraków, Poland: The title of the original: Systemy Domów Inteligentnych: Budynki Pasywne. Czy Reaktywne. Nowe Rozwiązania Inżynieryjne - Wsparcie Algorytmiczne, Zagadnienia aktualne poruszane przez młodych naukowców In Polish
30. Wilson, L.A., Holldobler, B.: The Ants. Springer, Berlin (1990)

Index

© Springer International Publishing AG 2018
U. Stańczyk et al. (eds.), *Advances in Feature Selection for Data and Pattern Recognition*, Intelligent Systems Reference Library 138,
https://doi.org/10.1007/978-3-319-67588-6

Printed in the United States
By Bookmasters